THEORY OF GROUPS OF

FINITE ORDER

THEORY OF GROUPS OF FINITE ORDER

W. BURNSIDE, M.A., F.R.S.,
LATE FELLOW OF PEMBROKE COLLEGE, CAMBRIDGE;
PROFESSOR OF MATHEMATICS AT THE ROYAL NAVAL COLLEGE, GREENWICH.

HAWK PRESS

Published by

Hawk Press
4836/24, Ansari Road, Daryaganj
New Delhi - 110 002
Phones: +91-11-23278618, +91-11-43667199
E-mail: thehawkpress@gmail.com
www.thehawkpress.com

ISBN: 978-93-93971-81-4

PREFACE.

THE theory of groups of finite order may be said to date from the time of Cauchy. To him are due the first attempts at classification with a view to forming a theory from a number of isolated facts. Galois introduced into the theory the exceedingly important idea of a self-conjugate sub-group, and the corresponding division of groups into simple and composite. Moreover, by shewing that to every equation of finite degree there corresponds a group of finite order on which all the properties of the equation depend, Galois indicated how far reaching the applications of the theory might be, and thereby contributed greatly, if indirectly, to its subsequent developement.

Many additions were made, mainly by French mathematicians, during the middle part of the century. The first connected exposition of the theory was given in the third edition of M. Serret's "*Cours d'Algèbre Supérieure*," which was published in 1866. This was followed in 1870 by M. Jordan's "*Traité des substitutions et des équations algébriques.*" The greater part of M. Jordan's treatise is devoted to a developement of the ideas of Galois and to their application to the theory of equations.

No considerable progress in the theory, as apart from its applications, was made till the appearance in 1872 of Herr Sylow's memoir "*Théorèmes sur les groupes de substitutions*" in the fifth volume of the *Mathematische Annalen.* Since the date of this memoir, but more especially in recent years, the theory has advanced continuously.

In 1882 appeared Herr Netto's "*Substitutionentheorie und ihre Anwendungen auf die Algebra,*" in which, as in M. Serret's and M. Jordan's works, the subject is treated entirely from the point of view of groups of substitutions. Last but not least among the works which give a detailed account of the subject must be mentioned Herr Weber's "*Lehrbuch der Algebra,*" of which the first volume appeared in 1895 and the second in 1896. In the last section of the first volume some of the more important properties of substitution groups are given. In the first section of the second volume, however, the subject is approached from a more general point of view, and a theory of finite groups is developed which is quite independent of any special mode of representing them.

The present treatise is intended to introduce to the reader the main outlines of the theory of groups of finite order apart from any applications. The subject is one which has hitherto attracted but little attention in this country; it will afford me much satisfaction if, by means of this book, I shall succeed in arousing interest among English mathematicians in a branch of pure mathematics which becomes the more fascinating the more it is studied.

Cayley's dictum that "a group is defined by means of the laws of combination of its symbols" would imply that, in dealing purely with the theory of groups, no more concrete mode of representation should be used than is absolutely necessary. It may then be asked why, in a book which professes to leave all applications on one side, a considerable space is devoted to substitution groups; while other particular modes of representation, such as groups of linear transformations, are not even referred to. My answer to this question is that while, in the present state of our knowledge, many results in the pure theory are arrived at most readily by dealing with properties of substitution groups, it would be difficult to find a result that could be most directly obtained by the consideration of groups of linear transformations.

The plan of the book is as follows. The first Chapter has been devoted to explaining the notation of substitutions. As this notation may not improbably be unfamiliar to many English readers, some such introduction is necessary to make the illustrations used in the following chapters intelligible. Chapters II to VII deal with the more important properties of groups which are independent of any special form of representation. The notation and methods of substitution groups have been rigorously excluded in the proofs and investigations contained in these chapters; for the purposes of illustration, however, the notation has been used whenever convenient. Chapters VIII to X deal with those properties of groups which depend on their representation as substitution groups. Chapter XI treats of the isomorphism of a group with itself. Here, though the properties involved are independent of the form of representation of the group, the methods of substitution groups are partially employed. Graphical modes of representing a group are considered in Chapters XII and XIII. In Chapter XIV the properties of a class of groups, of great importance in analysis, are investigated as a general illustration of the foregoing theory. The last Chapter contains a series of results in connection with the classification of groups as simple, composite, or soluble.

A few illustrative examples have been given throughout the book. As far as possible I have selected such examples as would serve to complete or continue the discussion in the text where they occur.

In addition to the works by Serret, Jordan, Netto and Weber already referred to, I have while writing this book consulted many original memoirs. Of these I may specially mention, as having been of great use to me, two by Herr Dyck published in the twentieth and twenty-second volumes of the *Mathematische Annalen* with the title "*Gruppentheoretische Studien*"; three by Herr Frobenius in the *Berliner Sitzungsberichte* for 1895 with the titles, "*Ueber endliche Gruppen*," "*Ueber auflösbare Gruppen*," and "*Verallgemeinerung des Sylow'schen Satzes*"; and one by Herr Hölder in the forty-sixth volume of the *Mathematische Annalen* with the title "*Bildung zusammengesetzter Gruppen*." Whenever a result is taken from an original memoir I have given a full reference; any omission to do so that may possibly occur is due to an oversight on my part.

To Mr A. R. Forsyth, Sc.D., F.R.S., Fellow of Trinity College, Cambridge, and Sadlerian Professor of Mathematics, and to Mr G. B. Mathews, M.A., F.R.S., late Fellow of St John's College, Cambridge, and formerly Professor of Mathematics in the University of North Wales, I am under a debt of gratitude for the care and patience with which they have read the proof-sheets. Without the assistance they have so generously given me, the errors and obscurities, which I can hardly hope to have entirely escaped, would have been far more numerous. I wish to express my grateful thanks also to Prof. O. Hölder of Königsberg who very kindly read and criticized parts of the last chapter. Finally I must thank the Syndics of the University Press of Cambridge for the assistance they have rendered in the publication of the book, and the whole Staff of the Press for the painstaking and careful way in which the printing has been done.

W. BURNSIDE.

July, 1897.

CONTENTS.

CHAPTER I.

ON SUBSTITUTIONS.

CHAPTER II.

THE DEFINITION OF A GROUP.

CHAPTER III.

ON THE SIMPLER PROPERTIES OF A GROUP WHICH ARE INDEPENDENT OF ITS MODE OF REPRESENTATION.

CHAPTER IV.

ON ABELIAN GROUPS.

CHAPTER V.

ON GROUPS WHOSE ORDERS ARE POWERS OF PRIMES.

CHAPTER VI.

ON SYLOW'S THEOREM.

CHAPTER VII.

ON THE COMPOSITION-SERIES OF A GROUP.

CHAPTER VIII.

ON SUBSTITUTION GROUPS: TRANSITIVE AND INTRANSITIVE GROUPS.

CHAPTER IX.

ON SUBSTITUTION GROUPS: PRIMITIVE AND IMPRIMITIVE GROUPS.

CHAPTER X.

ON SUBSTITUTION GROUPS: TRANSITIVITY AND PRIMITIVITY: (CONCLUDING PROPERTIES).

CHAPTER XI.

ON THE ISOMORPHISM OF A GROUP WITH ITSELF.

CHAPTER XII.

ON THE GRAPHICAL REPRESENTATION OF A GROUP.

CHAPTER XIII.

ON THE GRAPHICAL REPRESENTATION OF GROUPS: GROUPS OF GENUS ZERO AND UNITY: CAYLEY'S COLOUR GROUPS.

CHAPTER XIV.

ON THE LINEAR GROUP.

CHAPTER XV.

ON SOLUBLE AND COMPOSITE GROUPS.

CHAPTER I.

ON SUBSTITUTIONS.

1. AMONG the various notations used in the following pages, there is one of such frequent recurrence that a certain readiness in its use is very desirable in dealing with the subject of this treatise. We therefore propose to devote a preliminary chapter to explaining it in some detail.

2. Let $a_1, a_2, \ldots, a_n$ be a set of n distinct letters. The operation of replacing each letter of the set by another, which may be the same letter or a different one, when carried out under the condition that no two letters are replaced by one and the same letter, is called a *substitution* performed on the n letters. Such a substitution will change any given arrangement

$$a_1,\ a_2,\ \ldots,\ a_n$$

of the n letters into a definite new arrangement

$$b_1,\ b_2,\ \ldots,\ b_n$$

of the same n letters.

3. One obvious form in which to write the substitution is

$$\begin{pmatrix} a_1, a_2, \ldots, a_n \\ b_1, b_2, \ldots, b_n \end{pmatrix}$$

thereby indicating that each letter in the upper line is to be replaced by the letter standing under it in the lower. The disadvantage of this form is its unnecessary complexity, each of the n letters occurring twice in the expression for the substitution; by the following process, the expression of the substitution may be materially simplified.

Let p be any one of the n letters, and q the letter in the lower line standing under p in the upper. Suppose now that r is the letter in the lower line that stands under q in the upper, and so on. Since the number of letters is finite, we must arrive at last at a letter s in the upper line under which p stands. If the set of n letters is not thus exhausted, take any letter p' in the upper line, which has not yet occurred, and let q', r', ...

follow it as q, r, ... followed p, till we arrive at s' in the upper line with p' standing under it. If the set of n letters is still not exhausted, repeat the process, starting with a letter p'' which has not yet occurred. Since the number of letters is finite, we must in this way at last exhaust them; and the n letters are thus distributed into a number of sets

$$\begin{array}{lllll} p, & q, & r, & \ldots, & s; \\ p', & q', & r', & \ldots, & s'; \\ p'', & q'', & r'', & \ldots, & s''; \\ \ldots & \ldots & \ldots & \ldots & ; \end{array}$$

such that the substitution replaces each letter of a set by the one following it in that set, the last letter of each set being replaced by the first of the same set.

If now we represent by the symbol

$$(pqr \ldots s)$$

the operation of replacing p by q, q by r, ..., and s by p, the substitution will be completely represented by the symbol

$$(pqr \ldots s)(p'q'r' \ldots s')(p''q''r'' \ldots s'') \ldots .$$

The advantage of this mode of expressing the substitution is that each of the letters occurs only once in the symbol.

4. The separate components of the above symbol, such as $(pqr \ldots s)$ are called the *cycles* of the substitution. In particular cases, one or more of the cycles may contain a single letter; when this happens, the letters so occurring singly are unaltered by the substitution. The brackets enclosing single letters may clearly be omitted without risk of ambiguity, as also may the unaltered letters themselves. Thus the substitution

$$\begin{pmatrix} a, b, c, d, e \\ c, b, d, a, e \end{pmatrix}$$

may be written $(acd)(b)(e)$, or $(acd)be$, or simply (acd). If for any reason it were desirable to indicate that substitutions of the five letters a, b, c, d, e were under consideration, the second of these three forms would be used.

5. The form thus obtained for a substitution is not unique. The symbol $(qr \ldots sp)$ clearly represents the same substitution as $(pqr \ldots s)$, if the letters that occur between r and s in the two symbols are the same and occur in the same order; so that, as regards the letters inside the bracket, any one may be chosen to stand first so long as the cyclical order is preserved unchanged.

Moreover the order in which the brackets are arranged is clearly immaterial, since the operation denoted by any one bracket has no effect on the letters contained in the other brackets. This latter property is characteristic of the particular expression that has been obtained for a substitution; it depends upon the fact that the expression contains each of the letters once only.

6. When we proceed to consider the effect of performing two or more substitutions successively, it is seen at once that the order in which the substitutions are carried out in general affects the result. Thus to give a very simple instance, the substitution (ab) followed by (ac) changes a into b, since b is unaltered by the second substitution. Again, (ab) changes b into a and (ac) changes a into c, so that the two substitutions performed successively change b into c. Lastly, (ab) does not affect c and (ac) changes c into a. Hence the two substitutions performed successively change a into b, b into c, c into a, and affect no other symbols. The result of the two substitutions performed successively is therefore equivalent to the substitution (abc); and it may be similarly shewn that (ac) followed by (ab) gives (acb) as the resulting substitution. To avoid ambiguity it is therefore necessary to assign, once for all, the meaning to be attached to such a symbol as $s_1 s_2$, where s_1 and s_2 are the symbols of two given substitutions. We shall always understand by the symbol $s_1 s_2$ *the result of carrying out first the substitution* s_1 *and then the substitution* s_2. Thus the two simple examples given above may be expressed in the form

$$(ab)(ac) = (abc),$$
$$(ac)(ab) = (acb),$$

the sign of equality being used to represent that the substitutions are equivalent to each other.

If now

$$s_1 s_2 = s_4 \quad \text{and} \quad s_2 s_3 = s_5,$$

the symbol $s_1 s_2 s_3$ may be regarded as the substitution s_4 followed by s_3 or as s_1 followed by s_5. But if s_1 changes *any* letter a into b, while s_2 changes b into c and s_3 changes c into d, then s_4 changes a into c and s_5 changes b into d. Hence $s_4 s_3$ and $s_1 s_5$ both change a into d; and therefore, a being any letter operated upon by the substitutions,

$$s_4 s_3 = s_1 s_5.$$

Hence the meaning of the symbol $s_1 s_2 s_3$ is definite; it depends only on the component substitutions s_1, s_2, s_3 and their sequence, and it is independent of the way in which they are associated when their sequence is assigned. And the same clearly holds for the symbol representing the successive performance of any number of substitutions. To avoid circumlocution, it is convenient to speak of the substitution $s_1 s_2 \ldots s_n$ as the *product* of the substitutions s_1, s_2, $\ldots$, s_n in the sequence given. The product of a number of substitutions, thus defined, always obeys the associative law but does not in general obey the commutative law of algebraical multiplication.

7. The substitution which replaces every symbol by itself is called the *identical substitution.* The *inverse* of a given substitution is that substitution which, when performed after the given substitution, gives as result the identical substitution. Let s_{-1} be the substitution inverse to s, so that, if

$$s = \begin{pmatrix} a_1, a_2, \ldots, a_n \\ b_1, b_2, \ldots, b_n \end{pmatrix},$$

then

$$s_{-1} = \begin{pmatrix} b_1, b_2, \ldots, b_n \\ a_1, a_2, \ldots, a_n \end{pmatrix}.$$

Let s_0 denote the identical substitution which can be represented by

$$\begin{pmatrix} a_1, a_2, \ldots, a_n \\ a_1, a_2, \ldots, a_n \end{pmatrix}.$$

Then

$$s s_{-1} = s_0 \quad \text{and} \quad s_{-1} s = s_0,$$

so that s is the substitution inverse to s_{-1}.

Now if

$$ts = t's,$$

then

$$tss_{-1} = t'ss_{-1},$$

or

$$ts_0 = t's_0.$$

But ts_0 is the same substitution as t, since s_0 produces no change; and therefore

$$t = t'.$$

In exactly the same way, it may be shewn that the relation

$$st = st'$$

involves

$$t = t'.$$

8. The result of performing r times in succession the same substitution s is represented symbolically by s^r. Since, as has been seen, products of substitutions obey the associative law of multiplication, it follows that

$$s^\mu s^\nu = s^{\mu+\nu} = s^\nu s^\mu.$$

Now since there are only a finite number of distinct substitutions that can be performed on a given finite set of symbols, the series of substitutions s, s^2, s^3, ... cannot be all distinct. Suppose that s^{m+1} is the first of the series which is the same as s, so that

$$s^{m+1} = s.$$

Then

$$s^m ss_{-1} = ss_{-1},$$

or

$$s^m = s_0.$$

There is no index μ smaller than m for which this relation holds. For if

$$s^{\mu} = s_0,$$

then

$$s^{\mu+1} = ss_0 = s,$$

contrary to the supposition that s^{m+1} is the first of the series which is the same as s.

Moreover the $m-1$ substitutions s, s^2, ..., s^{m-1} must be all distinct. For if

$$s^{\mu} = s^{\nu}, \quad \nu < \mu < m,$$

then

$$s^{\mu-\nu} s^{\nu} (s^{\nu})_{-1} = s^{\nu} (s^{\nu})_{-1},$$

or

$$s^{\mu-\nu} = s_0,$$

which has just been shewn to be impossible.

The number m is called the *order* of the substitution s. In connection with the order of a substitution, two properties are to be noted. First, if

$$s^n = s_0,$$

it may be shewn at once that n is a multiple of m the order of s; and secondly, if

$$s^{\alpha} = s^{\beta},$$

then

$$\alpha - \beta \equiv 0 \text{ (mod. } m).$$

If now the equation

$$s^{\mu+\nu} = s^{\mu} s^{\nu}$$

be assumed to hold, when either or both of the integers μ and ν is a negative integer, a definite meaning is obtained for the symbol s^{-r}, implying the negative power of a substitution; and a definite meaning is also obtained for s^0. For

$$s^{\mu} s^{-\nu} = s^{\mu-\nu} = s^{\mu-\nu} s^{\nu} (s^{\nu})_{-1} = s^{\mu} (s^{\nu})_{-1},$$

so that

$$s^{-\nu} = (s^{\nu})_{-1}.$$

Similarly it can be shewn that

$$s^0 = s_0.$$

Since every power of s_0 is the same as s_0, and since wherever s_0 occurs in the symbol $s_1 s_2 \dots s_n$ of a compound substitution it may be omitted without affecting the result, it is clear that no ambiguity will result from replacing s_0 everywhere by 1; in other words, we may use 1 to represent the identical substitution which leaves every letter unchanged. But when this is done, it must of course be remembered that the equation

$$s^m = 1$$

is not a reducible algebraical equation, which is capable of being written in the form

$$(s-1)(s^{m-1} + s^{m-2} + \dots + 1) = 0.$$

Indeed the symbol $s + s'$, where s and s' are any two substitutions, has no meaning.

9. If the cycles of a substitution

$$s = (pqr \dots s)(p'q' \dots s')(p''q'' \dots s'') \dots$$

contain m, m', m'', ... letters respectively, and if

$$s^{\mu} = 1,$$

μ must be a common multiple of m, m', m'', ... For s^{μ} changes p into a letter μ places from it in the cyclical set p, q, r, ..., s; and therefore, if it changes p into itself, μ must be a multiple of m. In the same way, it must be a multiple of m', m'', ... Hence the order of s is the least common multiple of m, m', m'', ...

In particular, when a substitution consists of a single cycle, its order is equal to the number of letters which it interchanges. Such a substitution is called a *circular substitution*.

A substitution, all of whose cycles contain the same number of letters, is said to be *regular* in the letters which it interchanges; the order of such a substitution is clearly equal to the number of letters in one of its cycles.

10. Two substitutions, which contain the same number of cycles and the same number of letters in corresponding cycles, are called *similar.* If s, s' are similar substitutions, so also clearly are s^r, s'^r; and the orders of s and s' are the same.

Let now

$$s = (a_p a_q \dots a_s)(a_{p'} a_{q'} \dots a_{s'}) \dots$$

and

$$t = \begin{pmatrix} a_1, a_2, \dots, a_n \\ b_1, b_2, \dots, b_n \end{pmatrix}$$

be any two substitutions. Then

$$\begin{aligned} t^{-1}st &= \begin{pmatrix} b_1, b_2, \dots, b_n \\ a_1, a_2, \dots, a_n \end{pmatrix} (a_p a_q \dots a_s)(a_{p'} a_{q'} \dots a_{s'}) \dots \begin{pmatrix} a_1, a_2, \dots, a_n \\ b_1, b_2, \dots, b_n \end{pmatrix} \\ &= (b_p b_q \dots b_s)(b_{p'} b_{q'} \dots b_{s'}) \dots, \end{aligned}$$

the latter form of the substitution being obtained by actually carrying out the component substitutions of the earlier form. Hence s and $t^{-1}st$ are similar substitutions.

Since

$$s_2 s_1 = s_1^{-1} s_1 s_2 s_1,$$

it follows that $s_1 s_2$ and $s_2 s_1$ are similar substitutions and therefore that they are of the same order. Similarly it may be shewn that $s_1 s_2 s_3 \dots s_n$, $s_2 s_3 \dots s_n s_1$, $\dots$, $s_n s_1 \dots s_2 s_3$ are all similar substitutions.

It may happen in particular cases that s and $t^{-1}st$ are the same substitution. When this is so, t and s are *permutable*, that is, st and ts are equivalent to one another; for if

$$s = t^{-1}st,$$

then

$$ts = st.$$

This will certainly be the case when none of the symbols that are interchanged by t are altered by s; but it may happen when s and t operate on the same symbols. Thus if

$$s = (ab)(cd), \quad t = (ac)(bd),$$

then

$$st = (ad)(bc) = ts.$$

Ex. 1. Shew that every regular substitution is some power of a circular substitution.

Ex. 2. If s, s' are permutable regular substitutions of the same mn letters of orders m and n, these numbers being relatively prime, shew that ss' is a circular substitution in the mn letters.

Ex. 3*. If

$$\begin{aligned} s &= (123)(456)(789), \\ s_1 &= (147)(258)(369), \\ s_2 &= \quad\quad\;\; (456)(798), \end{aligned}$$

shew that s is permutable with both s_1 and s_2, and that it can be formed by a combination of s_1 and s_2.

Ex. 4. Shew that the only substitutions of n given letters which are permutable with a circular substitution of the n letters are the powers of the circular substitution.

Ex. 5. Determine all the substitutions of the ten symbols involved in

$$s = (abcde)(\alpha\beta\gamma\delta\epsilon)$$

which are permutable with s.

The determination of all the substitutions which are permutable with a given substitution will form the subject of investigation in Chapter X.

*It is often convenient to use digits rather than letters for the purpose of illustration.

11. A circular substitution of order two is called a *transposition.* It may be easily verified that

$$(pqr\dots s) = (pq)(pr)\dots(ps),$$

so that every circular substitution can be represented as a product of transpositions; and thence, since every substitution is the product of a number of circular substitutions, every substitution can be represented as a product of transpositions. It must be remembered, however, that, in general, when a substitution is represented in this way, some of the letters will occur more than once in the symbol, so that the order in which the constituent transpositions occur is essential. There is thus a fundamental difference from the case when the symbol of a substitution is the product of circular substitutions, no two of which contain a common letter.

Since

$$(p'q') = (pp')(pq')(pp'),$$

every transposition, and therefore every substitution of n letters, can be expressed in terms of the $n-1$ transpositions

$$(a_1a_2),\ (a_1a_3),\ \dots,\ (a_1a_n).$$

The number of different ways in which a given substitution may be represented as a product of transpositions is evidently unlimited; but it may be shewn that, however the representation is effected, the number of transpositions is either always even or always odd. To prove this, it is sufficient to consider the effect of a transposition on the square root of the discriminant of the n letters, which may be written

$$D = \prod_{r=1}^{r=n-1} \left\{ \prod_{s=r+1}^{s=n} (a_r - a_s) \right\}.$$

The transposition (a_ra_s) changes the sign of the factor $a_r - a_s$. When q is less than either r or s, the transposition interchanges the factors $a_q - a_r$ and $a_q - a_s$; and when q is greater than either r or s, it interchanges the factors $a_r - a_q$ and $a_s - a_q$. When q lies between r and s, the pair of factors

$a_r - a_q$ and $a_q - a_s$ are interchanged and are both changed in sign. Hence the effect of the single transposition on D is to change its sign. Since any substitution can be expressed as the product of a number of transpositions, the effect of any substitution on D must be either to leave it unaltered or to change its sign. If a substitution leaves D unaltered it must, when expressed as a product of transpositions in any way, contain an even number of transpositions; and if it changes the sign of D, every representation of it, as a product of transpositions, must contain an odd number of transpositions. Hence no substitution is capable of being expressed both by an even and by an odd number of transpositions.

A substitution is spoken of as *odd* or *even*, according as the transpositions which enter into its representation are odd or even in number.

Further, an even substitution can always be represented as a product of circular substitutions of order three. For any even substitution of n letters can be represented as the product of an even number of the $n - 1$ transpositions

$$(a_1a_2),\ (a_1a_3),\ \ldots,\ (a_1a_n),$$

in appropriate sequence and with the proper number of occurrences; and the product of any consecutive pair of these $(a_1a_r)(a_1a_s)$ is the circular substitution $(a_1a_ra_s)$.

Now

$$\begin{aligned}(a_1a_2a_s)(a_1a_2a_r)(a_1a_2a_s)^2 &= (a_1a_2a_s)(a_1a_2a_r)(a_1a_sa_2)\\ &= (a_1a_ra_s),\end{aligned}$$

so that every circular substitution of order three displacing a_1, and therefore every even substitution of n letters, can be expressed in terms of the $n - 2$ substitutions

$$(a_1a_2a_3),\ (a_1a_2a_4),\ \ldots,\ (a_1a_2a_n)$$

and their powers.

Ex. 1. Shew that every even substitution of n letters can be expressed in terms of

$$(a_1a_2a_3),\ (a_1a_4a_5),\ \ldots,\ (a_1a_{n-1}a_n),$$

when n is odd; and in terms of

$$(a_1 a_2 a_3),\ (a_1 a_4 a_5),\ \ldots,\ (a_1 a_{n-2} a_{n-1}),\ (a_1 a_2 a_n),$$

when n is even.

Ex. 2. If $n+1$ is odd, shew that every even substitution of $mn+1$ letters can be expressed in terms of

$$(a_1 a_2 \ldots a_{n+1}),\ (a_1 a_{n+2} \ldots a_{2n+1}),\ \ldots,\ (a_1 a_{(m-1)n+2} \ldots a_{mn+1});$$

and if $n+1$ is even, that every substitution of $mn+1$ letters can be expressed in terms of this set of m circular substitutions.

CHAPTER II.

THE DEFINITION OF A GROUP.

12. IN the present chapter we shall enter on our main subject and we shall begin with definitions, explanations and examples of what is meant by a group.

Definition. Let

$$A,\ B,\ C,\ \ldots$$

represent a set of operations, which can be performed on the same object or set of objects. Suppose this set of operations has the following characteristics.

(α) The operations of the set are all distinct, so that no two of them produce the same change in every possible application.

(β) The result of performing successively any number of operations of the set, say A, B, ..., K, is another definite operation of the set, which depends only on the component operations and the sequence in which they are carried out, and not on the way in which they may be regarded as associated. Thus A followed by B and B followed by C are operations of the set, say D and E; and D followed by C is the same operation as A followed by E.

(γ) A being any operation of the set, there is always another operation A_{-1} belonging to the set, such that A followed by A_{-1} produces no change in any object.

The operation A_{-1} is called the inverse of A.

The set of operations is then said to form a *Group*.

From the definition of the inverse of A given in (γ), it follows directly that A is the inverse of A_{-1}. For if A changes any object Ω into Ω', A_{-1} must change Ω' into Ω. Hence A_{-1} followed by A leaves Ω', and therefore every object, unchanged.

The operation resulting from the successive performance of the operations A, B, ..., K in the sequence given is denoted by the symbol $AB \ldots K$; and if Ω is any object on which the operations may be performed, the result of carrying out this compound operation on Ω is denoted by $\Omega \cdot AB \ldots K$.

If the component operations are all the same, say A, and r in number, the abbreviation A^r will be used for the resultant operation, and it will be called the rth *power* of A.

Definition. Two operations, A and B, are said to be *permutable* when AB and BA are the same operation.

13. If AB and AC are the same operation, so also are $A_{-1}AB$ and $A_{-1}AC$. But the operation $A_{-1}A$ produces no change in any object and therefore $A_{-1}AB$ and B, producing the same change in every object, are the same operation. Hence B and C are the same operation.

This is expressed symbolically by saying that, if

$$AB = AC,$$

then

$$B = C;$$

the sign of equality being used to imply that the symbols represent the same operation.

In a similar way, if

$$BA = CA,$$

it follows that

$$B = C.$$

From conditions (β) and (γ), AA_{-1} must be a definite operation of the group. This operation, by definition, produces no change in any possible object, and it must, by condition (α), be unique. It is called the *identical* operation. If it is represented by A_0 and if A be any other operation, then

$$A_0A = A = AA_0,$$

and for every integer r,

$$A_0^r = A_0.$$

Hence A_0 may, without ambiguity, be replaced by 1, wherever it occurs.

14. The number of distinct operations contained in a group may be either finite or infinite. When the number is infinite, the group may contain operations which produce an infinitesimal change in every possible object or operand.

Thus the totality of distinct displacements of a rigid body evidently forms a group, for they satisfy conditions (α), (β) and (γ) of the definition. Moreover this group contains operations of the kind in question, namely infinitesimal twists; and each operation of the group can be constructed by the continual repetition of a suitably chosen infinitesimal twist.

Next, the set of translations, that arise by shifting a cube parallel to its edges through distances which are any multiples of an edge, forms a group containing an infinite number of operations; but this group contains no operation which effects an infinitesimal change in the position of the cube.

As a third example, consider the set of displacements by which a complete right circular cone is brought to coincidence with itself. It consists of rotations through any angle about the axis of the cone, and rotations through two right angles about any line through the vertex at right angles to the axis. Once again this set of displacements satisfies the conditions (α), (β) and (γ) of the definition and forms a group.

This last group contains infinitesimal operations, namely rotations round the axis through an infinitesimal angle; and every finite rotation round the axis can be formed by the continued repetition of an infinitesimal rotation. There is however in this case no infinitesimal displacement of the group by whose continued repetition a rotation through two right angles about a line through the vertex at right angles to the axis can be constructed. Of these three groups with an infinite number of operations, the first is said to be a *continuous* group, the second a *discontinuous* group, and the third a *mixed* group.

Continuous groups and mixed groups lie entirely outside the plan of the present treatise; and though, later on, some of the properties of discontinuous groups with an infinite number of operations will be considered, such groups will be approached from a point of view suggested by the treatment of groups containing a finite number of operations. It is not therefore necessary here to deal in detail with the classification of infinite groups which is indicated by the three examples given above; and we pass on at once to the case of groups which contain a finite number only of distinct operations.

15. Definition. If the number of distinct operations contained in a group be finite, the number is called the *order* of the group.

Let S be an operation of a group of finite order N. Then the infinite series of operations

$$S,\ S^2,\ S^3,\ \ldots$$

must all be contained in the group, and therefore a finite number of them only can be distinct. If S^{m+1} is the first of the series which is the same as S, and if S_{-1} is the operation inverse to S, then

$$S^{m+1}S_{-1} = SS_{-1} = 1,$$

or

$$S^m = 1.$$

Exactly as in § 8, it may be shewn that, if

$$S^\mu = 1,$$

μ must be a multiple of m, and that the operations $S,\ S^2,\ \ldots,\ S^{m-1}$ are all distinct.

Since the group contains only N distinct operations, m must be equal to or less than N. It will be seen later that, if m is less than N, it must be a factor of N.

The integer m is called the *order* of the operation S. The order m' of the operation S^x is the least integer for which

$$S^{xm'} = 1,$$

that is, for which

$$xm' \equiv 0 \text{ (mod. } m).$$

Hence, if g is the greatest common factor of x and m,

$$m' = \frac{m}{g};$$

and, if m is prime, all the powers of S, whose indices are less than m, are of order m.

Since

$$S^x S^{m-x} = S^m = 1, \quad (x < m),$$

and

$$S^x (S^x)_{-1} = 1,$$

it follows that

$$(S^x)_{-1} = S^{m-x}.$$

If now a meaning be attached to S^{-x}, by assuming that the equation

$$S^{x+y} = S^x S^y$$

holds when either x or y is a negative integer, then

$$S^{m-x} = S^m S^{-x} = S^{-x},$$

and

$$(S^x)_{-1} = S^{-x},$$

so that S^{-x} denotes the inverse of the operation S^x.

Ex. If S_a, S_b, ..., S_c, S_d are operations of a group, shew that the operation inverse to $S_a^\alpha S_b^\beta \dots S_c^\gamma S_d^\delta$ is $S_d^{-\delta} S_c^{-\gamma} \dots S_b^{-\beta} S_a^{-\alpha}$.

16. If

$$1, S_1, S_2, \dots, S_{N-1}$$

are the N operations of a group of order N, the set of N operations

$$S_r, S_r S_1, S_r S_2, \dots, S_r S_{N-1}$$

are (§ 13) all distinct; and their number is equal to the order of the group. Hence every operation of the group occurs once and only once in this set.

Similarly every operation of the group occurs once and only once in the set

$$S_r, S_1 S_r, S_2 S_r, \dots, S_{N-1} S_r.$$

Every operation of the group can therefore be represented as the product of two operations of the group, and either the first factor or the second factor can be chosen at will.

A relation of the form

$$S_p = S_q S_r$$

between three operations of the group will not in general involve any necessary relation between the order of S_p and the orders of S_q and S_r. If however the two latter are permutable, the relation requires that, for all values of x,

$$S_p^x = S_q^x S_r^x;$$

and in that case the order of S_p is the least common multiple of the orders of S_q and S_r.

Suppose now that S, an operation of the group, is of order mn, where m and n are relatively prime. Then we may shew that, of the various ways in which S may be represented as the product of two operations of the group, there is just one in which the operations are permutable and of orders m and n respectively.

Thus let

$$S^n = M,$$

and

$$S^m = N,$$

so that M, N are operations of orders m and n. Since S^m and S^n are permutable, so also are M and N, and powers of M and N.

If x_0, y_0 are integers satisfying the equation

$$xn + ym = 1,$$

every other integral solution is given by

$$x = x_0 + tm, \quad y = y_0 - tn,$$

where t is an integer.

Now

$$M^x N^y = S^{xn+ym} = S;$$

and since x and m are relatively prime, as also are y and n, M^x and N^y are permutable operations of orders m and n, so that S is expressed in the desired form.

Moreover, it is the only expression of this form; for let

$$S = M_1 N_1,$$

where M_1 and N_1 are permutable and of orders m and n.

Then $S^n = M_1^n$, since $N_1^n = 1$.

Hence

$$M_1^n = M,$$

or

$$M_1^{xn} = M^x,$$

or

$$M_1^{1-ym} = M^x.$$

But $M_1^m = 1$, and therefore $M_1^{-ym} = 1$; hence

$$M_1 = M^x.$$

In the same way it is shewn that N_1 is the same as N^y. The representation of S in the desired form is therefore unique.

17. Two given operations of a group successively performed give rise to a third operation of the group which, when the operations are of known concrete form, may be determined by actually carrying out the two given operations. Thus the set of finite rotations, which bring a regular solid to coincidence with itself, evidently form a group; and it is a purely geometrical problem to determine that particular rotation of the group which arises from the successive performance of two given rotations of the group.

When the operations are represented by symbols, the relation in question is represented by an equation of the form

$$AB = C;$$

but the equation indicates nothing of the nature of the actual operations. Now it may happen, when the operations of two groups of equal order are represented by symbols,

$$\text{(i)}\quad 1,\ A,\ B,\ C,\dots$$
$$\text{(ii)}\quad 1,\ A',\ B',\ C',\dots$$

that, to every relation of the form

$$AB = C$$

between operations of the first group, there corresponds the relation

$$A'B' = C'$$

between operations of the second group. In such a case, although the nature of the actual operations in the first group may be entirely different from the nature of those in the second, the laws according to which the operations of each group combine among themselves are identical. The following series of groups of operations, of order six, will at once illustrate the possibility just mentioned, and will serve as concrete examples to familiarize the reader with the conception of a group of operations.

I. *Group of inversions.* Let P, Q, R be three circles with a common radical axis and let each pair of them intersect at an angle $\frac{1}{3}\pi$. Denote the operations of inversion with respect to P, Q, R by C, D, E; and denote successive inversions at P, R and at P, Q by A and B. The object of operation may be any point in the plane of the circles, except the two common points in which they intersect. Then it is easy to verify, from the geometrical properties of inversion, that the operations

$$1,\ A,\ B,\ C,\ D,\ E$$

are all distinct, and that they form a group. For instance, DE represents successive inversions at Q and R. But successive inversions at Q and R produce the same displacement of points as successive inversions at P and Q, and therefore

$$DE = B.$$

II. *Group of rotations.* Let POP', QOQ', ROR' be three concurrent lines in a plane such that each of the angles POQ and QOR is $\frac{1}{3}\pi$, and let IOI' be a perpendicular to their plane. Denote by A a rotation round II' through $\frac{2}{3}\pi$ bringing PP' to RR'; and by B a rotation round II' through $\frac{4}{3}\pi$ bringing PP' to $Q'Q$. Denote also by C, D, E rotations through two right angles round PP', QQ', RR'. The object of the rotations may be any point or set of points in space. Then it may again be verified, by simple geometrical considerations, that the operations

$$1,\ A,\ B,\ C,\ D,\ E$$

are distinct and that they form a group.

III. *Group of linear transformations of a single variable.* The operation of replacing x by a given function $f(x)$ of itself is sometimes represented by the symbol $(x, f(x))$. With this notation, if

$$A = \left(x, \frac{1}{1-x}\right), \quad B = \left(x, \frac{x-1}{x}\right), \quad C = \left(x, \frac{1}{x}\right), \quad D = (x, 1-x),$$
$$E = \left(x, \frac{x}{x-1}\right), \quad 1 = (x, x),$$

it may again be verified without difficulty that these six operations form a group.

IV. *Group of linear transformations of two variables.* With a similar notation, the six operations

$$A = \left(x, \frac{y}{x};\ y, \frac{1}{x}\right), \quad B = \left(x, \frac{1}{y};\ y, \frac{x}{y}\right), \quad C = (x, y;\ y, x),$$
$$D = \left(x, \frac{1}{x};\ y, \frac{y}{x}\right), \quad E = \left(x, \frac{x}{y};\ y, \frac{1}{y}\right), \quad 1 = (x, y;\ x, y)$$

form a group.

V. *Group of linear transformations to a prime modulus.*
The six operations defined by

$$A = (x, x+1), \quad B = (x, x+2), \quad C = (x, 2x),$$
$$D = (x, 2x+2), \quad E = (x, 2x+1), \quad 1 = (x, x),$$

where each transformation is taken to modulus 3, form a group.

VI. *Group of substitutions of* 3 *symbols.* The six substitutions

$$1, \quad A = (xyz), \quad B = (xzy), \quad C = x(yz), \quad D = y(zx), \quad E = z(xy)$$

are the only substitutions that can be formed with three symbols; they must therefore form a group.

VII. *Group of substitutions of* 6 *symbols.* The substitutions

$$1, \quad A = (xyz)(abc), \quad B = (xzy)(acb), \quad C = (xa)(yc)(zb),$$
$$D = (xb)(ya)(zc), \quad E = (xc)(yb)(za)$$

may be verified to form a group.

VIII. *Group of substitutions of* 6 *symbols.* The substitutions

$$1, \quad A = (xaybzc), \quad B = (xyz)(abc), \quad C = (xb)(yc)(za),$$
$$D = (xzy)(acb), \quad E = (xczbya)$$

form a group.

The operations in the first seven of these groups, as well as the objects of operation, are quite different from one group to another; but it may be shewn that the laws according to which the operations, denoted by the same letters in the different groups, combine together are identical for all seven. There is no difficulty in verifying that in each instance

$$A^3 = 1, \quad C^2 = 1, \quad B = A^2, \quad D = AC = CA^2, \quad E = A^2C = CA;$$

and from these relations the complete system, according to which the six operations in each of the seven groups combine together, may be at once constructed. This is given by the following multiplication table, where the left-hand vertical column gives the first factor and the top horizontal line the second factor in each product; thus the table is to be read $A1 = A$,

$AB = 1$, $AC = D$, and so on.

	1	A	B	C	D	E
1	1	A	B	C	D	E
A	A	B	1	D	E	C
B	B	1	A	E	C	D
C	C	E	D	1	B	A
D	D	C	E	A	1	B
E	E	D	C	B	A	1

But, though the operations of the seventh and eighth groups are of the same nature and though the operands are identical, the laws according to which the six operations combine together are quite distinct for the two groups. Thus, for the last group, it may be shewn that

$$B = A^2, \quad C = A^3, \quad D = A^4, \quad E = A^5, \quad A^6 = 1,$$

so that the operations of this group may, in fact, be represented by

$$1,\ A,\ A^2,\ A^3,\ A^4,\ A^5.$$

18. If we pay no attention to the nature of the actual operations and operands, and consider only the number of the former and the laws according to which they combine, the first seven groups of the preceding paragraph are identical with each other. From this point of view a group, abstractly considered, is completely defined by its multiplication table; and, conversely, the multiplication table must implicitly contain all properties of the group which are independent of any special mode of representation.

It is of course obvious that this table cannot be arbitrarily constructed. Thus, if

$$AB = P \quad \text{and} \quad BC = Q,$$

the entry in the table for PC must be the same as that for AQ. Except in the very simplest cases, the attempt to form a consistent multiplication table, merely by trial, would be most laborious.

The very existence of the table shews that the symbols denoting the different operations of the group are not all independent of each other; and since the number of symbols is finite, it follows that there must exist a set of symbols S_1, S_2, ..., S_n no one of which can be expressed in terms of the remainder, while every operation of the group is expressible in terms of the set. Such a set is called a set of *fundamental* or *generating* operations of the group. Moreover though no one of the generating operations can be expressed in terms of the remainder, there must be relations of the general form

$$S_m^a S_n^b \dots S_p^c = 1$$

among them, as otherwise the group would be of infinite order; and the number of these relations, which are independent of one another, must be finite. Among them there necessarily occur the relations

$$S_1^{a_1} = 1, \quad S_2^{a_2} = 1, \quad \dots, \quad S_n^{a_n} = 1$$

giving the orders of the fundamental operations.

We thus arrive at a virtually new conception of a group; it can be regarded as arising from a finite number of fundamental operations connected by a finite number of independent relations. But it is to be noted that there is no reason for supposing that such an origin for a group is unique; indeed, in general, it is not so. Thus there is no difficulty in verifying that the group, whose multiplication table is given in § 17, is completely specified either by the system of relations

$$A^3 = 1, \quad C^2 = 1, \quad (AC)^2 = 1,$$

or by the system

$$C^2 = 1, \quad D^2 = 1, \quad (CD)^3 = 1.$$

In other words, it may be generated by two operations of orders 2 and 3, or by two operations of order 2. So also the last group of § 17 is specified either by

$$A^6 = 1,$$

or by

$$B^3 = 1, \quad C^2 = 1, \quad BC = CB.$$

19. Definition. Let G and G' be two groups of equal order. If a correspondence can be established between the operations of G and G', so that to every operation of G there corresponds a single operation of G' and to every operation of G' there corresponds a single operation of G, while to the product AB of any two operations of G there corresponds the product $A'B'$ of the two corresponding operations of G', the groups G and G' are said to be *simply isomorphic**.

Two simply isomorphic groups are, abstractly considered, identical. In discussing the properties of groups, some definite mode of representation is, in general, indispensable; and as long as we are dealing with the properties of a group *per se*, and not with properties which depend on the form of representation, the group may, if convenient, be replaced by any group which is simply isomorphic with it. For the discussion of such properties, it would be most natural to suppose the group given either by its multiplication table or by its fundamental operations and the relations connecting them; and as far as possible we shall follow this course. Unfortunately, however, these purely abstract modes of representing a group are by no means the easiest to deal with. It thus becomes an important question to determine as far as possible what different concrete forms of representation any particular group may be capable of; and we shall accordingly end the present chapter with a demonstration of the following general theorem bearing on this question.

20. THEOREM. *Every group of finite order N is capable of representation as a group of substitutions of N symbols*†.

Let

$$1,\ S_1,\ S_2,\ \ldots,\ S_i,\ \ldots,\ S_{N-1}$$

be the N operations of the group; and form the complete multiplication

---

*We shall sometimes use the phrase that two groups are of the same *type* to denote that they are simply isomorphic.

†Dyck, "Gruppentheoretische Studien," *Math. Ann.*, Vol. XX (1882), p. 30.

table

$$\begin{array}{ccccccc}
1, & S_1, & S_2, & \ldots, & S_i, & \ldots, & S_{N-1}, \\
S_1, & S_1^2, & S_2S_1, & \ldots, & S_iS_1, & \ldots, & S_{N-1}S_1, \\
\cdots & \cdots & \cdots & \cdots & \cdots & \cdots & \cdots \\
S_i, & S_1S_i, & S_2S_i, & \ldots, & S_i^2, & \ldots, & S_{N-1}S_i, \\
\cdots & \cdots & \cdots & \cdots & \cdots & \cdots & \cdots \\
S_{N-1}, & S_1S_{N-1}, & S_2S_{N-1}, & \ldots, & S_iS_{N-1}, & \ldots, & S_{N-1}^2,
\end{array}$$

that results from multiplying the symbols in the first horizontal line by $1, S_1, S_2, \ldots, S_{N-1}$ in order. Each horizontal line in the table so obtained contains the same N symbols as the original line, but any given symbol occupies a different place in each line; for the supposition that the two symbols S_iS_p and S_iS_q were identical would involve

$$S_p = S_q,$$

which is not true.

It thus appears that the first line in the table, taken with the $(i+1)$th line, defines a substitution

$$\begin{pmatrix} 1, & S_1, & S_2, \ldots, & S_{N-1} \\ S_i, & S_1S_i, & S_2S_i, \ldots, & S_{N-1}S_i \end{pmatrix}$$

performed on the N symbols $1, S_1, S_2, \ldots, S_{N-1}$; and by taking the first line with each of the others, including itself, a set of N substitutions performed on these N symbols is obtained. Now this set of substitutions forms a group simply isomorphic to the given group. For if

$$S_pS_i = S_q \quad \text{and} \quad S_qS_j = S_r,$$

the substitutions

$$\begin{pmatrix} 1, & S_1, & S_2, \ldots, & S_{N-1} \\ S_i, & S_1S_i, & S_2S_i, \ldots, & S_{N-1}S_i \end{pmatrix}$$

and

$$\begin{pmatrix} 1, & S_1, & S_2, \ldots, & S_{N-1} \\ S_j, & S_1S_j, & S_2S_j, \ldots, & S_{N-1}S_j \end{pmatrix},$$

successively performed, change S_p into S_r.

But

$$S_p S_i S_j = S_r,$$

and therefore the product of the two substitutions, in the order given, is the substitution

$$\begin{pmatrix} 1\ , & S_1\ , & S_2\ , \ldots, & S_{N-1} \\ S_iS_j, & S_1S_iS_j, & S_2S_iS_j, \ldots, & S_{N-1}S_iS_j \end{pmatrix},$$

or the product of any two of the N substitutions is again one of the substitutions of the set. Moreover the above reasoning shews that the result of performing successively the substitutions corresponding to the operations S_i and S_j gives the substitution corresponding to the operation S_iS_j. Hence, since the number of substitutions is equal to the number of operations, the given group and the group of substitutions are simply isomorphic.

It may also be shewn that each of the N substitutions is regular (§ 9) in the N symbols. For the substitution

$$\begin{pmatrix} 1, & S_1\ , & S_2\ , \ldots, & S_{N-1} \\ S_i, & S_1S_i, & S_2S_i, \ldots, & S_{N-1}S_i \end{pmatrix}$$

changes S_r into S_rS_i, S_rS_i into $S_rS_i^2$, and so on. If then m is the order of S_i, the cycle of the substitution which contains S_r will be

$$(S_r,\ S_rS_i,\ S_rS_i^2,\ \ldots,\ S_rS_i^{m-1}).$$

Now S_r may be any symbol of the set; hence all the cycles of the substitution must contain the same number, m, of symbols.

The substitution is therefore regular in the N symbols, and this can only be the case if m is a factor of N. It follows at once, as was stated in § 15, that the order of any operation of a group of order N must be equal to or a factor of N.

All the substitutions in this form of representing a group being regular, the group itself is said to be expressed as a regular substitution group.

21. The form, in which it has just been shewn that every group can be represented, is by no means the only form of representation possessing this property. Thus Hurwitz* has shewn that every group can be expressed as a group of rational and reversible transformations which change a suitably chosen algebraic curve (or Riemann's surface) into itself. We shall see in Chapter XIV that every group can also be expressed as a group of linear transformations of a finite set of variables to a prime modulus.

*"Algebraische Gebilde mit eindeutigen Transformationen in sich," *Math. Ann.*, Vol. XLI (1893), p. 421.

CHAPTER III.

ON THE SIMPLER PROPERTIES OF A GROUP WHICH ARE INDEPENDENT OF ITS MODE OF REPRESENTATION.

22. IN this chapter we proceed to discuss some of the simplest of the properties of groups of finite order which are independent of their mode of representation.

If among the operations of a group G a certain set can be chosen which do not exhaust all the operations of the group G, yet which at the same time satisfy all the conditions of § 12 so that they form another group H, this group H is called a *sub-group* of the group G. Thus if S be any operation, order m, of G, the operations

$$1,\ S,\ S^2,\ \ldots,\ S^{m-1}$$

evidently form a group; and when the order of G is greater than m, this group is a sub-group of G. A sub-group of this nature, which consists of the different powers of a single operation, is called a *cyclical sub-group*; and a group, which consists of the different powers of a single operation, is called a *cyclical group*.

THEOREM I. *If H is a sub-group of G, the order n of H is a factor of the order N of G.*

Let

$$1,\ T_1,\ T_2,\ \ldots,\ T_{n-1}$$

be the n operations of H; and let S_1 be any operation of G which is not contained in H.

Then the operations

$$S_1,\ T_1S_1,\ T_2S_1,\ \ldots,\ T_{n-1}S_1$$

are all distinct from each other and from the operations of H.

For if

$$T_pS_1 = T_qS_1,$$

then

$$T_p = T_q,$$

contrary to supposition; and if

$$T_p = T_q S_1,$$

then

$$S_1 = T_q^{-1} T_p,$$

and S_1 would be contained among the operations of H.

If the $2n$ operations thus obtained do not exhaust all the operations of G, let S_2 be any operation of G not contained among them.

Then it may be shewn, by repeating the previous reasoning, that the n operations

$$S_2,\ T_1S_2,\ T_2S_2,\ \ldots,\ T_{n-1}S_2$$

are all different from each other and from the previous $2n$ operations. If the group G is still not exhausted, this process may be repeated; so that finally the N operations of G can be exhibited in the form

$$\begin{array}{ccccc}
1, & T_1, & T_2, & \ldots, & T_{n-1}, \\
S_1, & T_1S_1, & T_2S_1, & \ldots, & T_{n-1}S_1, \\
S_2, & T_1S_2, & T_2S_2, & \ldots, & T_{n-1}S_2, \\
\cdots & \cdots & \cdots & \cdots & \cdots \\
S_{m-1}, & T_1S_{m-1}, & T_2S_{m-1}, & \ldots, & T_{n-1}S_{m-1}.
\end{array}$$

Hence $N = mn$, and n is therefore a factor of N.

When N is a prime p, the group G can have no sub-group other than one of order unity consisting of the identical operation alone. Every operation S of the group, other than the identical operation, is of order p, and the group consists of the operations

$$1,\ S,\ S^2,\ \ldots,\ S^{p-1}.$$

A group whose order is prime is therefore necessarily cyclical.

23. THEOREM II. *The operations common to two groups G_1, G_2 themselves form a group g, whose order is a factor of the orders of G_1 and G_2.*

For if S, T are any two operations common to G_1 and G_2, ST is also common to both groups; and hence the common operations satisfy conditions (α) and (β) of the definition in § 12. But their orders are finite and they must therefore satisfy also condition (γ), and form a group g. Moreover g is a sub-group of both G_1 and G_2, and therefore by Theorem I its order is a factor of the orders of both these groups.

If G_1 and G_2 are sub-groups of a third group G, then g is also clearly a sub-group of G.

The set of operations, that arise by combining in every way the operations of the groups G_1 and G_2, evidently satisfy the conditions of § 12 and form a group; but this will not necessarily or generally be a group of finite order. If however G_1 and G_2 are sub-groups of a group G of finite order, the group g' that arises from their combination will necessarily be of finite order; it may either coincide with G or be a sub-group of G. In either case, the order of g' will be a multiple of the orders of G_1 and G_2.

It is convenient here to explain a notation that enables us to avoid an otherwise rather cumbrous phraseology. Let S_1, S_2, S_3, ... be a given set of operations, and G_1, G_2, ... a set of groups. Then the symbol

$$\{S_1, S_2, S_3, \dots, G_1, G_2, \dots\}$$

will be used to denote the group that arises by combining in every possible way the given operations and the operations of the given groups.

Thus, for instance, the group g' above would be represented by

$$\{G_1, G_2\};$$

the cyclical group that arises from the powers of an operation S by

$$\{S\};$$

and, as a further example, the sixth group of § 17 may be represented by

$$\{(xy), (xz)\}.$$

24. Definition. If S and T are any two operations of a group, the operations S and $T^{-1}ST$ are called *conjugate* operations; while $T^{-1}ST$ is spoken of as the result of *transforming* the operation S by T.

The two operations S and $T^{-1}ST$ are identical only when S and T are permutable. For if

$$S = T^{-1}ST,$$

then

$$TS = ST.$$

Two conjugate operations are always of the same order. For

$$\begin{aligned}(T^{-1}ST)^n &= T^{-1}ST \cdot T^{-1}ST \ldots T^{-1}ST \\ &= T^{-1}S^nT.\end{aligned}$$

Therefore, if

$$S^n = 1,$$
$$(T^{-1}ST)^n = T^{-1}T = 1;$$

and conversely, if

$$(T^{-1}ST)^n = 1,$$

then

$$S^n = T \cdot T^{-1}S^nT \cdot T^{-1} = T(T^{-1}ST)^nT^{-1} = TT^{-1} = 1.$$

The operations ST and TS are always conjugate and therefore of the same order; for

$$ST = T^{-1}T \cdot ST = T^{-1} \cdot TS \cdot T.$$

Ex. Shew that the operations $S_1S_2 \ldots S_{n-1}S_n$ and $S_rS_{r+1} \ldots S_nS_1 \ldots S_{r-1}$ are conjugate within the group $\{S_1, S_2, \ldots, S_n\}$.

Definition. An operation S of a group G, which is identical with all its conjugate operations, is called a *self-conjugate* operation. Such an operation must evidently be permutable with each of the operations of G.

In every group the identical operation is self-conjugate; and in a group, whose operations are all permutable, every operation is self-conjugate. A simple example of a group, which contains self-conjugate operations other than the identical operation, while at the same time its operations are not all self-conjugate, is given by

$$\{(1234), (13)\}.$$

It is easy to shew that the order of this group is 8, and that (13)(24) is a self-conjugate operation.

If all the operations of a group be transformed by a given operation, the set of transformed operations form a group. For if T_1 and T_2 are any two operations of the group, so that T_1T_2 is also an operation of the group, then

$$S^{-1}T_1S \cdot S^{-1}T_2S = S^{-1}T_1T_2S;$$

hence the product of any two operations of the transformed set is another operation belonging to the transformed set, and the set therefore forms a group. Moreover the preceding equation shews that the new group is simply isomorphic to the original group. If G is the given group, the symbol $S^{-1}GS$ will be used for the new group. When S belongs to the group G, the groups G and $S^{-1}GS$ are evidently the same.

Now unless S is a self-conjugate operation of G, the pairs of operations T and $S^{-1}TS$ will not all be identical when for T the different operations of G are put in succession. Hence the process of transforming all the operations of a group by one of themselves is equivalent to establishing a correspondence between the operations of the group, which exhibits it as simply isomorphic with itself.

Definitions. When H is a sub-group of G and S is any operation of G, the groups H and $S^{-1}HS$ are called *conjugate* sub-groups of G.

If H and $S^{-1}HS$ are identical, S is said to be permutable with the sub-group H. This does not necessarily involve that S is permutable with each of the operations of H.

If H and $S^{-1}HS$ are identical, whatever operation S is of G, H is said to be a *self-conjugate* sub-group of G.

A group is called *composite* or *simple*, according as it does or does not possess at least one self-conjugate sub-group other than that formed of the identical operation alone.

25. THEOREM III. *The operations of a group G, which are permutable with a given operation T, form a sub-group H; and the order of G divided by the order of H is the number of operations conjugate to T*.*

If R_1 and R_2 are any two operations permutable with T, so that

$$R_1T = TR_1 \quad \text{and} \quad R_2T = TR_2;$$

then

$$R_1R_2T = R_1TR_2 = TR_1R_2,$$

and therefore R_1R_2 is permutable with T. The operations permutable with T therefore form a group H. Let n be its order and

$$1,\ R_1,\ R_2,\ \ldots,\ R_{n-1}$$

its operations. Then if S is any operation of G not contained in H, the operations

$$S,\ R_1S,\ R_2S,\ \ldots,\ R_{n-1}S$$

all transform T into the same operation T'.

For

$$(R_iS)^{-1}TR_iS = S^{-1}R_i^{-1}TR_iS = S^{-1}TS.$$

Also the n operations thus obtained are the only operations which transform T into T'; for if

$$S'^{-1}TS' = T',$$

then

$$SS'^{-1}TS'S^{-1} = ST'S^{-1} = T;$$

and therefore $S'S^{-1}$ belongs to H. The number of operations which transform T into any operation conjugate to it is therefore equal to the number that transform T into itself, that is, to the order of H. If then N is the order of G, the operations of G may be divided into $\frac{N}{n}$ sets of n each, such that the operations of each set transform T into a distinct operation, those

*Among these T of course occurs.

of the first set, namely the operations of H, transforming T into itself. The number of operations conjugate to T, including itself, is therefore $\frac{N}{n}$.

Since

$$T = ST'S^{-1},$$

therefore

$$R_i^{-1}TR_i = R_i^{-1}ST'S^{-1}R_i;$$

hence

$$T' = S^{-1}TS = S^{-1}R_i^{-1}TR_iS = S^{-1}R_i^{-1}S \cdot T' \cdot S^{-1}R_iS,$$

so that every operation of the form $S^{-1}R_iS$ is permutable with T'. Hence if H is the group of operations permutable with T, and if

$$S^{-1}TS = T',$$

then $S^{-1}HS$ is the group of operations permutable with T'.

It is convenient to have a symbol to represent the set of operations

$$S,\ R_1S,\ R_2S,\ \ldots,\ R_{n-1}S,$$

where

$$1,\ R_1,\ R_2,\ \ldots,\ R_{n-1}$$

form a group H. We shall in future represent this set of operations by HS; and we shall use SH to represent the set

$$S,\ SR_1,\ SR_2,\ \ldots,\ SR_{n-1}.$$

THEOREM IV. *The operations of a group G which are permutable with a sub-group H form a sub-group I, which is either identical with H or contains H as a self-conjugate sub-group. The order of G divided by the order of I is the number of sub-groups conjugate to H*.*

If S_1, S_2 are any two operations of G which are permutable with H, then

$$S_1^{-1}HS_1 = H, \quad S_2^{-1}HS_2 = H,$$

*Among these sub-groups H itself occurs.

and therefore

$$S_2^{-1}S_1^{-1}HS_1S_2 = H,$$

so that S_1S_2 is permutable with H. The operations of G which are permutable with H therefore form a group I, which may be identical with H and, if not identical with H, must contain it. Also, if S is any operation of I,

$$S^{-1}HS = H,$$

and therefore H is a self-conjugate sub-group of I.

If now Σ is any operation of G not contained in I, it may be shewn, exactly as in the proof of Theorem III, that the operations $I\Sigma$ and no others transform H into a conjugate sub-group H' which is not identical with H; and therefore that the number of sub-groups in the conjugate set to which H belongs is the order of G divided by the order of I.

The operations of G which are permutable with H' may also be shewn to form the group $\Sigma^{-1}I\Sigma$.

It is perhaps not superfluous to point out that two distinct conjugate sub-groups may have some operations in common with one another.

26. Let S_1 be any operation of G, and

$$S_1,\ S_2,\ \ldots,\ S_m$$

the distinct operations obtained on transforming S_1 by every operation of G. The number, m, of these operations is, by Theorem III, a factor of N, the order of G. Moreover if, instead of transforming S_1, we transform any other operation of the set, S_r, by every operation of G, the same set of m distinct operations of G will result. Such a set of operations we call a *complete set* of conjugate operations. If T is any operation of G which does not belong to this complete set of conjugate operations, no operation that is conjugate to T can belong to the set. Hence the operations of G may be distributed into a number of distinct sets such that every operation belongs to one set and no operation belongs to more than one set; while any set forms by itself a complete set of conjugate operations. If $m_1, m_2, \ldots, m_s$ are the numbers of operations in the different sets, then

$$N = m_1 + m_2 + \ldots + m_s;$$

and, since the identical operation is self-conjugate, one at least of the m's must be unity.

Similarly, if H_1 is any sub-group of G, and

$$H_1,\ H_2,\ \ldots,\ H_p$$

the distinct sub-groups obtained on transforming H_1 by every operation of G, we call the set a *complete set* of conjugate sub-groups. If K is a sub-group of G not contained in the set, no sub-group conjugate to K can belong to the set. If the operation S_1 belongs to one or more of a complete set of conjugate sub-groups, $\Sigma^{-1}S_1\Sigma$ must also belong to one or more sub-groups of the set, Σ being any operation of G. Hence among the operations contained in the complete set of conjugate sub-groups, the complete set of conjugate operations

$$S_1,\ S_2,\ \ldots,\ S_m$$

occurs.

No sub-group of G can contain operations belonging to every one of the complete sets of conjugate operations of G. For if such a sub-group H existed, the complete set of conjugate sub-groups, to which H belongs, would contain all the operations of G. Let m be the order of H and n ($\geqslant m$) the order of the sub-group I formed of those operations of G which are permutable with H. Then H is one of $\dfrac{N}{n}$ conjugate sub-groups, each of which contains m operations. The identical operation is common to all these sub-groups, and they therefore cannot contain more than

$$1+\frac{N(m-1)}{n}$$

distinct operations in all. This number is less than N, and therefore the complete set of conjugate sub-groups cannot contain all the operations of G.

27. If a group contains self-conjugate operations, it must contain self-conjugate sub-groups. For the cyclical sub-group generated by any self-conjugate operation must be self-conjugate. The only exception to this statement is the case of the cyclical groups of prime order. Every operation of such a group is clearly self-conjugate; but since the cyclical sub-group generated by any operation coincides with the group itself, there can be no self-conjugate sub-group*.

If every operation of a group is not self-conjugate, or, in other words, if the operations of a group are not all permutable with each other, the totality of the self-conjugate operations forms a self-conjugate sub-group. For, if S_1 and S_2 are permutable with every operation of the group, so also is S_1S_2.

THEOREM V. *The operations common to a complete set of conjugate sub-groups form a self-conjugate sub-group.*

It is an immediate consequence of Theorem II that the operations common to a complete set of conjugate sub-groups form a sub-group. Also the set of conjugate sub-groups, when transformed by any operation of the group, is changed into itself. Hence their common sub-group must be self-conjugate.

It may of course happen that the identical operation is the only one which is common to every sub-group of the set.

Corollary. The operations permutable with each of a complete set of conjugate sub-groups form a self-conjugate sub-group.

For, if the operations permutable with the sub-group H form a sub-group I, the operations permutable with every sub-group of the conjugate set to which H belongs are the operations common to every sub-group of the conjugate set to which I belongs.

Further, the operations which are permutable with every operation of a complete set of conjugate sub-groups form a self-conjugate sub-group.

THEOREM VI. *If T_1, T_2, ..., T_r are a complete set of conjugate operations of G, the group $\{T_1, T_2, \ldots, T_r\}$, if it does not coincide with G, is a self-conjugate sub-group of G; and it is the self-conjugate sub-group of smallest order that contains T_1.*

Since the operations T_1, T_2, ..., T_r are merely rearranged in a new

*Strictly speaking, this statement should be qualified by the addition "except that formed by the identical operation alone." No real ambiguity however will be introduced by always leaving this exception unexpressed.

sequence when the set is transformed by any operation of G, it follows that

$$S^{-1}\{T_1, T_2, \ldots, T_r\}S = \{T_1, T_2, \ldots, T_r\},$$

whatever operation of G may be represented by S. Hence $\{T_1, T_2, \ldots, T_r\}$ is a self-conjugate sub-group. Also any self-conjugate sub-group of G that contains T_1 must contain T_2, T_3, ..., T_r; and therefore any self-conjugate sub-group of G which contains T_1 must contain $\{T_1, T_2, \ldots, T_r\}$.

In exactly the same way it may be shewn that, if

$$H_1,\ H_2,\ \ldots,\ H_s$$

are a complete set of conjugate sub-groups of G, the group $\{H_1, H_2, \ldots, H_s\}$, if it does not coincide with G, is the smallest self-conjugate sub-group of G which contains the sub-group H_1.

The theorem just proved suggests a process for determining whether any given group is simple or composite. To this end, the groups $\{T_1, T_2, \ldots, T_r\}$ corresponding to each set of conjugate operations in the group are formed. If any one of them differs from the group itself, it is a self-conjugate sub-group and the group is composite; but if each group so formed coincides with the original group, the latter is simple. If the order of T_1 contains more than one prime factor and if T_1^m is of prime order, it is easy to see that the distinct operations of the set T_1^m, T_2^m, ..., T_r^m form a complete set of conjugate operations, and that the group $\{T_1, T_2, \ldots, T_r\}$ contains the group $\{T_1^m, T_2^m, \ldots, T_r^m\}$. Hence practically it is sufficient to form the groups $\{T_1, T_2, \ldots, T_r\}$ for all conjugate sets of operations whose orders are prime.

With the notation of § 26 (p. 37), the order of any self-conjugate sub-group of G must be of the form $m_\alpha + m_\beta + \ldots$; for if the sub-group contains any given operation, it must contain all the operations conjugate with it. Moreover one at least of the numbers m_α, m_β, ... must be unity, since the sub-group must contain the identical operation. It may happen that the numbers m_r are such that the only factors of N of the form $m_\alpha + m_\beta + \ldots$, one of the m's being unity, are N itself and unity. When this is the case, G is necessarily a simple group. It must not however be inferred that, if N has factors of this form, other than N itself and unity, then G is necessarily composite.

If G_1 and G_2 are sub-groups of G, it has already been seen (§ 23) that the operations common to G_1 and G_2 form a sub-group g of G; and it is now obvious that, when G_1 and G_2 are self-conjugate sub-groups, so also is g. Moreover the group $\{G_1, G_2\}$ is a self-conjugate sub-group unless it coincides with G. For

$$S^{-1}\{G_1, G_2\}S = \{S^{-1}G_1S, S^{-1}G_2S\} = \{G_1, G_2\}.$$

Again, with the same notation, if T_1 is an operation of G not contained in the self-conjugate sub-group G_1, and if T_1, T_2, ..., T_r is a complete set of conjugate operations, the group $\{G_1, T_1, T_2, \ldots, T_r\}$ is a self-conjugate sub-group, unless it coincides with G.

Definitions. If G_1, a self-conjugate sub-group of G, is such that the group

$$\{G_1, T_1, T_2, \ldots, T_r\}$$

coincides with G, when T_1, $T2$, ..., T_r is any complete set of conjugate operations not contained in G_1, then G_1 is said to be a *maximum* self-conjugate sub-group of G. This does not imply that G_1 is the self-conjugate sub-group of G of absolutely greatest order; but that there is no self-conjugate sub-group of G, distinct from G itself, which contains G_1 and is of greater order than G_1.

If H is a sub-group of G, and if, for every operation S of G which does not belong to H, the group $\{H, S\}$ coincides with G, H is said to be a *maximum* sub-group of G.

28. Definition. When a correspondence can be established between the operations of a group G and the operations of a group G', whose order is smaller than the order of G, such that to each operation S of G there corresponds a single operation S' of G', while to the operation S_pS_q there corresponds the operation $S'_pS'_q$, the group G is said to be *multiply isomorphic* with the group G'.

THEOREM VII. *If a group G is multiply isomorphic with a group G', then* (i) *the operations of G, which correspond to the identical operation of G', form a self-conjugate sub-group of G;* (ii) *to each operation of G'*

there correspond the same number of operations of G; and (iii) *the order of G is a multiple of the order of G'.*

Let

$$S_0,\ S_1,\ S_2,\ \ldots,\ S_{n-1}$$

be the set of operations of G which correspond to the identical operation of G'. These operations must form a group, since to S_pS_q corresponds the operation $1 \cdot 1$, i.e. the identical operation of G'; and therefore S_pS_q must belong to the set.

Again, to the operation $T^{-1}S_pT$ of G corresponds the operation $T'^{-1} \cdot 1 \cdot T'$, that is, the identical operation of G'. Hence, whatever operation of G is taken for T,

$$T^{-1}\{S_0, S_1, \ldots, S_{n-1}\}T = \{S_0, S_1, \ldots, S_{n-1}\}.$$

The sub-group Γ of G formed of the operations

$$S_0,\ S_1,\ S_2,\ \ldots,\ S_{n-1}$$

is therefore self-conjugate.

Again, if T and T_1 are two operations of G which correspond to the operation T' of G', the operation $T^{-1}T_1$ corresponds to the identical operation of G', and therefore belongs to Γ. Hence the operations that correspond to T' are all contained in the set $T\Gamma$. The operations of this set are all distinct and equal in number to the order of Γ. Hence if n is the order of Γ, to each operation of G' there correspond n operations of G.

Finally, since to each operation of G there corresponds only one of G', while to each operation of G' there correspond n of G, the order of G is n times the order of G'.

To any sub-group of G' of order μ, there corresponds a sub group of G of order μn. For if $T'_pT'_q$ forms one of the set

$$1,\ T'_1,\ \ldots,\ T'_p,\ \ldots,\ T'_q,\ \ldots,\ T'_{\mu-1},$$

at least one, and therefore all, of the set $T_pT_q\Gamma$ must occur among

$$\Gamma,\ T_1\Gamma,\ \ldots,\ T_p\Gamma,\ \ldots,\ T_q\Gamma,\ \ldots,\ T_{\mu-1}\Gamma,$$

and hence these operations form a group. Moreover, if the sub-group of G' is self-conjugate, so also is the corresponding sub-group of G.

It should be noticed that no correspondence is thus established between a sub-group of G which does not contain Γ and any sub-group of G'.

29. The relation of multiple isomorphism between two groups can be presented in a manner rather different from that of the last paragraph. Let G be any composite group, and Γ a self-conjugate sub-group of G consisting of the operations

$$1,\ T_1,\ T_2,\ \ldots,\ T_{n-1}.$$

Then, as in § 22, the operations of G can be arranged in the scheme

$$\begin{array}{ccccc}
1, & T_1, & T_2, & \ldots, & T_{n-1}, \\
S_1, & T_1S_1, & T_2S_1, & \ldots, & T_{n-1}S_1, \\
\multicolumn{5}{c}{\dotfill} \\
S_i, & T_1S_i, & T_2S_i, & \ldots, & T_{n-1}S_i, \\
\multicolumn{5}{c}{\dotfill} \\
S_{m-1}, & T_1S_{m-1}, & T_2S_{m-1}, & \ldots, & T_{n-1}S_{m-1}.
\end{array}$$

Now Γ being a self-conjugate sub-group of G, it follows that

$$S_i^{-1}\Gamma S_i = \Gamma,$$

and therefore the two sets of operations ΓS_i and $S_i\Gamma$ coincide except as regards arrangement. Hence

$$T_\alpha S_i T_\beta S_j = T_\alpha T_{\beta'} S_i S_j = T_\gamma S_i S_j,$$

where

$$S_i T_\beta S_i^{-1} = T_{\beta'},$$

and

$$T_\alpha T_{\beta'} = T_\gamma;$$

so that both $T_{\beta'}$ and T_γ belong to Γ. Now all the operations of the set $\Gamma S_i S_j$ occur in a single line, say the $(k+1)$th, of the above scheme. Hence if any operation of the $(i+1)$th line be followed by any operation of the

$(j+1)$th line, the result is some operation of the $(k+1)$th line. If then we regard the set of operations contained in each line of the scheme as a single entity, they will by their laws of combination define a new group of order m. In fact, if we denote these entities by

$$S'_0,\ S'_1,\ \ldots,\ S'_{m-1},$$

a relation of the form

$$S'_i S'_j = S'_k$$

has been proved to hold for every pair. Moreover, these relations necessarily obey the associative law; for if

$$S'_j S'_p = S'_q,$$

then the relation

$$S'_k S'_p = S'_i S'_q$$

follows in consequence of the symbols of G itself obeying the associative law. The symbol S'_0, corresponding to the first line of the scheme, is clearly the symbol of the identical operation in the new group thus defined.

If now G and Γ coincide with the groups of the preceding paragraph which are represented by the same symbols, then G' of the preceding paragraph must be simply isomorphic with the group whose operations are

$$1,\ S'_1,\ S'_2,\ \ldots,\ S'_{m-1}.$$

It follows that a group G' with which a group G is multiply isomorphic, in such a way that to the identical operation of G' there corresponds a given self-conjugate sub-group Γ of G, is completely defined (as an abstract group) when G and Γ are given. This being so it is natural to use a symbol to denote directly the group thus defined in terms of G and Γ. Herr Hölder* has introduced the symbol

$$\frac{G}{\Gamma}$$

*"Zur Reduction der algebraischen Gleichungen," *Math. Ann.* XXXIV (1889), p. 31.

to represent this group; he calls it the *quotient* of G by Γ, and a *factor-group* of G. We shall in the sequel make constant use both of the symbol and of the phrase thus defined.

It may not be superfluous to notice that the symbol $\frac{G}{\Gamma}$ has no meaning*, unless Γ is a self-conjugate sub-group of G. Moreover, it may happen that G has two simply isomorphic self-conjugate sub-groups Γ and Γ'. When this is the case, there is no necessary relation between the factor-groups $\frac{G}{\Gamma}$ and $\frac{G}{\Gamma'}$ (except of course that their orders are equal); in other words, the type of the factor-group $\frac{G}{\Gamma}$ depends on the actual self-conjugate sub-group of G which is chosen for Γ and not merely on the type of Γ.

Further, though in relation to its definition by means of G and Γ we call $\frac{G}{\Gamma}$ a factor-group of G, we may without ambiguity, since the symbol represents a group of definite type, omit the word factor and speak of the group $\frac{G}{\Gamma}$.

It is also to be observed that G has not necessarily a sub-group simply isomorphic with $\frac{G}{\Gamma}$. This may or may not be the case.

30. If G is multiply isomorphic with G' so that the self-conjugate sub-group Γ of G corresponds to the identical operation of G', it was shewn, at the end of § 28, that to any self-conjugate sub-group of G' there corresponds a self-conjugate sub-group of G containing Γ. Hence, unless $\frac{G}{\Gamma}$ is a simple group, Γ cannot be a maximum self-conjugate sub-group of G. If g_1 is any self-conjugate sub-group of $\frac{G}{\Gamma}$, and G_1 the corresponding (necessarily self-conjugate) sub-group of G, containing Γ, we may form the factor-group $\frac{G}{G_1}$, and determine again whether this group is simple or composite. By contin-

*Herr Frobenius has extended the use of the symbol to the case in which Γ is any group, whether contained in G or not, with which every operation of G is permutable: "Ueber endliche Gruppen," *Berliner Sitzungsberichte*, 1895, p. 169. We shall always use the symbol in the sense defined in the text.

uing this process a maximum self-conjugate sub-group of G, containing Γ, must at last be reached.

31. Though $\frac{G}{\Gamma}$ is completely defined by G and Γ, where Γ is any given self-conjugate sub-group of G, the reader will easily verify that G is not in general determined when Γ and $\frac{G}{\Gamma}$ are given.

We shall have in the sequel to consider the solution of this problem in various particular cases. There is, however, in every case one solution of it which is immediately obvious. We may take any two groups G_1 and G_2, simply isomorphic with the given groups Γ and $\frac{G}{\Gamma}$, such that G_1 and G_2 have no common operation except identity, while each operation of one is permutable with each operation of the other. The group $\{G_1, G_2\}$, formed by combining these two, is clearly such that $\frac{\{G_1, G_2\}}{G_1}$ is simply isomorphic with $\frac{G}{\Gamma}$; it therefore gives a solution of the problem.

Definition. If two groups G_1, G_2 have no common operation except identity, and if each operation of G_1 is permutable with each operation of G_2, the group $\{G_1, G_2\}$ is called the *direct product* of G_1 and G_2.

Ex. If H, h are self-conjugate sub-groups of G, and if h is contained in H, so that $\frac{H}{h}$ is a self-conjugate sub-group of $\frac{G}{h}$, shew that the quotient of $\frac{G}{h}$ by $\frac{H}{h}$ is simply isomorphic with $\frac{G}{H}$.

32. If H is a self-conjugate sub-group of G, of order n, and if H' is a self-conjugate sub-group of G', of order n', and if $\frac{G}{H}$ and $\frac{G'}{H'}$ are simply isomorphic, a correspondence of the most general kind may be established between the operations of G and G'. To every operation of G (or G') there will correspond n' (or n) operations of G' (or G), in such a way that to the product of any two operations of G (or G') there corresponds a definite set of n' (or n) operations of G' (or G). Let

$$G = H,\ S_1H,\ S_2H,\ \ldots,\ S_{m-1}H,$$

and

$$G' = H',\ S'_1H',\ S'_2H',\ \ldots,\ S'_{m-1}H';$$

and in the simple isomorphism between $\dfrac{G}{H}$ and $\dfrac{G'}{H'}$, let S_r and S'_r $(r = 0, 1, \ldots, m-1)$ be corresponding operations. Then if we take the set S'_rH' as the n' operations of G' that correspond to any operation of the set S_rH of G, and the set S_rH as the n operations of G that correspond to any operation of the set S'_rH' of G', the correspondence is, in fact, established.

For, if h'_1 and h'_2 are any two operations of H', the set of operations $S'_rh'_1S'_sh'_2$ includes n' distinct operations only, namely those of the set $S'_rS'_sH'$. Hence to the product of any given operation of the set S_rH by any given operation of the set S_sH, there corresponds the set of n' operations $S'_rS'_sH'$; at the same time the product of the two given operations belongs $\left(\text{in consequence of the isomorphism between } \dfrac{G}{H} \text{ and } \dfrac{G'}{H'}\right)$ to the set S_rS_sH. The same statements clearly hold when we interchange accented and unaccented symbols.

We still speak of G and G' as isomorphic groups, and the correspondence between their operations is said to give an n-to-n' isomorphism of the two groups. We shall return to this general form of isomorphism in dealing with intransitive substitution groups.

33. Definition. Two groups G and G' are said to be *permutable* with each other when the distinct operations of the set $S_iS'_j$, where for S_i every operation of the group G is put in turn and for S'_j every operation of the group G', coincide with the distinct operations of the set S'_jS_i except possibly as regards arrangement.

If the two groups G and G' are permutable, the group $\{G, G'\}$ must be of finite order. For, by the definition, every operation

$$\ldots S_pS'_qS_rS'_t\ldots$$

can be reduced to the form $S_iS'_j$; and therefore the number of distinct operations of the group $\{G, G'\}$ cannot exceed the product of the orders of G and G'. Let g be the group formed of the common operations of G and G'. Divide the operations of these groups into the sets

$$g,\ \Sigma_1g,\ \Sigma_2g,\ \ldots,\ \Sigma_{m-1}g$$

and

$$g, \; g\Sigma'_1, \; g\Sigma'_2, \; \ldots, \; g\Sigma'_{m'-1}.$$

Then every operation of the set $S_i S'_j$ can clearly be expressed in the form

$$\Sigma_p \gamma \Sigma'_q,$$

where γ is some operation of g. And no two operations of this form can be identical, for if

$$\Sigma_p \gamma_1 \Sigma'_q = \Sigma_r \gamma_2 \Sigma'_t,$$

then

$$\gamma_2^{-1} \Sigma_r^{-1} \Sigma_p \gamma_1 = \Sigma'_t \Sigma_q'^{-1};$$

so that $\Sigma'_t \Sigma_q'^{-1}$ belongs to g. But this is only possible if

$$\Sigma'_t = \Sigma'_q,$$

which leads to

$$\Sigma_r = \Sigma_p,$$

and

$$\gamma_2 = \gamma_1.$$

The order of $\{G, G'\}$ is therefore the product of the orders of G and G', divided by the order of g.

If every operation of G is permutable with G', then g must be a self-conjugate sub-group of G. For G and G' are transformed, each into itself, by any operation of G; and therefore their common sub-group g must be transformed into itself by any operation of G.

Moreover those operations of G, which are permutable with every operation of G', form a self-conjugate sub-group of G. For if T is an operation of G, which is permutable with every operation S' of G', so that

$$T^{-1} S' T = S',$$

and if S is any operation of G, then

$$S^{-1} T^{-1} S \cdot S_{-1} S' S \cdot S_{-1} T S = S^{-1} S' S,$$

so that $S_{-1}TS$ is permutable with every operation of G'. Hence every operation of G, which is conjugate to T, is permutable with every operation of G'; and the operations of G, which are permutable with every operation of G', therefore form a self-conjugate sub-group.

If G is a simple group, g must consist of the identical operation only; and either all the operations of G, or none of them, must be permutable with every operation of G'.

A special case is that in which the two groups G' and G are respectively a self-conjugate sub-group Γ and any sub-group H of some third group; for then every operation of H is permutable with Γ. If H is a cyclical sub-group generated by an operation S of order n, and if S^m is the lowest power of S which occurs in Γ, then m must be a factor of n. For if m' is the greatest common factor of m and n, integers x and y can be found such that

$$xm + yn = m'.$$

Now

$$S^{m'} = S^{xm+yn} = S^{xm},$$

and therefore $S^{m'}$ belongs to Γ. Hence m' cannot be less than m, and therefore m is a factor of n. Moreover, since $\{S^m\}$ is a sub-group of Γ, the order of Γ must be divisible by $\frac{n}{m}$. Hence:—

THEOREM VIII. *If an operation S, of order n, is permutable with a group Γ, and if S^m is the lowest power of S which occurs in Γ; then m is a factor of n, and $\frac{n}{m}$ is a factor of the order of Γ.*

The operations of $\{\Gamma, S\}$ can clearly be distributed in the sets

$$\Gamma,\ \Gamma S,\ \Gamma S^2,\ \ldots,\ \Gamma S^{m-1};$$

and no two of the operations $S, S^2, \ldots, S^{m-1}$ are conjugate in $\{\Gamma, S\}$.

34. A still more special case, but it is most important, is that in which the two groups are both of them self-conjugate sub-groups of some third group. If in this case the two groups are G and H, while S and T are any operations of the two groups respectively, then

$$S^{-1}HS = H,$$

and

$$T^{-1}GT = G;$$

so that every operation of G is permutable with H and every operation of H is permutable with G.

Consider now the operation $S^{-1}T^{-1}ST$. Regarded as the product of S^{-1} and $T^{-1}ST$ it belongs to G, and regarded as the product of $S^{-1}T^{-1}S$ and T it belongs to H. Every operation of this form therefore belongs to the common group of G and H. If G and H have no common operation except identity, then

$$S^{-1}T^{-1}ST = 1,$$

or

$$ST = TS;$$

and S and T are permutable. Hence:—

THEOREM IX*. *If every operation of G transforms H into itself and every operation of H transforms G into itself, and if G and H have no common operation except identity; then every operation of G is permutable with every operation of H.*

Corollary. If every operation of G transforms H into itself and every operation of H transforms G into itself, and if either G or H is a simple group; then G and H have no common operation except identity, and every operation of G is permutable with every operation of H.

For, by § 33, if G and H had a common sub-group, it would be a self-conjugate sub-group of both of them; and neither of them could then be simple, contrary to hypothesis. Consequently, the only sub-group common to G and H is the identical operation.

35. If N and N' are the orders of two permutable groups G and G', the NN' equations expressing every operation of the form SS', where S belongs to G and S' to G', in the form $\Sigma'\Sigma$, where Σ and Σ' belong to G and G' respectively, are never all independent. In particular, if

$$S_1, S_2, \ldots, S_r$$

*Dyck, "Gruppentheoretische Studien," *Math. Ann.* Vol. XXII (1883), p. 97.

are a set of independent generating operations of G, and

$$m_1, m_2, \ldots, m_r$$

their orders, it is clearly sufficient that each operation of the form

$$S'S_p^\alpha \quad (\alpha = 1, 2, \ldots, m_p - 1;\ p = 1, 2, \ldots, r)$$

should be capable of expression in the form $\Sigma'\Sigma$.

When, in fact, these conditions are satisfied, it is always possible by a series of steps to express any operation $S'S$ in the form $\Sigma\Sigma'$.

Ex. 1. Shew that, in the group whose defining relations are

$$A^4 = 1, \quad B^3 = 1, \quad (AB)^2 = 1,$$

the three operations A^2, $B^{-1}A^2B$, BA^2B^{-1} are permutable and that they form a complete set of conjugate operations. Hence shew that $\{A^2, B\}$ is a self-conjugate sub-group, and that the order of the group is 24.

Ex. 2. Shew that the cyclical group generated by the substitution (1234567) is permutable with the group

$$\{(267)(345),\ (23)(47)\};$$

and that the order of the group resulting from their product is 168.

Ex. 3. If g_1 and g_2 are the orders of the groups G_1 and G_2, γ the order of their greatest common sub-group and g the order of $\{G_1, G_2\}$, shew that

$$g\gamma \geqslant g_1 g_2,$$

and that, if $g\gamma = g_1 g_2$, then G_1 and G_2 are permutable. (Frobenius.)

Ex. 4. If G_1 and G_2 are two sub-groups of G of orders g_1 and g_2, and S any operation of G, prove that the number of distinct operations of G contained in the set S_1SS_2, when for S_1 and S_2 are put in turn every pair of operations of G_1 and G_2 respectively, is $\dfrac{g_1 g_2}{\gamma}$; γ being the order of the greatest sub-group common to $S^{-1}G_1S$ and G_2.

If T is any other operation of G, shew also that the sets S_1SS_2 and S_1TS_2 are either identical or have no operation in common. (Frobenius.)

Ex. 5. If a group G of order mn has a sub-group H of order n, and if n has no prime factor which is less than m, shew that H must be a self-conjugate sub-group. (Frobenius.)

CHAPTER IV.

ON ABELIAN GROUPS.

36. WE shall now apply the general results, that have been obtained in the last chapter, to the study of two special classes of groups; in the present chapter we shall deal particularly with those groups whose operations are all permutable with each other.

Definition. A group, whose operations are all permutable with each other, is called an *Abelian** group.

It is to be expected (and it will be found) that the theory of Abelian groups is much simpler than that of groups in general; for the process of multiplication of the operations of such groups is commutative as well as associative.

Every sub-group of an Abelian group is itself an Abelian group, since its operations are necessarily all permutable. For the same reason, every operation and every sub-group of an Abelian group is self-conjugate both in the group itself and in any sub-group in which it is contained.

If G is an Abelian group and H any sub-group of G, then since H is necessarily self-conjugate, there exists a factor-group $\frac{G}{H}$, and this again must be an Abelian group. (The reader must not however infer that, if H

*On Abelian groups, the reader may consult Frobenius and Stickelberger, "Ueber Gruppen vertauschbarer Elemente," *Crelle's Journal*, Vol. LXXXVI (1879), p. 217; and a very complete discussion in the second volume of Herr Weber's recently published *Lehrbuch der Algebra*. In the proof of the existence of a set of independent generating operations (§ 41) we have directly followed Herr Weber.

The name "Abelian group" has been applied by M. Jordan (*Traité des substitutions etc.* pp. 171 et seq.) to an entirely different class of groups, whose operations are not permutable. Most writers, we believe, have used the phrase in the sense defined in the text.

The connection of Abel's name with groups of permutable operations is due to his having been the first to investigate, with complete generality, the application of such groups to the theory of equations, "Mémoire sur une classe particulière d'équations résolubles algébriquement," *Crelle's Journal*, Vol. IV (1829), p. 131; or Collected Works, 1881 edition, Vol. I, p. 478.

and $\frac{G}{H}$ are both Abelian, then G is also Abelian. It is indeed clear that this is not necessarily the case.)

37. Let now G be any Abelian group, and let p^m be the highest power of a prime p that divides its order. We shall first shew that G has a single sub-group of order p^m, consisting of all the operations of G whose orders are powers of p.

If S_1, S_2 are any two operations of G, it follows from § 33, that, because S_1 and S_2 are permutable, the order of $\{S_1, S_2\}$ is equal to or is a factor of the product of the orders of S_1 and S_2. So again, if S_3 is an operation of G not contained in $\{S_1, S_2\}$, the order of $\{S_1, S_2, S_3\}$ is equal to or is a factor of the product of the orders of $\{S_1, S_2\}$ and S_3. Now by continually including a fresh operation, not contained in the group already arrived at, we must in this way after a finite number of steps arrive at the group G, whose order is divisible by p. Hence, among the operations S_1, S_2, S_3, $\ldots$, there must be at least one whose order is divisible by p, and some power of this, say S, will be an operation of order p. Now, if m is greater than unity, the order of the factor-group $\frac{G}{\{S\}}$, which is also Abelian, is divisible by p, and therefore this factor-group must have an operation of order p. Hence G will (§ 28) contain a sub-group of order p^2. If m is greater than 2, the same reasoning may be repeated to shew that G has a sub-group of order p^3, and so on. Hence, finally, G has a sub-group of order p^m. Let H be this sub-group; and suppose if possible that G contains an operation T, whose order is a power of p, which does not belong to H. Then $\{H, T\}$ is a sub-group of G, whose order is a power of p, greater than p^m; and this is impossible (§ 22). The sub-group H must therefore contain all the operations of G whose orders are powers of p. Hence:—

THEOREM I. *If p^m is the highest power of a prime p that divides the order of an Abelian group G, then G contains a single sub-group of order p^m, which consists of all the operations of G whose orders are powers of p.*

38. Let the order of G be

$$N = p_1^{m_1} p_2^{m_2} \ldots p_n^{m_n},$$

where $p_1, p_2, \ldots, p_n$ are distinct primes; and let

$$H_1,\ H_2,\ \ldots,\ H_n$$

be the sub-groups of G of orders

$$p_1^{m_1},\ p_2^{m_2},\ \ldots,\ p_n^{m_n}.$$

Since the orders of H_1 and H_2 are relatively prime, they can have no common operation except identity; and therefore the order of $\{H_1, H_2\}$ is $p_1^{m_1}p_2^{m_2}$. This sub-group contains all the operations of G whose orders are relatively prime to $\frac{N}{p_1^{m_1}p_2^{m_2}}$. For if T were an operation of G of order $p_1^{\alpha}p_2^{\beta}$, not contained in $\{H_1, H_2\}$, then $\{H_1, H_2, T\}$ would be a sub-group of G, of order $p_1^{n_1}p_2^{n_2}$, where $n_1 > m_1$ if $\alpha > 0$ and $n_2 > m_2$ if $\beta > 0$; and this is impossible (§ 22).

This process may clearly be continued to shew that, if $N = \mu\nu$, where μ and ν are relatively prime, then G contains a single sub-group of order μ, consisting of all the operations of G whose orders are relatively prime to ν. Moreover G itself is the direct product (§ 31) of $H_1, H_2, \ldots, H_n$.

39. The first problem of pure group-theory that presents itself in connection with Abelian groups is the determination of all distinct Abelian groups of given order N. Let H_1 and H_1' be two distinct Abelian groups of order $p_1^{m_1}$, i.e. two groups which are not simply isomorphic. Then two Abelian groups of order N, whose sub-groups of order $p_1^{m_1}$ are simply isomorphic with H_1 and H_1' respectively, are necessarily distinct. Since then G is the direct product of $H_1, H_2, \ldots, H_n$, the general problem for any composite order N will be completely solved when we have determined all distinct types of Abelian groups of order $p_1^{m_1}, p_2^{m_2}, \ldots, p_n^{m_n}$. We may therefore, for the purpose of this problem, confine our attention to those Abelian groups whose orders are powers of primes.

40. Suppose then that G is an Abelian group whose order is the power of a prime. If among the operations of G we choose at random a set

$$P,\ P',\ P'',\ \ldots$$

from which the group can be generated, they will not in general be independent of each other.

As an instance, we may take the group whose multiplication table is:—

	1	P_1	P_2	P_3	P_4	P_5	P_6	P_7
1	1	P_1	P_2	P_3	P_4	P_5	P_6	P_7
P_1	P_1	P_2	P_3	1	P_5	P_6	P_7	P_4
P_2	P_2	P_3	1	P_1	P_6	P_7	P_4	P_5
P_3	P_3	1	P_1	P_2	P_7	P_4	P_5	P_6
P_4	P_4	P_5	P_6	P_7	1	P_1	P_2	P_3
P_5	P_5	P_6	P_7	P_4	P_1	P_2	P_3	1
P_6	P_6	P_7	P_4	P_5	P_2	P_3	1	P_1
P_7	P_7	P_4	P_5	P_6	P_3	1	P_1	P_2

Here P_1 and P_5 are two operations from which every operation of the group may be generated: an inspection of the table will shew that they are connected by the relation

$$P_1^2 P_5^2 = 1.$$

On the other hand, if we choose P_1 and P_4 as generating operations, we find that every operation of the group can be expressed in the form

$$P_1^{\alpha} P_4^{\beta}, \quad (\alpha = 0, 1, 2, 3; \ \beta = 0, 1),$$

while the only conditions to which the permutable operations P_1 and P_4 are submitted are

$$P_1^4 = 1, \quad P_4^2 = 1.$$

The question then arises as to whether G can be generated by a set of permutable and independent operations, i.e. by a set of permutable operations which are connected by no relations except those that give their orders. That this question is always to be answered in the affirmative may be proved in the following manner.

41. Let the order of G be p^m; and let P_1 be an operation of G whose order p^{m_1} is not less than that of any other operation of the group. Then every operation of the group satisfies the equation

$$S^{p^{m_1}} = 1.$$

If $m_1 = m$, the order of $\{P_1\}$ is equal to the order of G; the latter is then a cyclical group generated by the operation P_1.

If $m_1 < m$, G must contain other operations besides those of $\{P_1\}$. Denoting $\{P_1\}$ by G_1, let Q be any operation of G not contained in G_1, and let Q^a be the lowest power of Q that is contained in G_1. Then (§ 33) a must be a power of p; and when for Q each operation of G that is not contained in G_1 is taken in turn, a must have some maximum value, say p^{m_2}. Since no operation of G is of greater order than p^{m_1} it follows that $m_2 \leqslant m_1$. We may suppose then Q to be an operation of G, such that $Q^{p^{m_2}}$ is the lowest power of Q which is contained in G_1.

Then

$$Q^{p^{m_2}} = P_1^{\lambda_1},$$

and

$$Q^{p^{m_1}} = 1 = P_1^{\lambda_1 p^{m_1 - m_2}},$$

so that λ_1 is divisible by p^{m_2} and may be expressed in the form xp^{m_2}. The case $\lambda_1 = 0$ forms no exception to this statement, since λ_1 is congruent to zero, (mod. p^{m_1}); but we actually take $x = 0$ in this case.

If now we write

$$QP_1^{-x} = P_2,$$

then

$$P_2^{p^{m_2}} = 1,$$

and $P_2^{p^{m_2}}$ is the lowest power of P_2 which is contained in G_1. Let the sub-group $\{P_1, Q\}$ be denoted by G_2. Then G_2 is generated by the two independent operations P_1 and P_2 of orders p^{m_1} and p^{m_2}; and every operation S of G is such that

$$S^{p^{m_2}} = P_1^{\alpha_1},$$

where α_1 is divisible by p^{m_2}.

This process may now be continued step by step. That it will ultimately lead to a representation of G as generated by a set of independent operations may be shewn by induction. To this end, we will suppose that a sub-group G_n has been arrived at which is generated by the n independent operations

$$P_1,\ P_2,\ \ldots,\ P_n,$$

of orders

$$p^{m_1},\ p^{m_2},\ \ldots,\ p^{m_n},$$

where

$$m_1 \geqslant m_2 \geqslant \ldots \geqslant m_n.$$

We will also suppose that, if T is any operation of G,

$$T^{p^{m_n}} = P_1^{\alpha_1} P_2^{\alpha_2} \ldots P_{n-1}^{\alpha_{n-1}},$$

where α_1, α_2, …, α_{n-1} are all divisible by p^{m_n}. In the special case $n = 2$ these suppositions have been justified.

Let Q' be any operation of G, and let Q'^a be the lowest power of Q' that occurs in G_n. Then when for Q' each operation of G that is not contained in G_n is taken in turn, a will have a certain maximum value, say $p^{m_{n+1}}$; and from the suppositions made above, m_{n+1} is not greater than m_n. We may therefore write

$$Q'^{p^{m_{n+1}}} = P_1^{\beta_1} P_2^{\beta_2} \ldots P_{n-1}^{\beta_{n-1}} P_n^{\beta_n},$$

and hence

$$Q'^{p^{m_n}} = P_1^{\beta_1 p^{m_n - m_{n+1}}} P_2^{\beta_2 p^{m_n - m_{n+1}}} \ldots P_{n-1}^{\beta_{n-1} p^{m_n - m_{n+1}}} P_n^{\beta_n p^{m_n - m_{n+1}}}.$$

Now P_1, P_2, …, P_n are independent; it follows therefore from the second assumption made above, that the exponents of P_1, P_2, …, P_n in this last equation are divisible by p^{m_n}; and hence that β_1, β_2, …, β_n are divisible by $p^{m_{n+1}}$.

If we now suppose that Q' itself is chosen so that $Q'^{p^{m_{n+1}}}$ is the lowest power of Q' that occurs in G_n, then

$$Q'^{p^{m_{n+1}}} = P_1^{x_1 p^{m_{n+1}}} P_2^{x_2 p^{m_{n+1}}} \ldots P_n^{x_n p^{m_{n+1}}};$$

hence, if

$$Q' P_1^{-x_1} P_2^{-x_2} \ldots P_n^{-x_n} = P_{n+1},$$

then

$$P_{n+1}^{p^{m_{n+1}}} = 1;$$

and $P_{n+1}^{p^{m_{n+1}}}$ is the lowest power of P that occurs in G_n. Hence finally $\{P_1, P_2, \ldots, P_n, Q'\}$ is generated by the $n+1$ independent operations P_1, P_2, ..., P_{n+1} of orders p^{m_1}, p^{m_2}, ..., $p^{m_{n+1}}$ $(m_1 \geqslant m_2 \geqslant \ldots \geqslant m_{n+1})$; and S being any operation of G,

$$S^{p^{m_{n+1}}} = P_1^{\gamma_1} P_2^{\gamma_2} \ldots P_n^{\gamma_n},$$

where γ_1, γ_2, ..., γ_n are divisible by $p^{m_{n+1}}$.

This completes the proof by induction, and we may now state the result in the form of the following theorem:—

THEOREM II. *An Abelian group of order p^m, where p is a prime, can be generated by r $(\leqslant m)$ independent operations P_1, P_2, ..., P_r, of orders p^{m_1}, p^{m_2}, ..., p^{m_r}, where*

$$m_1 + m_2 + \ldots + m_r = m,$$

and $m_1 \geqslant m_2 \geqslant \ldots \geqslant m_r$. Moreover if μ is any positive integer such that $m_s \geqslant \mu \geqslant m_{s+1}$, while Q is any operation of the group, then

$$Q^{p^\mu} = P_1^{\alpha_1} P_2^{\alpha_2} \ldots P_s^{\alpha_s},$$

where α_1, α_2, ..., α_s are divisible by p^μ.

The actual existence of the set of independent generating operations is demonstrated by the above inductive proof. The other inferences in the theorem may be established as follows. A set of independent permutable operations of orders p^{m_1}, p^{m_2}, ..., p^{m_r} generate a group of order $p^{m_1+m_2+\ldots+m_r}$, and therefore

$$m_1 + m_2 + \ldots + m_r = m.$$

Since each number in this equation is a positive integer, the number of terms on the left-hand side cannot be greater than m. Hence

$$r \leqslant m;$$

and, if $r = m$, every operation of the group except identity is of order p.

Finally, from the above inductive proof it follows that, if Q is any operation of the group, then

$$Q^{p^{m_{s+1}}} = P_1^{\beta_1} P_2^{\beta_2} \dots P_s^{\beta_s},$$

where $\beta_1, \beta_2, \dots, \beta_s$ are all divisible by $p^{m_{s+1}}$.

Hence

$$\begin{aligned} Q^{p^\mu} &= P_1^{\beta_1 p^{\mu - m_{s+1}}} P_2^{\beta_2 p^{\mu - m_{s+1}}} \dots P_s^{\beta_s p^{\mu - m_{s+1}}} \\ &= P_1^{\alpha_1} P_2^{\alpha_2} \dots P_s^{\alpha_s}, \end{aligned}$$

where $\alpha_1, \alpha_2, \dots, \alpha_s$ are all divisible by p^μ.

42. It is clear, from the synthetic process by which it has been proved that an Abelian group of order p^m can be generated by a set of independent operations, that a considerable latitude exists in the choice of the actual generating operations; and the question arises as to the relations between the orders of the distinct sets of independent generating operations.

The discussion of this question is facilitated by a consideration of certain special sub-groups of G. If A and B are two operations of G, and if the order of A is not less than that of B, the order of AB is equal to, or is a factor of, the order of A. Hence the totality of those operations of G whose orders do not exceed p^μ, or in other words of those operations which satisfy the relation

$$S^{p^\mu} = 1,$$

form a sub-group G_μ. The order of G_μ clearly depends on the orders of the various operations of G and in no way on a special choice of generating operations. Now if

$$P_1^{\alpha_1} P_2^{\alpha_2} \dots P_r^{\alpha_r}$$

belongs to G_μ, then

$$P_1^{\alpha_1 p^\mu} P_2^{\alpha_2 p^\mu} \dots P_r^{\alpha_r p^\mu} = 1.$$

Hence if m_{s+1} is the first of the series

$$m_1, m_2, \dots, m_r,$$

which is less than μ, then $\alpha_{s+1}, \ldots, \alpha_r$ may have any values whatever; but α_t $(t = 1, 2, \ldots, s)$ must be a multiple of $p^{m_t-\mu}$.

It follows from this that G_μ is generated by the r independent operations

$$P_1^{p^{m_1-\mu}},\ P_2^{p^{m_2-\mu}},\ \ldots,\ P_s^{p^{m_s-\mu}},\ P_{s+1},\ \ldots,\ P_r.$$

If then the order of G_μ is p^ν, we have

$$\nu = \mu s + \sum_{s+1}^{r} m_t.$$

The order of G_1, the sub-group formed of all operations of G whose order is p, is clearly p^r.

43. Suppose now that by a fresh choice of independent generating operations, it were found that G could be generated by the r' independent operations

$$P_1',\ P_2',\ \ldots,\ P_{r'}'$$

of orders

$$p^{m_1'},\ p^{m_2'},\ \ldots,\ p^{m_{r'}'},$$

where

$$m_1' \geqslant m_2' \geqslant \ldots \geqslant m_{r'}'.$$

If $m_{s'+1}'$ is the first of this series which is less than μ, the order of G_μ will be $p^{\nu'}$, where

$$\nu' = \mu s' + \sum_{s'+1}^{r'} m_t'.$$

The order of G_μ is independent of the choice of generating operations; so that for all values of μ

$$\nu = \nu'.$$

Hence, by taking $\mu = 1$,

$$r = r',$$

or the number of independent generating operations is independent of their choice.

If now

$$m_t = m'_t \quad (t = s+1, s+2, \ldots, r),$$

and

$$m_s > m'_s,$$

and if we choose μ so that

$$m_s \geqslant \mu > m'_s;$$

then

$$\nu = \mu s + \sum_{s+1}^{r} m_t,$$

and

$$\nu' = \mu(s-1) + m'_s + \sum_{s+1}^{r} m_t.$$

The condition

$$\nu = \nu'$$

gives

$$\mu = m'_s,$$

in contradiction to the assumption just made.

Similarly we can prove that the assumption $m'_s > m_s$ cannot be maintained; hence

$$m_s = m'_s;$$

and therefore, however the independent generating operations of G are chosen, their number is always r, and their orders are

$$p^{m_1},\ p^{m_2},\ \ldots,\ p^{m_r}.$$

44. If G' is a second Abelian group of order p^m, simply isomorphic with G, and if

$$P'_1,\ P'_2,\ \ldots,\ P'_{r'}$$

of orders

$$p^{m'_1},\ p^{m'_2},\ \ldots,\ p^{m'_{r'}},$$

where

$$m'_1 \geqslant m'_2 \geqslant \ldots \geqslant m'_{r'},$$

are a set of independent generating operations of G', exactly the same process as that of the last paragraph may be used to shew that

$$r = r',$$

and

$$m_s = m'_s \quad (s = 1, 2, \ldots, r).$$

In fact, since corresponding operations of two simply isomorphic groups have the same order, the order of G_μ must be equal to the order of G'_μ; and this is the condition that has been used to obtain the result of the last paragraph.

Two Abelian groups of order p^m cannot therefore be simply isomorphic unless the series of integers m_1, m_2, ..., m_r is the same for each. On the other hand when this condition is satisfied, it is clear that the two groups are simply isomorphic, since by taking P_s and P'_s $(s = 1, 2, \ldots, r)$ as corresponding operations, the isomorphism is actually established.

The number of distinct types of Abelian groups of order p^m, where p is a prime, i.e. the number of such groups no one of which is simply isomorphic with any other, is therefore equal to the number of partitions of m. When the prime p is given, each type of group may be conveniently, and without ambiguity, represented by the symbol of the corresponding partition. Thus the typical group G that we have been dealing with would be represented by the symbol $(m_1, m_2, \ldots, m_r)$.

45. Having thus determined all distinct types of Abelian groups of order p^m, a second general problem in this connection is the determination of all possible types of sub-group when the group itself is given. This will be facilitated by the consideration of a second special class of sub-groups in addition to the sub-groups G_μ already dealt with.

If S and S' are any two operations of G, then

$$S^{p^\mu} S'^{p^\mu} = (SS')^{p^\mu};$$

and therefore the totality of the distinct operations obtained by raising every operation of G to the power p^μ will form a sub-group H_μ.

If

$$m_s \geqslant \mu > m_{s+1},$$

then (Theorem II, § 41)

$$S^{p^\mu} = P_1^{\alpha_1 p^\mu} P_2^{\alpha_2 p^\mu} \dots P_s^{\alpha_s p^\mu},$$

S being any operation of G. Hence H_μ is generated by the s independent operations

$$P_1^{p^\mu}, P_2^{p^\mu}, \dots, P_s^{p^\mu};$$

and the order of H_μ is $p^{\sum_1^s m_t - \mu s}$.

46. Let now Γ of type $(n_1, n_2, \dots, n_s)$ be any sub-group of G. The order of the group Γ_1, formed of all the operations of Γ which satisfy the equation

$$S^p = 1,$$

is p^s (§ 42). This group must be identical with or be a sub-group of G_1 whose order is p^r. Hence

$$s \leqslant r,$$

i.e. the number of independent generating operations of any sub-group of G is equal to or is less than the number of independent generating operations of G itself.

If p^n is the order of Γ, there must be one or more sub-groups of G of order p^{n+1} which contain Γ; for $\frac{G}{\Gamma}$ has operations of order p. Hence Γ must be contained in one or more sub-groups of G of order p^{m-1}. We may begin then by considering all possible types of sub-groups of G of order p^{m-1}. Let g be such a sub-group and $(m'_1, m'_2, \dots, m'_{r'})$ its type; and suppose that

$$m_t = m'_t \quad (t = 1, 2, \dots, k-1),$$

and

$$m_k \neq m'_k.$$

Then if

$$m_k < m'_k,$$

the order of h_{m_k}, the sub-group of g which results by raising all its operations to the power p^{m_k}, is greater than the order of H_{m_k}. This is impossible, since h_{m_k} must coincide with H_{m_k} or with one of its sub-groups. Hence

$$m'_k < m_k.$$

Now

$$\sum_1^{r'} m'_t = \sum_1^{r} m_t - 1;$$

therefore

$$m'_k = m_k - 1,$$

and

$$m'_t = m_t \quad (t = k+1, \ldots, r).$$

In the particular case in which m_r is unity if we take k equal to r, m'_r is zero; and in this case g would have $r-1$ generating operations.

It is easy to see that sub-groups of order p^{m-1} and of all the types just determined actually exist. For

$$P_1,\ P_2,\ \ldots,\ P_{k-1},\ P_{k^p},\ P_{k+1},\ \ldots,\ P_r$$

are a set of independent operations which generate a sub-group of the type $(m_1, m_2, \ldots, m_{k-1}, m_k - 1, m_{k+1}, \ldots, m_r)$.

47. Returning to the general case, we will assume for all orders not less than p^n the existence of a sub-group Γ of type $(n_1, n_2, \ldots, n_s)$, when the inequalities

$$\sum_1^t n_u \leqslant \sum_1^t m_u \quad (t = 1, 2, \ldots, r)$$

are satisfied, where if $t > s$, n_t is zero. When n is equal to $m-1$, the truth of this assumption has been established.

If Γ' of type $(n'_1, n'_2, \ldots, n'_{s'})$ is a sub-group of order p^{n-1}, it must be contained in some sub-group of order p^n, say Γ of type $(n_1, n_1, \ldots, n_s)$. Then, by the preceding discussion,

$$\sum_1^t n'_u \leqslant \sum_1^t n_u \quad (t = 1, 2, \ldots, s);$$

and therefore

$$\sum_1^t n'_u \leqslant \sum_1^t m_u \quad (t = 1, 2, \ldots, r),$$

so that the conditions assumed for sub-groups of order not less than p^n are necessary for sub-groups of order p^{n-1}. Moreover if the conditions

$$\sum_1^t n'_u \leqslant \sum_1^t m_u \quad (t = 1, 2, \ldots, r)$$

are satisfied, and if n'_k is the first of the series

$$n'_1, \ n'_2, \ \ldots,$$

which is less than the corresponding term of the series

$$m_1, \ m_2, \ \ldots,$$

we may take

$$\begin{aligned} n_u &= n'_u, \quad u \neq k, \\ n_k &= n'_k + 1; \end{aligned}$$

and then the conditions

$$\sum_1^t n_u \leqslant \sum_1^t m_u \quad (t = 1, 2, \ldots, r);$$

are satisfied. But these conditions being satisfied it follows, from the assumption made, that G contains a sub-group of order p^n and type $(n_1, n_2, \ldots, n_s)$ and therefore also a sub-group of order p^{n-1} and type $(n'_1, n'_2, \ldots, n'_{s'})$. If the conditions assumed are sufficient for the existence of a sub-group of order p^n, they are thus proved to be sufficient for the existence of one of order p^{n-1}. Hence since they are sufficient in the case of sub-groups of order p^{m-1}, they are so generally.

We may summarize the results obtained in the last three paragraphs as follows:—

THEOREM III. *The number of distinct types of Abelian groups of order p^m, where p is a prime, is equal to the number of partitions of m; and each type may be completely represented by the symbol $(m_1, m_2, \ldots, m_r)$ of the corresponding partition. If the numbers in the partition are written in descending order, a group of type $(m_1, m_2, \ldots, m_r)$ will have a sub-group of type $(n_1, n_2, \ldots, n_s)$, when the conditions*

$$s \leqslant r,$$
$$\sum_1^t n_u \leqslant \sum_1^t m_u \quad (t = 1, 2, \ldots, r),$$

are satisfied; and the type of every sub-group must satisfy these conditions.

48. It will be seen later that the Abelian group of order p^m and type $(1, 1, 1, \ldots, \text{with } m \text{ units})$ is of special importance in the general theory, and we shall here discuss one or two of its simpler properties.

Since the generating operations of the group are all of order p, every operation except identity is of order p; and therefore the type of any sub-group of order p^s is $(1, 1, 1, \ldots, \text{with } s \text{ units})$. If the group be denoted by G, every sub-group G_μ coincides with G; while of the sub-groups H_μ, the first coincides with G and all the rest consist of the identical operation only.

In choosing a set of independent generating operations, we may take for the first, P_1, any one of the $p^m - 1$ operations of the group, other than identity. The sub-group $\{P_1\}$ is of order p; and therefore G has $p^m - p$ operations which are not contained in $\{P_1\}$. If we choose any one of these, P_2, it is necessarily independent of P_1, and may be taken as a second generating operation. The sub-group $\{P_1, P_2\}$ is of order p^2 and type $(1, 1)$; and G has $p^m - p^2$ operations which are not contained in this sub-group. If P_3 be any one of these, no power of P_3 other than identity is contained in $\{P_1, P_2\}$; and P_1, P_2, P_3 are therefore three independent operations which generate a sub-group of order p^3. This process may clearly be continued till all m generating operations have been chosen. If then the position which each generating operation occupies in the set of m, when they are written in order, be taken into account, there are

$$(p^m - 1)(p^m - p)(p^m - p^2) \ldots (p^m - p^{m-1})$$

distinct ways in which a set may be chosen. If on the other hand the sets of generating operations which consist of the same operations written in different orders be regarded as identical, the number of distinct sets is

$$\frac{(p^m-1)(p^m-p)\dots(p^m-p^{m-1})}{m!}.$$

49. No operation P of the group can belong to two distinct sub-groups of order p except the identical operation. Hence since every sub-group of order p contains $p-1$ operations besides identity, G must contain $\frac{p^m-1}{p-1}$ sub-groups of order p.

Let $N_{m,r}$ be the number of sub-groups of G of order p^r, so that

$$N_{m,1}=\frac{p^m-1}{p-1}.$$

There are, in G, p^m-p^r operations not contained in any given sub-group of order p^r. If P occurs among these operations, so also do P^2, $P^3, \dots, P^{p-1}$. Hence there are $\frac{p^m-p^r}{p-1}$ sub-groups of order p in G which are not contained in a given sub-group of order p^r. Each of these may be combined with the given sub-group to give a sub-group of order p^{r+1}. When every sub-group of order p^r is treated in this way, every sub-group of order p^{r+1} will be formed and each of them the same number, x, of times. Hence

$$xN_{m,r+1}=N_{m,r}\frac{p^m-p^r}{p-1}.$$

Now a sub-group of order p^{r+1} contains $N_{r+1,r}$ sub-groups of order p^r, and $\frac{p^{r+1}-p^r}{p-1}$ sub-groups of order p which are not contained in any given sub-group of order p^r. Hence

$$x=\frac{p^{r+1}-p^r}{p-1}N_{r+1,r},$$

and therefore

$$N_{m,r+1}=\frac{N_{m,r}}{N_{r+1,r}}\cdot\frac{p^{m-r}-1}{p-1}.$$

We will now assume that

$$N_{m,t} = \frac{(p^m-1)(p^{m-1}-1)\dots(p^{m-t+1}-1)}{(p-1)(p^2-1)\dots(p^t-1)},$$

for all values of m and for values of t not exceeding r. This has been proved for $r=1$. Then it follows, from the above relation, that

$$N_{m,r+1} = \frac{(p^m-1)(p^{m-1}-1)\dots(p^{m-r}-1)}{(p-1)(p^2-1)\dots(p^{r+1}-1)},$$

that is to say, if the result is true for values of t not exceeding r, it is also true when $t=r+1$. Hence the formula is true generally.

It may be noticed that

$$N_{m,t} = N_{m,m-t}.$$

50. Ex. 1. Shew that a group whose operations except identity are all of order 2 is necessarily an Abelian group.

Ex. 2. Prove that in a group of order 16, whose operations except identity are all of order 2, the 15 operations of order 2 may be divided into 5 sets of 3 each so that each set of 3 with identity forms a sub-group of order 4; and that this division into sets may be carried out in 56 distinct ways.

Ex. 3. If G is an Abelian group and H a sub-group of G, shew that G contains one or more sub-groups simply isomorphic with $\frac{G}{H}$.

Ex. 4. If the symbols in the successive rows of a determinant of n rows are derived from those of the first row by performing on them the substitutions of a regular Abelian group of order n, prove that the determinant is the product of n linear factors. (*Messenger of Mathematics*, Vol. XXIII p. 112.)

Ex. 5. Discuss the number of ways in which a set of independent generating operations of an Abelian group of order p^m and given type may be chosen. Shew that, for a group of type $(m_1, m_2, \dots, m_r)$, where $m_1 > m_2 > \dots > m_r$, the number of ways is of the form $p^\alpha(p-1)^r$; and in particular that for a group of order $p^{\frac{1}{2}n(n+1)}$ and type $(n, n-1, \ldots, 2, 1)$, the number of ways is $p^\nu(p-1)^n$, where $\nu = \frac{1}{6}n(n+1)(2n+1) - n$.

Ex. 6. Shew that for any Abelian group a set of independent generating operations

$$S_1,\ S_2,\ \ldots,\ S_{r-1},\ S_r,\ \ldots,\ S_n$$

can be chosen such that, for each value of r, the order of S_r is equal to, or is a factor, of the order of S_{r-1}.

CHAPTER V.

ON GROUPS WHOSE ORDERS ARE THE POWERS OF PRIMES.

51. HAVING in the last chapter dealt in some detail with Abelian groups of order p^m, where p is a prime, we shall now investigate some of the more important properties of groups, which have the power of a prime for their order but are not necessarily Abelian. Besides illustrating and leading to many interesting applications of the general theorems of Chapter III, the discussion of groups, whose order is the power of a prime, will be found in many ways to facilitate the subsequent discussion of other groups, whose order is not thus limited.

52. If G is a group whose order is p^m, where p is a prime, the order of every sub-group of G must also be a power of p; and therefore (§ 25) the total number of operations of G which are conjugate with any given operation must be a power of p. The identical operation of G is self-conjugate. Hence if the operations of G, other than the identical operation, are distributed in conjugate sets containing p^α, p^β, ... operations, the total number of operations of G is $1 + p^\alpha + p^\beta + \ldots$; and therefore

$$p^m = 1 + p^\alpha + p^\beta + \ldots.$$

This equation can only be true if $p^{r_1} - 1$ of the indices α, β, ... are zero, r_1 being some integer not less than unity. Hence G must contain p^{r_1} self-conjugate operations, which form (§ 27) a self-conjugate sub-group*. Hence:—

THEOREM I. *Every group whose order is the power of a prime contains self-conjugate operations, other than the identical operation; and no such group can be simple.*

53. Let H_1 be the sub-group of G of order p^{r_1}, formed of its self-conjugate operations. Then $\frac{G}{H_1}$ is a group of order p^{m-r_1}, and it must contain self-conjugate operations other than identity, forming a self-conjugate sub-group of order p^{r_2}. Corresponding to this self-conjugate sub-group

*Sylow, *Math. Ann.* V. (1872), p. 588.

of $\frac{G}{H_1}$, G must have a self-conjugate sub-group H_2 of order $p^{r_1+r_2}$, which contains H_1. If $r_1 + r_2 < m$, $\frac{G}{H_2}$ is a group of order $p^{m-r_1-r_2}$; and it again will have self-conjugate operations forming a sub-group of order p^{r_3}. Corresponding to this, G will have a self-conjugate sub-group H_3 of order $p^{r_1+r_2+r_3}$, which contains H_2. If the process thus indicated is continued, till G itself is arrived at, a series of groups

$$H_1,\ H_2,\ \ldots,\ H_n,\ G$$

is formed, each of which is self-conjugate in G and contains all that precede it*. Moreover $\frac{H_{r+1}}{H_r}$ is the group formed of the self-conjugate operations of $\frac{G}{H_r}$, and it is therefore an Abelian group.

The series of sub-groups thus arrived at is of considerable importance. From the mode of formation, it is clear that a second group G' of order p^m cannot be simply isomorphic with G, unless $\frac{H_{r+1}}{H_r}$ and $\frac{H'_{r+1}}{H'_r}$ are simply isomorphic for each value of r. It by no means however follows that, when this latter condition is satisfied, G and G' are always simply isomorphic. This is indeed not necessarily the case; instances to the contrary will be found among the groups of order p^3 and p^4 in §§ 69–72. While in general there is no limitation on the type of any Abelian factor-group $\frac{H_{r+1}}{H_r}$ formed from two consecutive terms of the series, it is to be noted that the last factor-group $\frac{G}{H_n}$ cannot be cyclical†, a restriction that can be established as follows. If $\frac{G}{H_n}$ were cyclical, then, because $\frac{H_n}{H_{n-1}}$ is the group formed

*Let P be an operation or sub-group of G, and let H be a sub-group of G which contains P. If every operation of H is permutable with P, we shall speak of P as "self-conjugate in H." The phrase is to be regarded as emphasizing the relation of P to a sub-group of G which contains it; and not as implying that H is the greatest, or the only, sub-group of G, within which P is a self-conjugate operation or sub-group.

†Young, *American Journal*, vol. XV (1893), p. 132.

of the self-conjugate operations of $\frac{G}{H_{n-1}}$, the operations of $\frac{G}{H_{n-1}}$ could be arranged in the sets

$$\frac{H_n}{H_{n-1}},\quad P\frac{H_n}{H_{n-1}},\quad P^2\frac{H_n}{H_{n-1}},\quad \ldots,\quad P^{p^{\alpha-1}}\frac{H_n}{H_{n-1}},$$

p^α being the order of $\frac{G}{H_n}$ and P^{p^α} being an operation of $\frac{H_n}{H_{n-1}}$. But since P is permutable with its own powers and with every operation of $\frac{H_n}{H_{n-1}}$, these operations would form an Abelian group. Now the self-conjugate operations of $\frac{G}{H_{n-1}}$ form the group $\frac{H_n}{H_{n-1}}$, and therefore $\frac{G}{H_{n-1}}$ cannot be Abelian. Hence the assumption, that $\frac{G}{H_n}$ is cyclical, is incorrect.

One immediate consequence of this result is that a group G of order p^2 is necessarily Abelian. For if it were not, its self-conjugate operations would form a sub-group H of order p, and $\frac{G}{H}$ being of order p would be cyclical.

A second consequence is that if G, of order p^m, is not Abelian, the order of the sub-group H_1, formed of the self-conjugate operations of G, cannot exceed p^{m-2}.

54. It was proved in the last chapter that an Abelian group has one or more sub-groups the order of which is any given factor of the order of the group itself. Hence, since $\frac{H_{r+1}}{H_r}$ is Abelian, H_{r+1} must have one or more sub-groups which contain H_r, of any given order greater than that of H_r and less than its own.

In particular, if p^s and p^t are the orders of H_r and H_{r+1}, then $\frac{H_{r+1}}{H_r}$ must have a series of sub-groups of orders

$$p^{t-s-1},\ p^{t-s-2},\ \ldots,\ p,\ 1,$$

each of which contains those that follow it, while each one is of necessity self-conjugate in $\frac{G}{H_r}$. Hence between H_r and H_{r+1}, a series of groups h_{s+1},

$h_{s+2}, \ldots, h_{t-1}$ of orders $p^{s+1}, p^{s+2}, \ldots, p^{t-1}$ can be chosen, so that each of the series

$$H_r, h_{s+1}, h_{s+2}, \ldots, h_{t-1}, H_{r+1}$$

is self-conjugate in G and contains all that precede it. We can therefore always form, in at least one way, a series of sub-groups of G of orders

$$1, p, p^2, \ldots, p^{m-1}, p^m,$$

such that each is self-conjugate in G and contains all that precede it.

It will be shewn in the next paragraph that every sub-group of order p^s of G is contained self-conjugately in some sub-group of order p^{s+1}, and that every sub-group of order p^s of a group of order p^{s+1} is self-conjugate. Assuming this result, it follows at once that a series of sub-groups of orders

$$1, p, p^2, \ldots, p^{m-1}, p^m$$

can always be formed, which shall contain any one given sub-group and which shall be such that each group of the series contains the previous one self-conjugately. It will not however, in general, be the case, as in the previous series, that each group of the series is self-conjugate in G.

55. Let G_s be any sub-group of G of order p^s, and let H_{r+1} be the first of the series of sub-groups

$$H_1, H_2, \ldots, H_n$$

which is not contained in G_s. Then $\dfrac{G_s}{H_r}$ is a sub-group of $\dfrac{G}{H_r}$ which does not contain all the operations of $\dfrac{H_{r+1}}{H_r}$. Every operation of $\dfrac{H_{r+1}}{H_r}$ is self-conjugate in $\dfrac{G}{H_r}$; and therefore $\dfrac{G_s}{H_r}$ is self-conjugate in $\left\{\dfrac{G_s}{H_r}, \dfrac{H_{r+1}}{H_r}\right\}$, a group of order greater than its own. Hence G_s must be contained self-conjugately in some group G_{s+t} of order p^{s+t}, where t is not less than unity. Moreover since $\dfrac{G_{s+t}}{G_s}$ must contain operations of order p, there must be one or more groups of order p^{s+1} which contain G_s self-conjugately. Hence:—

THEOREM II. *If G_s of order p^s is a sub-group of G, which is of order p^m, then G must contain a sub-group of order p^{s+t}, $t \nless 1$, within which G_s is self-conjugate. In particular, every sub-group of order p^{m-1} of G is a self-conjugate sub-group**.

Suppose now that G_{s+t}, of order p^{s+t}, is the greatest sub-group of G which contains a given sub-group G_s, of order p^s, self-conjugately; so that G_s is one of p^{m-s-t} conjugate sub-groups. Suppose also that G_{s+t+u} of order p^{s+t+u}, $(u \nless 1)$, is the greatest sub-group of G that contains G_{s+t} self-conjugately. Every operation of G_{s+t+u} transforms G_{s+t} into itself; and no operation of G_{s+t+u} that is not contained in G_{s+t} transforms G_s into itself. Hence, in G_{s+t+u}, G_s is one of p^u conjugate sub-groups and each of these is self-conjugate in G_{s+t}.

The p^{m-s-t} sub-groups conjugate with G_s may therefore be divided into $p^{m-s-t-u}$ sets of p^u each, $(u \nless 1)$, such that any operation of one of the sets transforms each sub-group of that set into itself.

Similarly if G_r, of order p^r, is the greatest sub-group of G that contains a given operation P self-conjugately, and if G_{r+s}, $s \nless 1$, is the greatest sub-group that contains G_r self-conjugately, then G_r must contain self-conjugately p^s operations of the conjugate set to which P belongs, and therefore any two of these p^s operations are permutable. Hence the p^{m-r} conjugate operations of the set to which P belongs can be divided into p^{m-r-s} sets of p^s each, $(s \nless 1)$, such that all the operations of any one set are permutable with each other. In particular, if P is one of a set of p conjugate operations, all the operations of the set are permutable.

56. If P is an operation of G, of order p^n, which is conjugate to one of its own powers P^α, there must be some other operation Q such that

$$Q^{-1}PQ = P^\alpha.$$

From this equation it follows that

$$Q^{-1}P^xQ = (Q^{-1}PQ)^x = P^{x\alpha}.$$

*Frobenius, "Ueber endliche Gruppen," *Berliner Sitzungsberichte* (1895), p. 173: Burnside, "Notes on the theory of groups of finite order," *Proc. London Mathematical Society*, Vol. XXVI (1895), p. 209.

Again

$$Q^{-2}PQ^2 = Q^{-1}P^{\alpha}Q \ = P^{\alpha^2},$$
$$Q^{-3}PQ^3 = Q^{-1}P^{\alpha^2}Q = P^{\alpha^3},$$

and so on. Hence

$$Q^{-y}P^xQ^y = P^{x\alpha^y}.$$

If Q^β is the lowest power of Q which is permutable with P, then in $\{P, Q\}$ P will be one of β conjugate operations, and therefore β must be a power of p, say p^r, less than p^n. Now

$$Q^{-p^r}PQ^{p^r} = P^{\alpha^{p^r}};$$

and therefore

$$\alpha^{p^r} \equiv 1 \text{ (mod. } p^n),$$

while

$$\alpha^{p^{r-1}} \not\equiv 1 \text{ (mod. } p^n).$$

First, we will suppose that p is an odd prime. Then since

$$x^{p^r} = x \text{ (mod. } p),$$

whatever integer x may be, we may assume that $\alpha = 1 + kp^s$, where k is not a multiple of p; and then

$$\alpha^{p^r} - 1 + kp^{r+s} + \ldots$$
$$\alpha^{p^{r-1}} = 1 + kp^{r+s-1} + \ldots$$

Hence

$$r + s = n,$$

and

$$\alpha \equiv 1 \text{ (mod. } p^{n-r}).$$

In particular, we see that no operation of order p can be conjugate to one of its powers. Hence if P and P' are two conjugate operations of order p, $\{P\}$ and $\{P'\}$ have no operation in common except identity. Also, if $\{P\}$ be a self-conjugate sub-group of order p, each of its operations is self-conjugate.

If p is 2, we must take

$$\alpha = \pm 1 + k2^s,$$

where k is odd, and we are led by the same process to the result

$$\alpha \equiv \pm 1 \ (\text{mod. } 2^{n-r}).$$

57. If P is an operation of G which belongs to the sub-group H_{r+1} (§ 53) and to no previous sub-group in the set

$$1,\ H_1,\ H_2,\ \ldots,\ H_n,\ G,$$

then $\{P, H_r\}$ is a sub-group of H_{r+1}; and therefore every operation of $\frac{\{P, H_r\}}{H_r}$ is self-conjugate in $\frac{G}{H_r}$. The set of operations PH_r is therefore transformed into itself by every operation of G, and hence every operation that is conjugate to P is contained in the set PH_r. Suppose now, if possible, that all the operations conjugate with P are contained in the set PH_{r-1}. Then every operation of G transforms this set into itself, and therefore every operation of $\frac{\{P, H_{r-1}\}}{H_{r-1}}$ is self-conjugate in $\frac{G}{H_{r-1}}$. This however cannot be the case, since by supposition P does not belong to H_r. Hence among the operations conjugate to P, there must be some which belong to PH_r and not to PH_{r-1}. In particular, if P is one of p conjugate operations, no operation conjugate to P can belong to the set PH_{r-1}. For there must be an operation Q such that $Q^{-1}PQ$ belongs to PH_r and not to PH_{r-1}; and this being the case, no one of the operations $Q^{-x}PQ^x$, $(x = 1, 2, \ldots, p-1)$, can belong to PH_{r-1}. But these operations with P constitute the conjugate set.

It follows from the above reasoning that, if P belongs to H_{r+1}, then each operation of the set $P^{-\alpha}Q^{-1}P^{\alpha}Q$, where Q is any operation of G, belongs to H_r. If Q belongs to H_{s+1}, $(s < r)$, the operation $P^{-\alpha}Q^{-1}P^{\alpha}Q$, regarded as the product of $P^{-\alpha}Q^{-1}P^{\alpha}$ and Q, must similarly belong to H_s. Hence unless every operation of H_{s+1} is permutable with P, there must be operations conjugate to P, which belong to the set PH_s. Moreover those operations of G which transform P into operations of the set PH_s form a sub-group. For if

$$S^{-1}PS = Ph_s,$$

and

$$S'^{-1}PS' = Ph'_s,$$

then

$$S'^{-1}S^{-1}PSS' = Ph'_s \cdot S'^{-1}h_sS,$$

and when both h_s and h'_s belong to H_s, so also does $h'_sS'^{-1}h_sS'$.

58. In illustration of the foregoing paragraph, we will consider a group G of order p^{r+s} where p is an odd prime in which the group H_1, is of type (r), while $\frac{G}{H_1}$ is an Abelian group of type $(1, 1, \dots$ with s units).

Let Q be an operation of order p^r that generates H_1; and let $P_1, P_2, \dots, P_s$ be s operations no one of which is a power of any other, such that, with Q, they generate G.

If P_1 and P_2 are not permutable, then

$$P_2^{-1}P_1P_2 = P_1Q^{\alpha},$$

and

$$P_2^{-p}P_1P_2^{p} = P_1Q^{p\alpha}.$$

Now P_2^p belongs to H_1, and therefore

$$Q^{p\alpha} = 1,$$

so that

$$\alpha = kp^{r-1}.$$

Hence the only operations which are conjugate to P_1 are

$$P_1,\ P_1Q^{p^{r-1}},\ P_1Q^{2p^{r-1}},\ \dots,\ P_1Q^{(p-1)p^{r-1}};$$

and P_1 is one of a set of p conjugate operations, which are transformed among themselves by P_2. Similarly P_2 is one of a set of p conjugate operations which are transformed among themselves by P_1. Hence both P_1 and P_2 are permutable with each of the operations P_3, P_4, ..., P_s. Again, P_3 is not self-conjugate, and there must therefore be another, say P_4, of the set P_3, P_4, ..., P_s which, with its powers, transforms P_3 into a set of p conjugate operations. Then P_3 will similarly transform P_4; and P_3, P_4 will be permutable with each of the set P_5, ..., P_s, P_1, P_2. Hence finally the set of s operations P_1, P_2, ..., P_s must be divisible into sets of two, such that each pair are permutable with all the remaining operations, but are not permutable with each other. The group can therefore only exist if s is even*.

The sub-group $\{P_1, P_2, Q\}$ is a group which satisfies the same conditions as G, when $s = 2$. Its order is p^{r+2}, and it contains a self-conjugate operation of order p^r. Now we shall see in §§ 65, 66 that such a group is necessarily of one of two types†, the generating operations of which satisfy one or the other of the sets of equations

$$Q^{p^r} = 1, \quad P_1^p = 1, \quad P_2^p = 1, \quad P_2^{-1}P_1P_2 = P_1Q^{p^{r-1}} :$$

or

$$Q^{p^r} = 1, \quad P_1^p = Q, \quad P_2^p = 1, \quad P_2^{-1}P_1P_2 = P_1Q^{p^{r-1}}.$$

The same is true for the sub-groups $\{P_3, P_4, Q\}$, etc.; and all the operations of any one of these sub-groups are permutable with each operation of $\{P_1, P_2, Q\}$.

---

*Young, "On groups whose order is the power of a prime," *American Journal of Mathematics*, Vol. XV (1893), p. 171.

†Let

$$S_1^{\alpha_1} = 1, \quad S_2^{\alpha_2} = 1, \quad \ldots, \quad S_n^{\alpha_n} = 1,$$
$$f_1(S_i) = 1, \quad f_2(S_i) = 1, \quad \ldots, \quad f_k(S_i) = 1,$$

where $f_j(S_i)$ is an abbreviation for an expression of the form $S_p^a S_q^b \ldots S_r^c$, be a set of relations, such as was considered in § 18, which completely specify a group. We may then, without altering the sense in which the word "type" has been used (§§ 19, 44), speak of *the type of group* defined by these relations. It is essential however that the relations should completely specify the group, as otherwise they will define more than one type. For instance, it is clear, from § 56, that the equations

$$Q^p = 1, \quad P^{p^2} = 1, \quad Q^{-1}PQ = P^\alpha,$$

where p is a given prime, but α is not given, will define more than one type of group. Any data in fact, which completely specify a group, may be said to define a type of group. Thus in dealing with Abelian groups of order p^m, a type is defined by a partition of m.

Hence finally, since the equations to be satisfied by each pair of operations, such as P_1 and P_2, may be chosen in two distinct ways, there are in all $2^{\frac{1}{2}s}$ distinct types of group of order p^{r+s}, for which H_1 is a cyclical group of order p^r, and $\frac{G}{H_1}$ is an Abelian group of type $(1, 1, \ldots, \text{to } s \text{ units})$.

59. It has been seen in § 55 that every sub-group G' of order p^{m-1} of a group G of order p^m is self-conjugate. Suppose now that G contains two such sub-groups G' and G''. Then since G' and G'' are permutable with each other, while the order of $\{G', G''\}$ is p^m, the order of the greatest group g' common to them must (§ 33) be p^{m-2}; and since g' is the greatest common sub-group of two self-conjugate sub-groups of G, it must itself be a self-conjugate sub-group of G. The factor group $\frac{G}{g'}$ of order p^2 contains the two distinct sub-groups $\frac{G'}{g'}$ and $\frac{G''}{g'}$, which are of order p and permutable with each other. Hence $\frac{G}{g'}$ must be an Abelian group of type $(1, 1)$, and it therefore contains (§ 49) $p+1$ sub-groups of order p. Hence, besides G', G must contain p other sub-groups of order p^{m-1} which have in common with G' the sub-group g'. If the $p+1$ sub-groups thus obtained do not exhaust the sub-groups of G of order p^{m-1}, let G''' be a new one. Then, as before, G' and G''' must have a common sub-group g'', of order p^{m-2}, which is self-conjugate in G. If g'' were the same as g', $\frac{G'''}{g'}$ would be contained in $\frac{G}{g'}$, which by supposition is not the case. It may now be shewn as above that there are, in addition to G', p sub-groups of order p^{m-1} which have in common with G' the group g''. These are therefore necessarily distinct from those before obtained. This process may clearly be repeated till all the sub-groups of order p^{m-1} are exhausted. Hence finally, if the number of sub-groups of G, of order p^{m-1} be r_{m-1}, we have $r_{m-1} \equiv 1 \pmod{p}$.

60. The self-conjugate operations of a group G of order p^m, whose orders are p, form with identity a self-conjugate sub-group whose order is some power of p; and therefore their number must be congruent to

-1, (mod. p). On the other hand, if P is any operation of G of order p which is not self-conjugate, the number of operations in the conjugate set to which P belongs is a power of p. Hence the total number of operations of G, of order p, is congruent to -1, (mod. p). Now if r_1 is the total number of sub-groups of G of order p, the number of operations of order p is $r_1(p-1)$, since no two of these sub-groups can have a common operation, except identity. It follows that

$$r_1(p-1) \equiv -1, \text{ (mod. } p),$$

and therefore

$$r_1 \equiv 1, \text{ (mod. } p).$$

If now G_s is any sub-group of G of order p^s, and if G_{s+t} is the greatest sub-group of G in which G_s is contained self-conjugately, then every sub-group of G which contains G_s self-conjugately is contained in G_{s+t}. But every sub-group of order p^{s+1}, which contains G_s, contains G_s self-conjugately; and therefore every sub-group of order p^{s+1}, which contains G_s, is itself contained in G_{s+t}. By the preceding result, the number of sub-groups of $\dfrac{G_{s+t}}{G_s}$ of order p is congruent to unity, (mod. p). Hence the number of sub-groups of G of order p^{s+1}, which contain G_s of order p^s, is congruent to unity, (mod. p).

61. Let now r_s represent the total number of sub-groups of order p^s contained in a group G of order p^m. If any one of them is contained in a_x sub-groups of order p^{s+1}, and if any one of the sub-groups of order p^{s+1} contains b_y sub-groups of order p^s, then

$$\sum_{x=1}^{x=r} a_x = \sum_{y=1}^{y=r_{s+1}} b_y;$$

for the numbers on either side of this equation are both equal to the number of sub-groups of order p^{s+1}, when each of the latter is reckoned once for every sub-group of order p^s that it contains. It has however been shewn, in the two preceding paragraphs, that for all values of x and y

$$a_x \equiv 1, \quad b_y \equiv 1 \text{ (mod. } p).$$

Hence

$$r_s \equiv r_{s+1} \text{ (mod. } p).$$

Now it has just been proved that

$$r_1 \equiv 1 \quad \text{and} \quad r_{m-1} \equiv 1 \text{ (mod. } p);$$

and therefore finally, for all values of s,

$$r_s \equiv 1 \text{ (mod. } p).$$

We may state the results thus obtained as follows:—

THEOREM III. *The number of sub-groups of any given order p^s of a group of order p^m is congruent to unity,* (mod. p)*.

Corollary. The number of self-conjugate sub-groups of order p^s of a group of order p^m is congruent to unity, (mod. p).

This is an immediate consequence of the theorem, since the number of sub-groups in any conjugate set is a power of p.

62. Having shewn that the number of sub-groups of G of order p^s is of the form $1 + kp$, we may now discuss under what circumstances it is possible for k to be zero, so that G then contains only one sub-group G_s of order p^s.

If this is the case, and if P is any operation of G not contained in G_s, the order of P must be not less than p^{s+1}; for if it were less, G would have some sub-group of order p^s containing P and this would necessarily be different from G_s. If the order of P is p^{s+t}, then $\{P^{p^t}\}$ is a cyclical sub-group of order p^s; and it must coincide with G_s. Hence, if G_s is the only sub-group of order p^s, it must be cyclical.

Suppose now that G contains operations of order p^r $(r > s)$, but no operations of order p^{r+1}; and let P be an operation of G of order p^r. Then $\{P\}$ must be contained self-conjugately in a non-cyclical sub-group of order p^{r+1}.

---

*Frobenius, "Verallgemeinerung des Sylow'schen Satzes," *Berliner Sitzungsberichte*, (1895), p. 989.

We will take first the case in which p is an odd prime. Then (§ 55) G must contain an operation P' which does not belong to $\{P\}$, such that

$$P'^{-1}PP' = P^{\alpha}, \quad P'^{p} = P^{\beta}.$$

If α were unity, $\{P, P'\}$ would be an Abelian group of order p^{r+1} containing no operation of order p^{r+1}. Its type would therefore be $(r, 1)$, and it would necessarily contain an operation of order p not occurring in $\{P\}$. It has been shewn that this is impossible if G_s is the only sub-group of order p^s, and therefore α cannot be unity.

We may then without loss of generality (§ 56) assume that

$$\alpha = 1 + p^{r-1}.$$

Moreover if β were not divisible by p, the order of P' would be p^{r+1}, contrary to supposition. Hence we must have the relations

$$P'^{-1}PP' = P^{1+p^{r-1}}, \quad P'^{p} = P^{\gamma p}.$$

By successive applications of the first of these equations, we get

$$P'^{-y}P^{x}P'^{y} = P^{x(1+yp^{r-1})},$$

for all values of x and y; and from this it immediately follows that

$$\begin{aligned}(P^{x}P')^{p} &= P'^{p}P^{x\{p+\frac{1}{2}p(p+1)p^{r-1}\}}\\ &= P^{(x+\gamma)p}.\end{aligned}$$

Hence if

$$x = p^{r-1} - \gamma,$$

the order of $P^{x}P'$, an operation not contained in $\{P\}$, is p. This is impossible if G_s is the only sub-group of order p^s. If then $r < m$, G must contain operations of order greater than p^r; and G is therefore a cyclical group. Hence:—

THEOREM IV. *If G, of order p^m, where p is an odd prime, contains only one sub-group of order p^s, G must be cyclical.*

63. When $p = 2$, the result is not so simple.

Let Q be an operation that transforms $\{P\}$ into itself, and suppose that the lowest power of Q that occurs in $\{P\}$ is Q^4. Then $\{P, Q\}$ must be defined by

$$Q^{-1}PQ = P^{\alpha}, \quad Q^4 = P^{\beta},$$

where

$$\alpha = \pm 1 + k2^{r-2}.$$

Moreover β must be a multiple of 4, as otherwise the order of Q would be greater than 2^r. A simple calculation now gives

$$(P^x Q^2)^2 = P^{\beta + x(1+\alpha^2)};$$

and since

$$\beta \equiv 0 \text{ (mod. 4)},$$

and

$$\alpha^2 \equiv 1 \text{ (mod. } 2^{r-1}),$$

x can always be chosen so that

$$\beta + x(1 + \alpha^2) \equiv 0 \text{ (mod. } 2^r).$$

When x is thus chosen, $P^x Q^2$ is an operation of order 2 which is not contained in $\{P\}$. But this is inconsistent with G_s being the only sub-group of order 2^s; hence $\{P\}$ must contain Q^2.

If now $\{P\}$ is not a self-conjugate sub-group, there must (§ 55) be some operation P' of order p^r, conjugate with P and not contained in $\{P\}$, which transforms $\{P\}$ into itself; and then $\{P, P'\}$ is defined by

$$P'^{-1}PP' = P^{\alpha}, \quad P'^2 = P^{2\beta},$$

where α is -1 or $\pm 1 + 2^{r-1}$, and β is odd.

If

$$\alpha = 1 + 2^{r-1},$$

then

$$(P^x P')^2 = P^{2\{\beta + x(1+2^{r-2})\}};$$

and x can be chosen so that $P^x P'$ is of order 2. Hence this case cannot occur.

If α is either -1 or $-1+2^{r-1}$, the defining relations of $\{P, P'\}$ are not self-consistent unless r is 2; for they lead to

$$P'^{-1}P^{2\beta}P' = P^{-2\beta},$$

and

$$P'^{-1}P^{2\beta}P' = P^{2\beta}.$$

If r is 2, and if P'' is an operation that transforms P into P', so that

$$P''^{-1}PP'' = P',$$

then

$$(PP'')^2 = PP''^2P' = P^{-1}P',$$

since

$$P^2 = P'^2 = P''^2.$$

Hence PP'' would be of order 8; and therefore this case cannot occur.

Hence, finally, $\{P\}$ must be self-conjugate.

If $r+1 < m$, $\{P\}$ must be transformed into itself by some operation Q' not contained in $\{P,\ Q\}$, so that

$$Q^{-1}PQ = P^{-1+k2^{r-1}},$$

and

$$Q'^{-1}PQ' = P^{-1+k'2^{r-1}},$$

where k and k' are each either zero or unity. If both are zero or both unity, QQ' is permutable with P; and if one is zero and the other unity,

$$(QQ')^{-1}P(QQ') = P^{1+2^{r-1}}.$$

In either case, the group contains an operation of order 2 that does not belong to $\{P\}$, and this is inadmissible. Hence, lastly, r must not be less than $m-1$.

Suppose now that r is equal to $m-1$, and that of the operations of G, not contained in $\{P\}$, Q has as small an order as possible, say 2^t $(t \not< s+1)$. Then Q^2 of order 2^{t-1} is contained in $\{P\}$; without loss of generality we may assume

$$Q^2 = P^{2^{m-t}}.$$

Moreover

$$Q^{-1}PQ = P^{-1} \quad \text{or} \quad P^{-1+2^{m-2}},$$

and hence

$$Q^{-1}P^{2^{m-t}}Q = P^{-2^{m-t}}.$$

Now Q is permutable with $P^{2^{m-t}}$, one of its own powers, and therefore $P^{2^{m-t}}$ is an operation of order 2. Hence t is 2 and s is unity. If then s is greater than unity, G must contain operations of order 2^m and must therefore be cyclical. Moreover if s is unity, the relations

$$Q^{-1}PQ = P^{-1+2^{m-2}}, \quad Q^2 = P^{2^{m-2}},$$

lead to

$$(PQ)^2 = 1,$$

and they are therefore inadmissible.

On the other hand, the relations

$$Q^{-1}PQ = P^{-1}, \quad Q^2 = P^{2^{m-2}},$$

give

$$(P^xQ)^2 = P^{2^{m-2}};$$

and every operation of G, not contained in $\{P\}$, is of order 4. Also these relations are clearly self-consistent, and they define a group of order 2^m. Hence finally:—

THEOREM V. *If a group G, of order 2^m, has a single sub-group of order 2^s, $(s > 1)$, it must be cyclical; if it has a single sub-group of order 2, it is either cyclical or of the type defined by*

$$P^{2^{m-1}} = 1, \quad Q^2 = P^{2^{m-2}}, \quad Q^{-1}PQ = P^{-1}.$$

64. We shall now proceed to discuss, in application of the foregoing theorems and for the importance of the results themselves, the various types of groups of order p^m which contain self-conjugate cyclical sub-groups of orders p^{m-1} and p^{m-2} respectively. It is clear from Theorem V that the case $p = 2$ requires independent investigation; we shall only deal in detail with the case in which p is an odd prime, and shall state the results for the case when $p = 2$.

The types of Abelian groups of order p^m which contain operations of order p^{m-2} are those corresponding to the symbols (m), $(m-1, 1)$, $(m-2, 2)$, and $(m-2, 1, 1)$. We will assume that the groups which we consider in the following paragraphs are not Abelian.

65. We will first consider a group G, of order p^m, which contains an operation P of order p^{m-1}. The cyclical sub-group $\{P\}$ is self-conjugate and contains a single sub-group $\{P^{p^{m-2}}\}$ of order p. By Theorem IV, since G is not cyclical, it must contain an operation Q', of order p, which does not occur in $\{P\}$. Since $\{P\}$ is self-conjugate and the group is not Abelian, Q' must transform P into one of its own powers. Hence

$$Q'^{-1}PQ' = P^{\alpha},$$

and since Q'^p is permutable with P it follows, from § 56, that

$$\alpha = 1 + kp^{m-2}.$$

Since the group is not Abelian, k cannot be zero; but it may have any value from 1 to $p-1$. If now

$$kx \equiv 1 \text{ (mod. } p),$$

then

$$Q'^{-x}PQ'^{x} = P^{1+p^{m-2}};$$

and therefore, writing Q for Q'^x, the group is defined by

$$P^{p^{m-1}} = 1, \quad Q^p = 1, \quad Q^{-1}PQ = P^{1+pm-2}.$$

These relations are clearly self-consistent, and they define a group of order p^m.

There is therefore a single type of non-Abelian group of order p^m which contains operations of order p^{m-1}, because, for any such group, a pair of generating operations may be chosen which satisfy the above relations.

From the relation

$$Q^{-1}PQ = P^{1+p^{m-2}},$$

it follows by repetition and multiplication that

$$Q^{-y}P^xQ^y = P^{x(1+yp^{m-2})},$$

and therefore that

$$(P^xQ^y)^z = P^{xz\{1+\frac{1}{2}(z+1)yp^{m-2}\}(1-yzp^{m-2})}Q^{yz},$$

and

$$(P^xQ^y)^p = P^{xp}.$$

Hence G contains p cyclical sub-groups of order p^{m-1} of which P and PQ^y $(y = 1, 2, \dots, p-1)$ may be taken as the generating operations. Since Q and P^p are permutable, G also contains an Abelian non-cyclical sub-group $\{Q, P^p\}$ of order p^{m-1} It is easy to verify that the $1+p$ sub-groups thus obtained exhaust the sub-groups of order p^{m-1}; and that, for any other order p^s, there are also exactly $p+1$ sub-groups of which p are cyclical and one is Abelian of type $(s-1, 1)$.

The reader will find it an instructive exercise to verify the results of the corresponding case where p is 2; they may be stated thus. There are four distinct types of non-Abelian group of order 2^m, which contain operations of order 2^{m-1}, when $m > 3$. Of these, one is the type given in Theorem V, and the remaining three are defined by

$$P^{2^{m-1}} = 1, \quad Q^2 = 1, \quad QPQ = P^{1+2^{m-2}};$$
$$P^{2^{m-1}} = 1, \quad Q^2 = 1, \quad QPQ = P^{-1+2^{m-2}};$$
$$P^{2^{m-1}} = 1, \quad Q^2 = 1, \quad QPQ = P^{-1}.$$

When $m = 3$, there are only two distinct types. In this case, the second and the fourth of the above groups are identical, and the third is Abelian.

66. Suppose next that G, a group of order p^m, has a self-conjugate cyclical sub-group $\{P\}$ of order p^{m-2}, and that no operation of G is of higher order than p^{m-2}. We may at once distinguish two cases for separate discussion; viz. (i) that in which P is a self-conjugate operation, (ii) that in which P is not self-conjugate.

Taking the first case, there can be no operation Q' in G such that Q'^{p^2} is the lowest power of Q' contained in $\{P\}$, for if there were, $\{Q', P\}$ would be Abelian and, its order being p^m, it would necessarily coincide with G. Hence any operation Q', not contained in $\{P\}$, generates with P an Abelian group of type $(m-2, 1)$, and we may choose P and Q as independent generators of this sub-group, the order of Q being p. If now R' is any operation of G not contained in $\{Q, P\}$, R'^p must occur in this sub-group, and therefore

$$R'^p = P^\alpha Q^\beta.$$

If β were not zero, R^{p^2} would be the lowest power of R that occurs in $\{P\}$, and we have just seen that this cannot be the case. Hence

$$R'^p = P^\alpha,$$

and $\{R', P\}$ is again an Abelian group of type $(m-2, 1)$. If P and R are independent generators of this group, the latter cannot occur in $\{Q, P\}$. Now since Q is not self-conjugate,

$$R^{-1}QR = QP^\beta;$$

and since R^p is permutable with Q

$$P^{p\beta} = 1,$$

so that

$$\beta \equiv 0 \ (\text{mod. } p^{m-3}).$$

Hence

$$R^{-1}QR = QP^{kp^{m-3}},$$

where k is not a multiple of p. If finally, P^k be taken as a generating operation in the place of P, the group is defined by

$$P^{p^{m-2}} = 1, \quad Q^p = 1, \quad R^p = 1, \quad R^{-1}QR = QP^{p^{m-3}},$$
$$PQ = QP, \quad PR = RP.$$

There is therefore a single type of group of order p^m, which contains a self-conjugate operation of order p^{m-2} and no operation of order p^{m-1}.

67. Taking next the second case, we suppose that G contains a cyclical self-conjugate sub-group $\{P\}$ of order p^{m-2}, but no self-conjugate operation of this order.

If G contains an operation Q' such that Q'^{p^2} is the lowest power of Q' occurring in $\{P\}$, it follows (§ 56), since $\{P\}$ is self-conjugate, that

$$Q'^{-1}PQ' = P^{1+kp^{m-4}}.$$

Moreover, since the order of Q' cannot exceed p^{m-2}, we have

$$Q'^{p^2} = P^{\alpha p^2}.$$

A simple calculation, similar to that of § 65, will now shew that

$$(P^x Q')^{p^2} = P^{(x+\alpha)p^2},$$

while

$$(P^x Q')^p = Q'^p P^{x\{p+\frac{1}{2}k(p+1)p^{m-3}\}}.$$

Hence $P^{-\alpha}Q'$ is an operation of order p^2, none of whose powers, except identity, is contained in $\{P\}$. If this operation be represented by Q, G is defined by

$$P^{p^{m-2}} = 1, \quad Q^{p^2} = 1, \quad Q^{-1}PQ = P^{1+kp^{m-4}}.$$

If k is a multiple of p^2, the group is Abelian.

If k is $k'p$, and if Q^α is taken for a new generating operation, where α is chosen so that

$$\alpha k' \equiv 1 \text{ (mod. } p),$$

we have a single type defined by

$$P^{p^{m-2}} = 1, \quad Q^{p^2} = 1, \quad Q^{-1}PQ = P^{1+p^{m-3}}.$$

If k is not a multiple of p, we may take it equal to $k_1 + k_2p$, where k_1 and k_2 are both less than p, and k_1 is not zero. (This case cannot occur if $m < 5$.) Then

$$Q^{-x_1-x_2p}PQ^{x_1+x_2p} = P^{1+k_1x_1p^{m-4}+(k_1x_2+k_2x_1)p^{m-3}},$$

when $m > 5$; while if $m = 5$, the coefficient of p^{m-3}, viz. of p^2, in the index of P must be increased by $\frac{1}{2}x_1(x_1 - 1)k_1^2$.

If now we choose x_1 so that

$$k_1x_1 = 1 + y_1p,$$

and then x_2 so that

$$y_1 + k_1x_2 + k_2x_1 \equiv 0 \text{ (mod. } p),$$

(with the suitable modification when $m = 5$), then $Q^{x_1+x_2p}$ transforms P into $P^{1+p^{m-4}}$. We therefore again in this case get a single type of group, which, if $Q^{x_1+x_2p}$ is represented by R, is defined by

$$P^{p^{m-2}} = 1, \quad R^{p^2} = 1, \quad R^{-1}PR = P^{1+p^{m-4}}.$$

As already stated, this type exists only when $m > 4$.

If every operation of G is such that its pth power occurs in $\{P\}$, P must be one of a set of p conjugate operations; for if P is one of p^2 conjugate operations, there must be an operation Q which transforms P into $P^{1+kp^{m-4}}$, where k is not a multiple of p, and then Q^{p^2} is the lowest power of Q which is permutable with P. Since P is one of p conjugate operations, it must be self-conjugate in an Abelian group $\{Q, P\}$ of type $(m-2, 1)$. Again, since P and $P^{1+p^{m-3}}$ are conjugate operations, P must be contained in a group of order p^{m-1}, defined by

$$P^{p^{m-2}} = 1, \quad R^p = 1, \quad R^{-1}PR = P^{1+p^{m-3}}.$$

If Q is not a self-conjugate operation, it must be one of p conjugate operations and (§ 57) these must be

$$Q, \quad QP^{p^{m-3}}, \quad QP^{2p^{m-3}}, \ldots, \quad QP^{(p-1)p^{m-3}}.$$

Now if

$$R^{-1}QR = QP^{\alpha p^{m-3}},$$

then

$$R^{-1}P^{-\alpha}QR = P^{-\alpha}Q.$$

Hence, if Q is not self-conjugate, $P^{-\alpha}Q$ is a self-conjugate operation of order p^{m-2}: and the group is that determined in the last paragraph. Therefore Q must be self-conjugate, and the group is defined by

$$P^{p^{m-2}} = 1, \quad Q^p = 1, \quad R^p = 1, \quad R^{-1}PR = P^{1+p^{m-3}},$$
$$PQ = QP, \quad RQ = QR.$$

It is clear from these relations that the group thus arrived at is the direct product of the groups $\{P, R\}$ and $\{Q\}$*.

It is to be expected, from the result of the corresponding case at the end of § 65, that the number of distinct types when $p = 2$ is much greater than when p is

*In each of the five distinct cases to which we have been led in the discussion contained in §§ 65–67, we have arrived at a set of defining relations, containing no indeterminate symbols, such that in each case a set of generating operations can be chosen to satisfy these relations. To justify the statement, in each particular case, that such a set of relations gives a distinct type of group, it is finally necessary to verify that the relations actually define a group of order p^m. In the cases dealt with in the text, this verification is implicitly contained in the process by which the relations have been arrived at. We have therefore omitted the direct verification, which moreover is extremely simple. We shall similarly omit the corresponding verifications in the discussion of groups of orders p^3 and p^4, as in none of these cases does it present any difficulty.

To illustrate the necessity of such a verification in general, we may consider a simple case. The relations

$$P^3 = 1, \quad Q^9 = 1, \quad P^{-1}QP = Q^\alpha,$$

where α is any given integer, certainly define a group whose order is equal to or is a factor of 27, since they indicate that $\{P\}$ and $\{Q\}$ are permutable. They will not however give a type of group of order 27, unless α is 1, 4 or 7. For instance, if $\alpha = 5$, an odd prime. There are, in fact, when $m > 5$, fourteen distinct types of groups of order 2^m, which contain a self-conjugate cyclical sub-group of order 2^{m-2} and no operation of order 2^{m-1}. They may be classified as follows.

Suppose first that the group has a self-conjugate operation of order 2^{m-2}. There is then a single type defined by the relations

$$\text{(i)}\quad A^{2^{m-2}} = 1, \quad B^2 = 1, \quad C^2 = 1, \quad CBC = BA^{2^{m-3}}.$$

Suppose next that the group G has no self-conjugate operation of order 2^{m-2}, and let $\{A\}$ be a self-conjugate cyclical sub-group of order 2^{m-2}. If $\frac{G}{\{A\}}$ is cyclical, there are, when $m > 5$, five distinct types. The common defining relations of these are

$$A^{2^{m-2}} = 1, \quad B^4 = 1, \quad B^{-1}AB = A^{\alpha};$$

and the five distinct types are

$$\text{(ii)}\quad \alpha = -1, \qquad \text{(iii)}\quad \alpha = 1 + 2^{m-3}, \qquad \text{(iv)}\quad \alpha = -1 + 2^{m-3},$$
$$\text{(v)}\quad \alpha = 1 + 2^{m-4}, \qquad \text{(vi)}\quad \alpha = -1 + 2^{m-4}.$$

If $m = 5$, then (iv) and (v) are identical, and (vi) is Abelian; so that there are only three distinct types. If $m = 4$, there is a single type; it is given by (ii).

When $\frac{G}{\{A\}}$ is not cyclical, the square of every operation of G is contained in $\{A\}$. If all the self-conjugate operations of G are not contained in $\{A\}$, there must be a self-conjugate operation B, of order 2, which does not occur in $\{A\}$. If C is any operation of G, not contained in $\{A, B\}$, then $\{A, C\}$ is a self-conjugate sub-group of order 2^{m-1}, which has no operation except identity in common

---

the relations involve

$$Q = P^{-3}QP^3 = Q^8, \quad \text{or} \quad Q^7 = 1.$$

Hence

$$Q = Q^{4\cdot 7 - 3\cdot 9} = 1,$$

and the relations hold only for a group of order 3.

Again, if $\alpha = 3$, the relations give

$$P^{-1}Q^3P = Q^9 = 1, \quad \text{or} \quad Q^3 = 1,$$

and as before they define a group of order 3.

with $\{B\}$. Hence G is a direct product of a group of order 2 and a group of order 2^{m-1}. There are therefore, for this case, four types (vii), (viii), (ix), (x), when $m > 4$, corresponding to the four groups of order 2^{m-1} of § 65. If $m = 4$, there are two types.

Next, let all the self-conjugate operations of G be contained in $\{A\}$; and suppose that A is one of two conjugate operations. Then G must contain an Abelian sub-group of type $(m-2, 1)$, in which A occurs; and it may be shewn that, when $m > 4$, there are two types defined by the relations

(xi) and (xii)
$$A^{2^{m-2}} = 1, \quad B^2 = 1, \quad BAB = A, \quad C^2 = 1,$$
$$CBC = BA^{2^{m-3}}, \quad CAC = A^{-1} \quad \text{or} \quad A^{-1+2^{m-3}}.$$

When $m = 4$, there is, for this case, no type.

Lastly, suppose that A is one of four conjugate operations. Then G must contain sub-groups of order 2^{m-1}, of the second and the third types of § 65, and a sub-group of order 2^{m-1} of either the first or fourth type (*l.c.*). In this last case, there are two distinct types defined by

(xiii) and (xiv)
$$A^{2^{m-2}} = 1, \quad B^2 = 1, \quad BAB = A^{1+2^{m-3}},$$
$$C^2 = 1, \quad CAC = A^{-1+2^{m-3}}, \quad CBC = B \quad \text{or} \quad BA^{2^{m-3}}.$$

These two types exist only when $m > 4$.

68. We shall now, as a final illustration, determine and tabulate all types of groups of orders p^2, p^3 and p^4. It has been already seen that when $p = 2$ the discussion must, in part at least, be distinct from that for an odd prime; for the sake of brevity we shall not deal in detail with this case, but shall state the results only and leave their verification as an exercise to the reader.

It has been shewn (§ 53) that all groups of order p^2 are Abelian; and hence the only distinct types are those represented by (2) and $(1, 1)$.

For Abelian groups of order p^3, the distinct types are (3), $(2, 1)$ and $(1, 1, 1)$.

If a non-Abelian group of order p^3 contains an operation of order p^2, the sub-group it generates is self-conjugate; hence (§ 65) in this case there is a single type of group defined by

$$P^{p^2} = 1, \quad Q^p = 1, \quad Q^{-1}PQ = P^{1+p}.$$

If there is no operation of order p^2, then since there must be a self-conjugate operation of order p, the group comes under the head discussed in § 66; there is again a single type of group defined by

$$P^p = 1, \quad Q^p = 1, \quad R^p = 1, \quad R^{-1}QR = QP,$$
$$R^{-1}PR = P, \quad Q^{-1}PQ = P.$$

These two types exhaust all the possibilities for non-Abelian groups of order p^3.

69. For Abelian groups of order p^4, the possible distinct types are (4), (3, 1), (2, 2), (2, 1, 1) and (1, 1, 1, 1).

For non-Abelian groups of order p^4 which contain operations of order p^3 there is a single type, namely that given in § 65 when m is put equal to 4.

For non-Abelian groups which contain a self-conjugate cyclical sub-group of order p^2 and no operation of order p^3, there are three distinct types, obtained by writing 4 for m in the group of § 66 and in the first and the last groups of § 67. The defining relations of these need not be here repeated, as they will be given in the summarizing table (§ 73).

It remains now to determine all distinct types of groups of order p^4, which contain no operation of order p^3 and no self-conjugate cyclical sub-group of order p^2. We shall first deal with groups which contain operations of order p^2.

Let S be an operation of order p^2 in a group G of order p^4. The cyclical sub-group $\{S\}$ must be self-conjugate in a non-cyclical sub-group $\{S, T\}$ of order p^3, defined by

$$S^{p^2} = 1, \quad T^p = 1, \quad T^{-1}ST = S^{1+kp}.$$

If R is any operation of G, not contained in $\{S, T\}$, then since $\{S\}$ is not self-conjugate, we must have (§ 57)

$$R^{-1}SR = S^{1+\alpha p}T^{\beta},$$

and therefore

$$R^{-1}S^pR = S^p.$$

Now

$$T^{-1}S^pT = S^p,$$

and therefore the pth power of every operation of order p^2 in G is a self-conjugate operation.

First let us suppose that G contains other self-conjugate operations besides those of $\{S^p\}$. Every such operation must occur in the group that contains $\{S\}$ self-conjugately; hence in this case T must be self-conjugate.

We now therefore have

$$S^{p^2}=1,\quad T^p=1,\quad T^{-1}ST=S,$$
$$R^{-1}SR=S^{1+\alpha p}T^\beta,\quad R^{-1}TR=T,$$
$$R^p=S^{\gamma p}T^\delta.$$

These equations give

$$(S^xR)^p=R^pS^{xp}=S^{(x+\gamma)p}T^\delta.$$

Hence, if $\delta=0$, $S^{-\gamma}R$ is an operation of order p. Denoting this by R' and $S^{\alpha p}T^\beta$ by T', the group is defined by

$$S^{p^2}=1,\quad T'^p=1,\quad R'^p=1,\quad R'^{-1}SR'=ST',$$
$$T'^{-1}ST'=S,\quad R'^{-1}T'R'=T'.$$

On the other hand, if δ is not zero, $S^{\frac{\delta\alpha}{\beta}-\gamma}R$ is an operation of order p^2 such that R transforms it into a power of itself. This is contrary to the supposition that the group contains no cyclical self-conjugate sub-group of order p^2. Hence δ cannot be different from zero: we therefore have only one type of group.

70. Next, let $\{S^p\}$ contain all the self-conjugate operations of G; and as before, let $\{S,T\}$ be the group that contains $\{S\}$ self-conjugately. If G contains an operation S' of order p^2 which does not occur in $\{S,T\}$, there must also be a non-cyclical sub-group $\{S',T'\}$ of order p^3 which contains $\{S'\}$ self-conjugately. Now $\{S,T\}$ and $\{S',T'\}$ must have a common sub-group of order p^2; since this is self-conjugate in G, it cannot be cyclical. The only non-cyclical sub-groups of orders p^2 that $\{S,T\}$ and $\{S',T'\}$ contain are $\{S^p,T\}$ and $\{S'^p,T'\}$. Hence these must be identical, and therefore T must occur in $\{S',T'\}$. If now $\{S,T\}$ and $\{S',T'\}$ were both Abelian, T would be permutable both with S and with S', and would therefore, contrary to supposition, be a self-conjugate operation. Hence either (i) G must contain a non-Abelian group $\{S,T\}$ of order p^3, in which S is an operation of order p^2; or (ii) the Abelian group $\{S,T\}$, in which S is an operation of order p^2, must contain all the operations of G of order p^2.

In the case (i), the group is defined by

$$S^{p^2}=1,\quad T^p=1,\quad T^{-1}ST=S^{1+p},$$
$$R^{-1}SR=S^{1+\alpha p}T^\beta,\quad R^{-1}TR=S^{\gamma p}T,$$
$$R^p=S^{\delta p}.$$

These equations give

$$(S^x R)^p = R^p S^{xp\{1+\frac{1}{6}(p-1)p(p+1)\beta(\gamma-2)\}},$$

and therefore, except when $p = 3$, $S^{-\delta}R$ is an operation of order p. Hence, if $p > 3$, we may, without loss of generality, put zero for δ. In this case a simple calculation gives

$$R^{-x}SR^x = S^{1+\alpha_x p}T^{\beta_x}, \quad R^{-x}TR^x = S^{\gamma_x p}T,$$

where

$$\alpha_x = \tfrac{1}{2}\beta\gamma x(x-1) + x\alpha, \quad \beta_x = x\beta, \quad \gamma_x = x\gamma.$$

Hence if $R^x = R'$, $S^y = S'$, $S^{zp}T = T'$, where x, y and z are not multiples of p, then

$$S'^{p^2} = 1, \quad T'^p = 1, \quad T'^{-1}S'T' = S'^{1+p},$$
$$R'^{-1}S'R' = S'^{1+(\alpha_x - z\beta_x)p}T'^{y\beta},$$
$$R'^{-1}T'R' = S'^{\frac{\gamma_x p}{y}}T', \quad R'^p = 1.$$

If now we take

$$z \equiv \frac{\alpha_x}{\beta_x}, \quad y \equiv \frac{1}{\beta_x} \pmod{p},$$

then

$$R'^{-1}S'R' = S'T', \quad R'^{-1}T'R' = S'^{x^2\beta\gamma p}T'.$$

Dropping the accents, the defining relations now become

$$S^{p^2} = 1, \quad T^p = 1, \quad R^p = 1, \quad T^{-1}ST = S^{1+p},$$
$$R^{-1}SR = ST, \quad R^{-1}TR = S^{\alpha p}T;$$

where α is either zero, unity, or any given non-residue. Since the sub-group $\{R, T, S^p\}$ contains all the operations of order p of the group, it follows at once that these three cases give three distinct types.

When $p = 3$, it will be found that the defining relations again give three distinct types. The operation R may always be chosen so that α and γ are zero. If it is so chosen, the three types correspond to the values 1, 0; -1, 1; and -1, -1; of β and δ.

In case (ii), the group is defined by

$$S^{p^2} = 1, \quad T^p = 1, \quad T^{-1}ST = S,$$
$$R^{-1}SR = S^{1+\alpha p}T^{\beta}, \quad R^{-1}TR = S^{\gamma p}T,$$
$$R^p = 1;$$

with the condition that all operations of G, not contained in $\{S, T\}$, are of order p.

The formulæ for $R^{-x}SR^x$ and $R^{-x}TR^x$ enable us to calculate directly the power of any given operation of G. Thus they give

$$(S^xR)^p = S^{px\{1+\frac{1}{6}\beta\gamma(p+1)p(p-1)+\frac{1}{2}\alpha p(p+1)\}}.$$

If $p > 3$, this gives

$$(S^xR)^p = S^{px},$$

so that, if x is not a multiple of p, the order of S^xR is p^2. Hence the type of group under consideration can only occur when $p = 3$.

In this case

$$(S^xR)^3 = S^{3x(1+\beta\gamma)}.$$

Hence if $p = 3$ and $\beta\gamma \equiv -1$ (mod. 3), we obtain a new type. A reduction similar to that in the previous case may now be effected; and taking unity for x, the group is defined by

$$S^9 = 1, \quad T^3 = 1, \quad R^3 = 1, \quad T^{-1}ST = S,$$
$$R^{-1}SR = ST, \quad R^{-1}TR = S^{-3}T.$$

71. It only remains to determine the distinct types which contain no operation of order p^2.

Suppose first that the self-conjugate operations of G form a group of order p^2. This must be generated by two independent operations P and Q of order p.

If now R is any other operation of the group, $\{P, Q, R\}$ must be an Abelian group of type $(1, 1, 1)$. If again S is any operation not contained in $\{P, Q, R\}$, it cannot be permutable with R; for if it were, R would be self-conjugate. There must therefore be a relation of the form (§ 57)

$$S^{-1}RS = RP^{\alpha}Q^{\beta}.$$

Since any operation of $\{P, Q\}$ may be taken for one of its generating operations, we may take $P^\alpha Q^\beta$ or P' for one. If then Q' is an independent operation of $\{P, Q\}$, G will be defined by

$$P'^p = 1, \quad Q'^p = 1, \quad R^p = 1, \quad S^p = 1, \quad S^{-1}RS = RP',$$

in addition to the relations expressing that P' and Q' are self-conjugate. There is then in this case a single type. That all the operations of G in this case are actually of order p follows from the obvious fact that every sub-group of order p^3 is either Abelian or of the second type of non-Abelian groups of order p^3.

72. Suppose, secondly, that the self-conjugate operations of G form a sub-group of order p, generated by P. There must then be some operation Q which belongs to a set of p conjugate operations; for if every operation of G which is not self-conjugate were one of a set of p^2 conjugate operations, the total number of operations in the group would be congruent to p (mod. p^2). It follows that Q must be self-conjugate in a group of order p^3; and since P is also self-conjugate in this group, it must be Abelian. Let P, Q, R be generators of this group and S any operation of G not contained in it. We may now assume that Q belongs to the sub-group H_2 (§ 53), and therefore that

$$S^{-1}QS = QP^\alpha,$$

while

$$S^{-1}RS = RQ^\beta P^\gamma.$$

If β were zero, $Q^{\frac{1}{\alpha}} R^{-\frac{1}{\gamma}}$ would be a self-conjugate operation not contained in $\{P\}$; and therefore β must be different from zero. We may now put

$$Q^\beta P^\gamma = Q', \quad P^{\alpha\beta} = P';$$

and the group is then defined by the relations

$$P'^p = 1, \quad Q'^p = 1, \quad R^p = 1, \quad S^p = 1,$$
$$S^{-1}RS = RQ', \quad S^{-1}Q'S = Q'P',$$

together with the relations expressing that P is self-conjugate. There is thus again in this case, at most, a single type. It remains to determine whether the operations are all actually of order p.

The defining relations give

$$S^{-1}P^{\alpha_x}Q^{\beta_x}R^{\gamma_x}S = P^{\alpha_{x+1}}Q^{\beta_{x+1}}R^{\gamma_{x+1}},$$

where

$$\alpha_{x+1} = \alpha_x + \beta_x, \quad \beta_{x+1} = \beta_x + \gamma_x, \quad \gamma_{x+1} = \gamma_x;$$

and therefore

$$S^{-x}P^{\alpha}Q^{\beta}R^{\gamma}S^{x} = P^{\alpha_x}Q^{\beta_x}R^{\gamma_x},$$

where

$$\alpha_x = \alpha + x\beta + \tfrac{1}{2}x(x-1)\gamma, \quad \beta_x = \beta + x\gamma, \quad \gamma_x = \gamma.$$

Hence

$$\begin{aligned}(P^{\alpha}Q^{\beta}R^{\gamma}S)^{p} &= P^{\sum_1^p \alpha_x}Q^{\sum_1^p \beta_x}R^{\sum_1^p \gamma_x} \\ &= P^{p\alpha+\frac{1}{2}p(p-1)\beta+\frac{1}{6}(p+1)p(p-1)\gamma}Q^{p\beta+\frac{1}{2}p(p-1)\gamma}R^{p\gamma}.\end{aligned}$$

If p is a greater prime than 3, the indices of P, Q, R are all multiples of p; hence $P^{\alpha}Q^{\beta}R^{\gamma}S$ is of order p, S being any operation not contained in G. If however $p = 3$, then

$$(P^{\alpha}Q^{\beta}R^{\gamma}S)^{3} = P^{\gamma},$$

so that, if γ is not a multiple of 3, $P^{\alpha}Q^{\beta}R^{\gamma}S$ is an operation of order 9. Hence this last type of group exists as a distinct type for all primes greater than 3; but for $p = 3$, it is not distinct from one of the previous types containing operations of order 9.

73. In tabulating, as follows, the types of group thus obtained, we give with each group G a symbol of the form

$$(a, b, \ldots)(a', b', \ldots)(a'', b'', \ldots)\ldots,$$

indicating the types of H_1, $\dfrac{H_2}{H_1}$, $\dfrac{H_3}{H_2}$, $\ldots$, where

$$H_1,\ H_2,\ H_3,\ \ldots,\ H_n,\ G$$

is the series of self-conjugate sub-groups defined in § 53. This symbol is to be read from the left so that $(a, b, \ldots)$ is the type of H_1.

Moreover in each group there is no operation of higher order than that denoted by P.

Table of groups of order p^n, p an odd prime.*

I. $n=2$, two types.

(i) (2); (ii) $(1,1)$.

II. $n=3$, five types.

(i) (3); (ii) $(2,1)$; (iii) $(1,1,1)$;

(iv) $P^{p^2}=1,\quad Q^p=1,\quad Q^{-1}PQ=P^{1+p},\quad (1)(11)$;

(v) $P^p=1,\quad Q^p=1,\quad R^p=1,\quad R^{-1}QR=QP,$

$$R^{-1}PR=P,\quad Q^{-1}PQ=P,\quad (1)(11).$$

III. $n=4$, fifteen types.

(i) (4); (ii) $(3,1)$; (iii) $(2,2)$;

(iv) $(2,1,1)$; (v) $(1,1,1,1)$;

(vi) $P^{p^3}=1,\quad Q^p=1,\quad Q^{-1}PQ=P^{1+p^2},\quad (2)(11)$;

(vii) $P^{p^2}=1,\quad Q^p=1,\quad R^p=1,\quad R^{-1}QR=QP^p,$

$$Q^{-1}PQ=P,\quad R^{-1}PR=P,\quad (2)(11);$$

(viii) $P^{p^2}=1,\quad Q^{p^2}=1,\quad Q^{-1}PQ=P^{1+p},\quad (11)(11)$;

(ix) $P^{p^2}=1,\quad Q^p=1,\quad R^p=1,\quad R^{-1}PR=P^{1+p},$

$$P^{-1}QP=Q,\quad R^{-1}QR=Q,\quad (11)(11),$$

this group (ix) being the direct product of $\{Q\}$ and $\{P,R\}$;

(x) $P^{p^2}=1,\quad Q^p=1,\quad R^p=1,\quad R^{-1}PR=PQ,$

$$Q^{-1}PQ=P,\quad R^{-1}QR=Q,\quad (11)(11);$$

(xi), (xii), and (xiii) $p>3$,

$$P^{p^2}=1,\quad Q^p=1,\quad R^p=1,\quad Q^{-1}PQ=P^{1+p},$$
$$R^{-1}PR=PQ,\quad R^{-1}QR=P^{\alpha p}Q,\quad (1)(1)(11),$$

where for (xi) $\alpha=0$, for (xii) $\alpha=1$, for (xiii) $\alpha=$ any non-residue, (mod. p);

*On groups of orders p^3 and p^4, the reader may consult, in addition to Young's memoir already referred to, Hölder, "Die Gruppen der Ordnungen p^3, pq^2, pqr, p^4," *Math. Ann.*, XLIII, (1893), in particular, pp. 371–410.

(xi), (xii), and (xiii) $p = 3$,

$$P^9 = 1, \quad Q^3 = 1, \quad R^3 = P^{3\alpha}, \quad Q^{-1}PQ = P^4,$$
$$R^{-1}PR = PQ^{\beta}, \quad R^{-1}QR = Q, \quad (1)(1)(11),$$

where for (xi) $\alpha = 0$, $\beta = 1$, for (xii) $\alpha = 1$, $\beta = -1$, for (xiii) $\alpha = -1$, $\beta = -1$.

(xiv) $$P^p = 1, \quad Q^p = 1, \quad R^p = 1, \quad S^p = 1, \quad S^{-1}RS = RP,$$
$$S^{-1}QS = Q, \quad S^{-1}PS = P, \quad R^{-1}QR = Q,$$
$$R^{-1}PR = P, \quad Q^{-1}PQ = P, \quad (11)(11),$$

this group (xiv) being the direct product of $\{Q\}$ and $\{P, R, S\}$;

(xv) $p > 3$,

$$P^p = 1,\ Q^p = 1,\ R^p = 1,\ S^p = 1,\ S^{-1}RS = RQ,\ S^{-1}QS = QP,$$
$$S^{-1}PS = P,\ R^{-1}QR = Q,\ R^{-1}PR = P,$$
$$Q^{-1}PQ = P,\ (1)(1)(11);$$

(xv) $p = 3$,

$$P^9 = 1,\ Q^3 = 1,\ R^3 = 1,\ Q^{-1}PQ = P,\ R^{-1}PR = PQ,$$
$$R^{-1}QR = P^{-3}Q,\ (1)(1)(11).$$

74. To complete the list, we add, as was promised in § 68, the types of non-Abelian groups of orders 2^3 and 2^4; the possible types of Abelian groups being the same as for an odd prime.

Non-Abelian groups of order 2^3; two types.

(i) identical with II (iv), writing 2 for p;

(ii) $P^4 = 1, \quad Q^4 = 1, \quad Q^{-1}PQ = P^{-1}, \quad Q^2 = P^2, \quad (1)(11)$.

Non-Abelian groups of order 2^4; nine types.

(i), (ii), (iii), (iv) and (v) identical with III (vi), (vii), (viii), (ix) and (x), writing 2 for p;

(vi) $$P^4 = 1, \quad Q^4 = 1, \quad R^2 = 1, \quad Q^{-1}PQ = P^{-1},$$
$$Q^2 = P^2, \quad R^{-1}QR = Q, \quad R^{-1}PR = P, \quad (11)(11),$$

this group (vi) being the direct product of $\{R\}$ and $\{P, Q\}$;

(vii) $P^8 = 1,\quad Q^2 = 1,\quad Q^{-1}PQ = P^{-1},\quad (1)(1)(11);$
(viii) $P^8 = 1,\quad Q^2 = 1,\quad Q^{-1}PQ = P^3,\quad (1)(1)(11);$
(ix) $P^8 = 1,\quad Q^4 = 1,\quad Q^{-1}PQ = P^{-1},\quad Q^2 = P^4,\quad (1)(1)(11).$

75. In the following examples, G is a non-Abelian group whose order is the power of a prime p; and the sub-groups $H_1, H_2, \ldots, H_n$ referred to are the series of self-conjugate sub-groups of § 53.

Ex. 1. If P and Q are any two operations of G, the totality of operations of the form $P^{-1}Q^{-1}PQ$ generate a self-conjugate sub-group identical with or contained in H_n.

Ex. 2. Shew that, if every sub-group of G is Abelian, $\dfrac{G}{H_1}$ is an Abelian group of type $(1, 1)$; and any two operations of G, neither of which belongs to H_1 and neither of which is a power of the other, generate G.

Ex. 3. If $\dfrac{G}{H_n}, \dfrac{H_n}{Hn-1}, \dfrac{H_{n-1}}{H_{n-2}}, \ldots$ are of types $(1, 1)$, (1), (1), $\ldots$, shew that G can be generated by two operations.

Ex. 4. If G, of order p^m, is not Abelian, and if every sub-group of G is self-conjugate, shew that p must be 2. (Dedekind.)

Ex. 5. If G is of order p^m, and H_1 of type $(1, 1, \ldots,$ with $m - 2$ units), and if $m > 5$, G is the direct product of two groups. If $m = 5$, there is one type for which G is not a direct product, viz.

$$Q_1^{p^2} = 1,\quad Q_2^{p^2} = 1,\quad P^p = 1,\quad Q_2^{-1}Q_1Q_2 = Q_1P,$$
$$Q_1^{-1}PQ_1 = P,\quad Q_2^{-1}PQ_2 = P.$$

Ex. 6. A group G, of order p^m, $(m > 4)$, where p is an odd prime, contains an operation P of order p^{m-2}, and no cyclical self-conjugate sub-group of order p^{m-2}. Shew that, if P is one of a set of p^2 conjugate operations, the defining relations of the group are of the form

$$P^{p^{m-2}} = 1,\quad Q^p = 1,\quad Q^{-1}PQ = P^{1+p^{m-3}},\quad R^p = P^{\alpha p}Q^{\beta},$$
$$R^{-1}QR = Q,\quad R^{-1}PR = P^{1+\beta p^{m-4}}Q;$$

and that, if P is one of a set of p conjugate operations, the defining relations are of the form

$$P^{p^{m-2}} = 1, \quad Q^p = 1, \quad Q^{-1}PQ = P, \quad R^p = Q^\alpha,$$
$$R^{-1}QR = P^{\beta p^{m-3}}Q, \quad R^{-1}PR = PQ.$$

Determine in each case the number of distinct types.

CHAPTER VI.

ON SYLOW'S THEOREM.

76. It has been proved (§ 22) that the order of any sub-group of a group G is a factor of the order of G; and it results at once from the investigation of §§ 45–47 that, in an Abelian group, there is always at least one sub-group whose order is any given factor of the order of the group. The latter result is not however generally true for groups which are not Abelian; and for factors of the order of a group which contain more than one distinct prime, no general law is known as to the existence or non-existence of corresponding sub-groups. If however p^m, where p is a prime, divides the order of the group, it may be shewn that the group will always contain a sub-group of order p^m. The special form of this theorem, that a group whose order is divisible by a prime p contains operations of order p, is due originally to Cauchy*. The more general result was first established by Sylow. He has shewn† that, if p^α is the highest power of a prime p which divides the order of a group, the group contains a single conjugate set of $kp+1$ sub-groups of order p^α.

For the special case $p = 2$, the following proof of Cauchy's theorem is perhaps worth giving for its simplicity. Arrange the operations of a group of even order in pairs S and S^{-1} of inverse operations. If at any stage no further pairs can be formed, the remaining operations must all, except the identical operation, be of order 2. The number of these remaining operations, which include the identical operation, is even. Hence there must be at least one operation of order 2, and the total number of such operations is odd.

We shall devote the present chapter to the proof of Sylow's theorem; and a consideration of some of its more immediate consequences. These constitute, as will be seen later on, a most important set of results.

77. We shall divide the proof of Sylow's theorem into two parts. First we shew that, if p^α is the highest power of a prime p which divides

---

*Cauchy, *Exercises d'analyse*, III, (1844), p. 250.

†Sylow, Théorèmes sur les groupes de substitutions, *Math. Ann.*, V (1872) pp. 584 et seq. Compare also Frobenius, Neuer Beweis des Sylow'schen Satzes, *Crelle*, C. (1886), p. 179.

the order of a group, the group must have a sub-group of order p^α; and secondly that the sub-groups of order p^α form a single conjugate set and that their number is congruent to unity, (mod. p).

Lemma. *If p^α is the highest power of a prime p by which the order of a group G is divisible, G must contain a sub-group whose order is divisible by p^α.*

If the group G is Abelian, this has been already proved in § 37. Suppose then that G of order N $(= p^\alpha m$, where m is prime to $p)$ is not Abelian; and let G' of order N' be the sub-group of G formed of its self-conjugate operations. If S is any operation of G which is not self-conjugate, and if n is the order of the greatest sub-group within which S is permutable, S forms one of a set of $\frac{N}{n}$ conjugate operations; hence, by equating the order of the group to the sum of the numbers of operations contained in the different conjugate sets, we obtain the equation

$$N = N' + \sum \frac{N}{n},$$

where the sign of summation is extended to all the different conjugate sets of those operations which are not self-conjugate.

If N' is not divisible by p, or in other words if G contains no self-conjugate operations of order p, this equation requires that at least one of the symbols n should be divisible by p^α; therefore in this case the lemma is true.

If N' is divisible by $p^{\alpha'}$, the group G', being Abelian, must contain a sub-group g' of order $p^{\alpha'}$; and this sub-group is self-conjugate in G. Consider now the group $\frac{G}{g'}$ of order $p^{\alpha-\alpha'}m$. If it has no self-conjugate operations of order p, it must (by the result just obtained) contain a sub-group whose order is divisible by $p^{\alpha-\alpha'}$; and hence G contains a sub-group whose order is divisible by p^α. If, on the other hand, $\frac{G}{g'}$ has a sub-group of self-conjugate operations of order $p^{\alpha''}$, then G must contain a self-conjugate sub-group of order $p^{\alpha'+\alpha''}$, say g''. We may now repeat the same reasoning with the group $\frac{G}{g''}$ of order $p^{\alpha-\alpha'-\alpha''}m$. In this way we must in any case be ultimately led to a sub-group of G whose order is divisible by p^α. The lemma is therefore established.

We have as an immediate inference:—

THEOREM I. *If p^α is the highest power of a prime p which divides the order of a group, the group contains a sub-group of order p^α.*

For the group contains a sub-group whose order is divisible by p^α; this sub-group must itself contain a sub-group whose order is divisible by p^α, and so on. Hence at last a sub-group must be arrived at whose order is p^α.

Corollary. Since it has been seen in § 53 that a group of order p^α contains sub-groups of every order p^β $(\beta < \alpha)$, it follows that, if the order of a group is divisible by p^β, the group must contain a sub-group of order p^β. In particular, a group whose order is divisible by a prime p has operations of order p.

78. THEOREM II. *If p^α is the highest power of a prime p which divides the order of a group G, the sub-groups of G of order p^α form a single conjugate set, and their number is congruent to unity,* (mod. p).

If H is a sub-group of G of order p^α, the only operations of G, which are permutable with H and have powers of p for their orders, are the operations of H itself. For if P is an operation of order p^r, not contained in H and permutable with it, and if p^s is the order of the greatest group common to $\{P\}$ and H, the order of $\{H, P\}$ is $p^{\alpha+r+s}$. But G can have no such sub-group, since $p^{\alpha+r-s}$ is not a factor of the order of G.

Suppose now that H' is any sub-group conjugate to H; and let p^β be the order of the group h common to H and H'. When H' is transformed by all the operations of H, the operations of h are the only ones which transform H' into itself. Hence the operations of H can be divided into $p^{\alpha-\beta}$ sets of p^β each, such that the operations of each set transform H' into a distinct sub-group. In this way, $p^{\alpha-\beta}$ sub-groups are obtained distinct from each other and from H and conjugate to H. If these sub-groups do not exhaust the set of sub-groups conjugate to H, let H'' be a new one. From H'' another set of $p^{\alpha-\beta'}$ sub-groups can be formed, distinct from each other and from H and conjugate to H. Moreover no sub-group of this latter set can coincide with one of the previous set. For if

$$P_1^{-1}H''P_1 = P_2^{-1}H'P_2,$$

where P_1 and P_2 are operations of H, then

$$H'' = P_3^{-1}H'P_3,$$

where P_3 $(= P_2P_1^{-1})$ is an operation of H; and this is contrary to the supposition that H'' is different from each group of the previous set. By continuing this process, it may be shewn that the number of sub-groups in the conjugate set containing H is

$$1 + p^{\alpha-\beta} + p^{\alpha-\beta'} + \ldots,$$

where no one of the indices $\alpha - \beta$, $\alpha - \beta'$, ... can be less than unity. The number of sub-groups in the conjugate set containing H is therefore congruent to unity, (mod. p).

If now G contains another sub-group H_1, of order p^α, it must belong to a different conjugate set. The number of sub-groups in the new set may be shewn, as above, to be congruent to unity, (mod. p). But on transforming H_1 by the operations of H, a set of $p^{\alpha-\gamma}$ conjugate sub-groups is obtained, where p^γ is the order of the sub-group common to H and H_1. A further sub-group of the set, if it exists, gives rise to $p^{\alpha-\gamma'}$ additional conjugate sub-groups, distinct from each other and from the previous $p^{\alpha-\gamma}$. Proceeding thus we shew that the number of sub-groups in the conjugate set is a multiple of p; and as it cannot be at once a multiple of p and congruent to unity, (mod. p), the set does not exist. The sub-groups of order p^α therefore form a single conjugate set and their number is congruent to unity, (mod. p).

Corollary I. If $p^\alpha m$ is the order of the greatest group I, within which the group H of order p^α is contained self-conjugately, the order of the group G must be of the form

$$p^\alpha m(1 + kp).$$

Corollary II. The number of groups of order p^α contained in G, i.e. the factor $1 + kp$ in the preceding expression for the order of the group, can be expressed in the form

$$1 + k_1p + k_2p^2 + \ldots + k_\alpha p^\alpha,$$

where k_rp^r is the number of groups having with a given group H of the set greatest common sub-groups of order $p^{\alpha-r}$.

This follows immediately from the arrangement of the set of groups given in the proof of the theorem. Thus each of the $p^{\alpha-\beta}$ groups, obtained on transforming H' by the operations of H, has in common with H a greatest common sub-group of order p^β. It may of course happen that any one or more of the numbers $k_1, k_2, \ldots, k_\alpha$ is zero. If no two sub-groups of the set have a common sub-group whose order is greater than p^r, then $k_1, k_2, \ldots, k_{\alpha-r-1}$ all vanish; and the number of sub-groups in the set is congruent to unity, (mod. $p^{\alpha-r}$). Conversely, if p^s is the highest power of p that divides kp, some two sub-groups of the set must have a common sub-group whose order is not less than $p^{\alpha-s}$; for if there were no such common sub-groups, the number of sub-groups in the set would be congruent to unity, (mod. p^{s+1}).

Corollary III. Every sub-group of G whose order is p^β, $(\beta < \alpha)$, must be contained in one or more sub-groups of order p^α.

For if the sub-group of order p^β is contained in no sub-group of order $p^{\beta+1}$, the only operations whose orders are powers of p that transform it into itself are its own. In this case, the preceding method may be used to shew that the number of sub-groups in the conjugate set to which the given sub-group belongs must be congruent to unity, (mod. p). But this is impossible, as the number of such sub-groups must, on the assumption made, be a multiple of $p^{\alpha-\beta}$. The sub-group of order p^β is therefore contained in one of order $p^{\beta+1}$, and hence repeating the same reasoning in one of order p^α.

79. We shall refer to Theorems I and II together as Sylow's theorem. In discussing in this and the following paragraphs some of the results that follow from Sylow's theorem, we shall adhere to the notation that has been used in establishing the theorem itself. Thus p^α will always denote the highest power of a prime p which divides the order of G; the sub-groups of G of order p^α will be denoted by $H, H_1, \ldots$, and the greatest sub-groups of G that contain these self-conjugately by $I, I_1, \ldots$. These latter form a single conjugate set of sub-groups of G, whose orders are $p^\alpha m$, the order of G itself being $p^\alpha m(1 + kp)$. Moreover the number of groups in this conjugate set is $1 + kp$.

Suppose now that S is any operation of G whose order is a power of p. When the $1 + kp$ sub-groups

$$H,\ H_1,\ H_2,\ \ldots,\ H_{kp}$$

are transformed by S, each one that contains S is transformed into itself, while the remainder are interchanged in sets, the number in any set being a power of p. Hence the number of these groups which contain S must be congruent to unity, (mod. p). In precisely the same way, it may be shewn that the number of sub-groups of order p^α, which contain a given sub-group of order p^β, is congruent to unity, (mod. p).

A sub-group (or operation), having a power of p for its order and occurring in $1 + lp$ sub-groups of order p^α, will not necessarily be one of the same number of conjugate sub-groups (or operations) in each of these $1+lp$ sub-groups. If then h is a sub-group, of order p^r, that occurs in $1 + lp$ sub-groups of order p^α, we may choose one of these, say H, in which h is one of as small a number as possible of conjugate sub-groups. Let this number be $p^{\alpha-r-s}$, so that, in H, h is self-conjugate in a group of order p^{r+s}. Then in G, the order of the greatest group i, in which h is self-conjugate, must be $p^{r+s}n$, where n is relatively prime to p. The number of sub-groups of i of order p^{r+s} must be congruent to unity, (mod. p). No two of these groups of order p^{r+s} can occur in the same group of order p^α; for if they did, they would generate a group of order p^{r+s+t}, $(t \nless 1)$, and this is impossible, p^{r+s} being the highest power of p that divides the order of i. Moreover the number of groups of order p^α in which any one of these groups of order p^{r+s} enters is congruent to unity, (mod. p). Hence the number of groups of order p^α, in which h is one of $p^{\alpha-r-s}$ (the least possible number) conjugate sub-groups, is congruent to unity, (mod. p). Suppose now that, in H', h is one of $p^{\alpha-r-s'}$ $(s' < s)$ conjugate sub-groups; so that in H' it is self-conjugate in a sub-group of order $p^{r+s'}$ and in no greater sub-group. Then the highest power of p that divides the order of the group common to I' and i is $p^{r+s'}$; and therefore, when H' is transformed by all the operations of i, the number of groups of order p^{α} formed is a multiple of $p^{s-s'}$. In each of these groups, h is one of $p^{\alpha-r-s'}$ conjugate sub-groups. If this does not exhaust all the groups of order p^{α} in which h is one of $p^{\alpha-r-s'}$ conjugate sub-groups, let H'' be another. Then from this another set of groups, whose number is a multiple of $p^{s-s'}$, may be formed, which are distinct from each other and from the previous set, such that in each of them h is one of $p^{\alpha-r-s'}$ conjugate sub-groups. This process may clearly be continued till all such groups are exhausted. Hence the number of groups of order p^{α}, in which h is one of $p^{\alpha-r-s'}$ $(s' < s)$ conjugate sub-groups, is a multiple of $p^{s-s'}$. If h enters as one of $p^{\alpha-r-s'}$ conjugate sub-groups in H', a sub-group conjugate to h must enter as one of $p^{\alpha-r-s'}$ conjugate sub-groups in H. Hence if p^{r+s}, $p^{r+s'}$, $p^{r+s''}$, ... are the orders of the greatest sub-groups that contain h self-conjugately in the various sub-groups of order p^{α} in which it appears, then H must contain sub-groups of the conjugate set (in G) to which h belongs in conjugate sets of p^{r+s}, $p^{r+s'}$, $p^{r+s''}$, ... only. The total number x of such sub-groups contained in H is clearly connected with the total number $1 + lp$ of sub-groups of order p^{α}, in which h appears, by the relation

$$x(1 + kp) = y(1 + lp),$$

y being the number of sub-groups in the conjugate set. For every sub-group of order p^{α} will contain x of the set: and the two sides of the equation represent two distinct ways of reckoning all the sub-groups of the set h contained in the sub-groups of the set H, when all repetitions are counted.

It is to be noticed that, with the above notation,

$$y = \frac{p^{\alpha-r-s}m(1 + kp)}{n};$$

and therefore

$$x = \frac{p^{\alpha-r-s}m(1 + lp)}{n}.$$

80. THEOREM III. *Let p^α be the highest power of a prime p which divides the order of a group G, and let H be a sub-group of G of order p^α. Let h be a sub-group common to H and some other sub-group of order p^α, such that no sub-group, which contains h and is of greater order, is common to any two sub-groups of order p^α. Then there must be some operation of G, of order prime to p, which is permutable with h and not with H**.

Suppose that H and H' are two groups of order p^α to which h is common; and let h_1 and h_1' be sub-groups of H and H', of greater order than h, in which h is self-conjugate. If h_1 and h_1' generate a group whose order is a power of p, it must occur in some group H'' of order p^α; and then H and H'' have a common group h_1, which contains h and is of greater order. This is contrary to supposition, and therefore the order of the group generated by h_1 and h_1' is not a power of p. Hence h is permutable with some operation whose order is prime to p.

Let p^r be the order of h, and $p^{r+s}n$ be the order of the greatest sub-group i of G that contains h self-conjugately. If i contained a self-conjugate sub-group of order p^{r+s}, h_1 and h_1' would be sub-groups of it and they would generate a group whose order is a power of p. This is not the case, and therefore i must contain $1 + k'p$ sub-groups of order p^{r+s}, so that we may write $m'(1 + k'p)$ for n; and then, in i, a sub-group of order p^{r+s} is self-conjugate in a sub-group of order $p^{r+s}m'$. No sub-group of i of order p^{r+t} $(t > 0)$ can occur in more than one sub-group of order p^α; and the $1+k'p$ sub-groups of i of order p^{r+s} belong therefore to $1+k'p$ distinct sub-groups of order p^α. Moreover h occurs in no sub-groups of order p^α other than these $1 + k'p$. For if h occurred in another sub-group H_1, it would in this sub-group be self-conjugate in a group of order $p^{r+s'}$ $(s' > 0)$; and this group would occur in i. This group would then be common to two sub-groups of order p^α, contrary to supposition.

An operation of i, which transforms one of its sub-groups of order p^{r+s} into another, must transform the sub-group of order p^α containing the one into that containing the other. Hence i must contain operations which are

*Frobenius, "Ueber endliche Gruppen," *Berliner Sitzungsberichte* (1895), p. 176: and Burnside, "Notes on the theory of groups of finite order," *Proc. London Mathematical Society*, Vol. XXVI (1895), p. 209.

not permutable with H. The greatest common sub-group of i and I is that sub-group of i of order $p^{r+s}m'$ which contains the sub-group of order p^{r+s} belonging to H self-conjugately. For every operation, that transforms this sub-group of order p^{r+s} into itself, must transform H into itself; and no operation can transform H into itself which transforms this sub-group of order p^{r+s} into another.

81. Let P be an operation, or sub-group, which is self-conjugate in H; and let Q be another operation, or sub-group, of H, which is conjugate to P in G, but not conjugate to P in I. Suppose first that, if possible, Q is self-conjugate in H. There must be an operation S which transforms P into Q and H into some other sub-group H', so that

$$S^{-1}PS = Q,$$
$$S^{-1}HS = H'.$$

Now in the sub-group which contains Q self-conjugately, the sub-groups of order p^α form a single conjugate set, and H must occur among them. Hence this sub-group must contain an operation T such that

$$T^{-1}QT = Q,$$

and

$$T^{-1}H'T = H.$$

It follows that

$$T^{-1}S^{-1}PST = Q,$$
$$T^{-1}S^{-1}HST = H,$$

or that, contrary to supposition, P and Q are conjugate in I. Hence:—

THEOREM IV. *Let G and H be defined as in the previous theorem, and let I be the greatest sub-group of G which contains H self-conjugately. Then if P and Q are two self-conjugate operations or sub-groups of H, which are not conjugate in I, they are not conjugate in G.*

Corollary. If H is Abelian, no two operations of H which are not conjugate in I can be conjugate in G. Hence the number of distinct sets of conjugate operations in G, which have powers of p for their orders, is the same as the number of such sets in I.

82. Suppose next that Q is not self-conjugate in H. Then every operation that transforms Q into P must transform H into a sub-group of order p^α in which P is not self-conjugate. Of the sub-groups of order p^α, to which P belongs and in which P is not self-conjugate, choose H' so that, in H', P forms one of as small a number of conjugate operations or sub-groups as possible. Let g be the greatest sub-group of H' that contains P self-conjugately. Among the sub-groups of order p^α that contain P self-conjugately, there must be one or more to which g belongs. Let H be one of these; and suppose that h and h' are the greatest sub-groups of H and H' respectively that contain g self-conjugately. The orders of both h and h' must (Theorem II, § 55) be greater than the order of g; and in consequence of the assumption made with respect to H', every sub-group, having a power of p for its order and containing h, must contain P self-conjugately.

Now consider the sub-group $\{h, h'\}$. Since it does not contain P self-conjugately, its order cannot be a power of p. Also if p^β is the highest power of p that divides its order, it must contain more than one sub-group of order p^β. For any sub-group of order p^β, to which h belongs, contains P self-conjugately; and any sub-group of order p^β, to which h' belongs, does not. Suppose now that S is an operation of $\{h, h'\}$, having its order prime to p and transforming a sub-group of $\{h, h'\}$ of order p^β, to which h belongs, into one to which h' belongs. Then S cannot be permutable with P; for if it were, P would be self-conjugate in each of these sub-groups of order p^β.

When P is an operation, we may reason in the same way with respect to $\{P\}$.

Since g is self-conjugate in both h and h', S must transform g into itself. Now P is self-conjugate in g, and therefore $S^{-r}PS^r$ is also self-conjugate in g for all values of r. If then S^t is the first power of S which is permutable with P, the series of groups P, $S^{-1}PS$, ..., $S^{-t+1}PS^{t-1}$ are all distinct and each is a self-conjugate sub-group of g. Every group in this series is therefore permutable with every other. Hence:—

THEOREM V. *If G and H are defined as in the two preceding theorems, and if P is a self-conjugate sub-group or operation of H, then either*

(i) *P must be self-conjugate in every sub-group of G, of order p^α, in which it enters, or* (ii) *there must be an operation S, of order q prime to p, such that the set of sub-groups $S^{-r}\{P\}S^r$ $(r = 0, 1, \ldots, q-1)$ are all distinct and permutable with each other.*

83. In illustration of Sylow's theorem and its consequences, we will now consider certain special types of group; and we will deal first with a *group whose order is the product of two different primes.*

If p_1 and p_2 $(p_1 < p_2)$ are distinct primes, a group of order p_1p_2 must, by Sylow's theorem, have a self-conjugate sub-group of order p_2. If p_2 is not congruent to unity, (mod. p_1), the group must also have a self-conjugate sub-group of order p_1. The two self-conjugate sub-groups of orders p_1 and p_2 can have no common operation except the identical operation; and therefore (Theorem IX, § 34) every operation of one must be permutable with every operation of the other. Hence if p_2 is not congruent to unity, (mod. p_1), a group of order p_1p_2 must be Abelian, and therefore also cyclical.

If p_2 is congruent to unity, (mod. p_1), there may be either 1 or p_2 conjugate sub-groups of order p_1.

If there is one, it is self-conjugate and the group is cyclical.

If there are p_2 sub-groups of order p_1, let P_1 denote an operation which generates one of them. Then if P_2 is an operation of order p_2,

$$P_1^{-1}\{P_2\}P_1 = \{P_2\};$$

therefore

$$P_1^{-1}P_2P_1 = P_2^\alpha,$$

so that

$$P_2 = P_1^{-p_1}P_2P_1^{p_1} = P_2^{\alpha^{p_1}}.$$

Hence

$$\alpha^{p_1} \equiv 1 \text{ (mod. } p_2).$$

If α were unity, P_2 would be permutable with P_1, and $\{P_1\}$ would not be one of p_2 conjugate sub-groups. Hence α must be a primitive root of the above congruence.

A group of order p_1p_2, which has p_2 conjugate sub-groups of order p_1, must therefore, when it exists, be defined by the relations

$$P_1^{p_1} = 1, \quad P_2^{p_2} = 1, \quad P_1^{-1}P_2P_1 = P_2^\alpha,$$

where α is a primitive root of the congruence*

$$\alpha^{p_1} \equiv 1 \ (\text{mod. } p_2).$$

It follows from § 33 that the order of the group defined by these relations cannot exceed p_1p_2. But also from the given relations it is clearly impossible to deduce new relations of the form

$$P_1^x P_2^y = P_1^{x'} P_2^{y'},$$

where x and x' are less than p_1, and y and y' are less than p_2; hence the order cannot be less than p_1p_2.

Again, let β be a primitive root of the congruence, distinct from α, so that

$$\beta^x \equiv \alpha \ (\text{mod. } p_2).$$

Then if, in the group defined by

$$P_1^{p_1} = 1, \quad P_2^{p_2} = 1, \quad P_1^{-1}P_2P_1 = P_2^{\beta},$$

we represent P_1^x by P_1', the defining relations become

$$P_1'^{p_1} = 1, \quad P_2^{p_2} = 1, \quad P_1'^{-1}P_2P_1' = P_2^{\alpha};$$

and the group is simply isomorphic with the previous group.

Hence finally, when p_2 is congruent to unity, (mod. p_1), there is a single type of group of order p_1p_2, which contains p_2 conjugate sub-groups of order p_1.

84. We will next deal with the problem of determining all distinct types of group of order 24.

---

*It should be noticed that, if α is not a root of the congruence, the relations define a group of order p_1. Thus from

$$P_1^{-1}P_2P_1 = P_2^{\alpha},$$

we have

$$P_1^{-p_1}P_2P_1^{p_1} = P_2^{\alpha^{p_1}}.$$

Hence

$$P_2^{\alpha^{p_1}-1} = 1, \quad \text{and} \quad P_2^{p_2} = 1,$$

so that $P_2 = 1$, and the group reduces to $\{P_1\}$.

A group of order 24 must contain either 1 or 3 sub-groups of order 8, and either 1 or 4 sub-groups of order 3. If it has one sub-group of order 8 and one sub-group of order 3, the group must, since each of these sub-groups is self-conjugate, be their direct product. We have seen (§§ 68, 74) that there are five distinct types of group of order 8; there are therefore five distinct types of group of order 24, which are obtained by taking the direct product of any group of order 8 and a group of order 3.

If there are 3 groups of order 8, some two of them must (Theorem II, Cor. II, § 78) have a common sub-group of order 4; and (Theorem III, § 80) this common sub-group must be a self-conjugate sub-group of the group of order 24. Moreover if, in this case, a sub-group of order 8 is Abelian, each operation of the self-conjugate sub-group of order 4 must (Theorem IV, Cor. § 81) be a self-conjugate operation of the group of order 24.

With the aid of these general considerations, it now is easy to determine for each type of group of order 8, the possible types of group of order 24, in addition to the five types already obtained.

(i) Suppose a group of order 8 to be cyclical, and let A be an operation that generates it. If $\{A\}$ is self-conjugate and B is an operation of order 3, then

$$B^{-1}AB = A^{\alpha},$$

and therefore

$$B^{-3}AB^3 = A^{\alpha^3}.$$

Hence

$$\alpha^3 \equiv 1 \pmod{8},$$

and therefore

$$\alpha \equiv 1 \pmod{8};$$

so that A and B are permutable. This is one of the types already obtained. Hence for a new type, $\{A\}$ cannot be self-conjugate, and A^2 must be a self-conjugate operation; B is therefore one of two conjugate operations, while $\{B\}$ is self-conjugate. Hence the only possible new type in this case is given by

$$A^{-1}BA = B^{-1}.$$

(ii) Next, let a group of order 8 be an Abelian group defined by

$$A^4 = 1, \quad B^2 = 1, \quad AB = BA.$$

If this is self-conjugate, then, by considerations similar to those of the preceding case, we infer that the group is the direct product of groups of orders 8 and 3. Hence there is not in this case a new type.

If the group of order 8 is not self-conjugate, the self-conjugate group of order 4 may be either $\{A\}$ or $\{A^2, B\}$. In either case, if C is an operation of order 3, it must be one of two conjugate operations while $\{C\}$ is self-conjugate. Hence there are two new types respectively given by

$$C^3 = 1, \quad BCB = C^{-1}, \quad A^{-1}CA = C;$$

and

$$C^3 = 1, \quad A^{-1}CA = C^{-1}, \quad BCB = C.$$

(iii) Let a group of order 8 be an Abelian group defined by

$$A^2 = 1, \quad B^2 = 1, \quad C^2 = 1, \quad AB = BA, \quad BC = CB, \quad CA = AB.$$

If it is self-conjugate, and if the group of order 24 is not the direct product of groups of orders 8 and 3, an operation D of order 3 must transform the 7 operations of order 2 among themselves; and it must therefore be permutable with one of them. Now the relations

$$D^{-1}AD = A, \quad D^{-1}BD = AB,$$

are not self-consistent, because they give

$$D^{-2}BD^2 = B.$$

Hence, since the group of order 8 is generated by A, B and any other operation of order 2 except AB, we may assume, without loss of generality, that

$$D^{-1}AD = A, \quad D^{-1}BD = C, \quad D^{-1}CD = A^x B^y C^z.$$

These relations give

$$B = D^{-3}BD^3 = D^{-1}A^x B^y C^z D = A^{x(1+z)} B^{yz} C^{y+z^2},$$

and therefore

$$y = z = 1.$$

Now if

$$D^{-1}CD = ABC,$$

and if

$$AB = B', \quad AC = C',$$

then

$$D^{-1}B'D = C', \quad D^{-1}C'D = B'C';$$

so that the two alternatives $x = 0$ and $x = 1$ lead to simply isomorphic groups.

Hence there is in this case a single type. It is the direct product of $\{A\}$ and $\{D, B, C\}$, where

$$D^{-1}BD = C, \quad D^{-1}CD = BC.$$

If the group of order 8 is not self-conjugate, the self-conjugate group of order 4 may be taken to be $\{A, B\}$; and D being an operation of order 3, there is a single new type given by

$$D^3 = 1, \quad CDC = D^{-1}, \quad ADA = D, \quad BDB = D.$$

(iv) Let a group of order 8 be a non-Abelian group defined by

$$A^4 = 1, \quad B^4 = 1, \quad A^2 = B^2, \quad B^{-1}AB = A^{-1};$$

and let C be an operation of order 3. If the group of order 8 is self-conjugate, and the group of order 24 is not a direct product of groups of orders 8 and 3, C must transform the 3 sub-groups of order 4, $\{A\}$, $\{B\}$ and $\{AB\}$, among themselves. Hence we may take

$$C^{-1}AC = B,$$

and

$$C^{-1}BC = AB \quad \text{or} \quad (AB)^3.$$

If C transforms B into $(AB)^3$, then

$$C^{-3}AC^3 = A^{-1},$$

and C cannot be an operation of order 3. Hence in this case there is only one new type, given by

$$C^3 = 1, \quad C^{-1}AC = B, \quad C^{-1}BC = AB.$$

If the sub-group of order 8 is not self-conjugate, the self-conjugate sub-group of order 4 is cyclical, and each of its operations must be permutable with C. Hence again we get a single new type, given by

$$C^3 = 1, \quad A^{-1}CA = C, \quad B^{-1}CB = C^{-1}.$$

(v) Lastly, let a sub-group of order 8 be a non-Abelian group defined by

$$A^4 = 1, \quad B^2 = 1, \quad BAB = A^{-1}.$$

This contains one cyclical and two non-cyclical sub-groups of order 4. If it is self-conjugate, the group of order 24 must therefore be the direct product of groups of orders 8 and 3; and there is no new type.

If the sub-group of order 8 is not self-conjugate, and the self-conjugate sub-group of order 4 is the cyclical group $\{A\}$, then A must be permutable with an operation C of order 3, and there is a single new type given by

$$C^3 = 1, \quad A^{-1}CA = C, \quad B^{-1}CB = C^{-1}.$$

If the self-conjugate sub-group of order 4 is not cyclical, it may be taken to be $\{1, A^2, B, A^2B\}$. If C is permutable with each operation of this sub-group, there is a single type given by

$$C^3 = 1, \quad A^{-1}CA = C^{-1}, \quad B^{-1}CB = C.$$

If C is not permutable with every operation of the self-conjugate sub-group, it must transform A^2, B, A^2B among themselves and we may take

$$C^{-1}A^2C = B, \quad C^{-1}BC = A^2B.$$

Now $\{C, A^2, B\}$ is self-conjugate, and therefore A must transform C into another operation of order 3 contained in this sub-group. Hence

$$A^{-1}CA = C^x A^{2y} B^z.$$

The only values of x, y, z which are consistent with the previous relation

$$A^2CA^2 = CA^2B,$$

are

$$x = 2, \quad y = z = 1.$$

The last new type is therefore defined by

$$A^4 = 1, \quad B^2 = 1, \quad BAB = A^{-1},$$
$$C^3 = 1, \quad C^{-1}A^2C = B, \quad C^{-1}BC = A^2B,$$
$$A^{-1}CA = C^2A^2B.$$

When B is eliminated between these relations, it will be found that the only independent relations remaining are

$$A^4 = 1, \quad C^3 = 1, \quad (AC)^2 = 1.$$

It is a good exercise to verify that these form a complete set of defining relations for the group. (Compare Ex. 1, § 35.)

There are therefore, in all, fifteen distinct types of group of order 24. The last of these is the only type, which has neither a self-conjugate sub-group of order 8, nor one of order 3. The reader should satisfy himself, as an exercise, that, in the ten cases where the group is not a direct product of groups of orders 8 and 3, the defining relations which we have given are self-consistent. This is of course an essential part of the investigation, and it may, in more complicated cases, involve some little difficulty. We have omitted the verification here, where it is very easy, for the sake of brevity.

It is to be noticed that the last type obtained gives an example, and indeed the simplest possible, of Theorem V, § 82. Thus in $\{A, B\}$ of order 8, A^2 is a self-conjugate operation and B is not. In the group of order 24, the operations A^2 and B are conjugate; and C is an operation, of order prime to 2, such that A^2, $C^{-1}A^2C$, $C^{-2}A^2C^2$ generate three mutually permutable sub-groups.

A discussion similar to that of the present section (but simpler, since in each case the number of types is smaller), will verify the following table*:—

Order	6	10	12	14	15	18	20	22	26	28	30
Number	2	2	5	2	1	5	5	2	2	4	4

This table, taken with the results of Chapter V, gives the number of distinct types of groups for all orders less than 32.

85. As a second example, we will discuss the various distinct types of group of order 60.

A group of order 60 must, by Sylow's theorem, contain either 1 or 6 cyclical sub-groups of order 5.

We will first suppose that a group G of order 60 contains a single cyclical sub-group of order 5, which is necessarily self-conjugate. There will then be four operations of order 5 in G; and we may deal with two sub-cases according as these operations are or are not self-conjugate.

*Miller, *Comptes Rendus*, CXXII (1896), p. 370.

(i) Suppose that each operation of order 5 is self-conjugate. There must then be either 1 or 3 sub-groups of order 4. If there is only one, it must be permutable with an operation of order 3; and then G contains a sub-group of order 12. If there are three, it follows, by Theorem II, Cor. II (§ 78), that some pair of them must have a common sub-group of order 2. But (Theorem III, § 80) this sub-group of order 2 must be permutable with some operation of order prime to 2, which is not permutable with a sub-group of order 4. This operation must be of order 3; hence in this case also there must be a sub-group of order 12. Thus then the group is, with either alternative, the direct product of a group of order 5 and a group of order 12. Now there are five distinct types of group of order 12; there are therefore five distinct types of group of order 60 which contain self-conjugate operations of order 5.

(ii) Suppose that S is an operation of order 5 which is not self-conjugate. If T is an operation which is not permutable with S, then

$$T^{-1}ST = S^{\alpha},$$

where α is not unity. Also, if T^x is the lowest power of T which is permutable with S, then

$$T^{-x}ST^x = S^{\alpha^x} = S,$$

and therefore

$$\alpha^x \equiv 1 \pmod{5}.$$

It follows that x must be either 2 or 4. If x is 2 for every operation T which is not permutable with S, then S and S^4 form a complete set of conjugate operations; as also do S^2 and S^3. If x is 4 for any operation T, the four operations S, S^2, S^3, S^4 form a single conjugate set.

First, let S be one of two conjugate operations; it must then be self-conjugate in a sub-group of order 30, and by Sylow's theorem this sub-group must contain a single sub-group of order 3. It will therefore be given by

$$\begin{aligned} &(\alpha) \quad A^2 = 1, \quad B^3 = 1, \quad S^5 = 1, \quad ABA = B, \\ \text{or} \quad &(\beta) \quad A^2 = 1, \quad B^3 = 1, \quad S^5 = 1, \quad ABA = B^2, \end{aligned}$$

according as B is or is not a self-conjugate operation; S in either case being permutable with both A and B.

If the sub-groups of order 4 are cyclical, G must contain an operation A_1 of order 4 whose square is A; and A_1 must transform S into its inverse and $\{B\}$

into itself. The latter condition clearly cannot be satisfied if the self-conjugate sub-group of order 30 is of type (β). Hence we have two types given by

$$A_1^4 = 1, \quad B^3 = 1, \quad S^5 = 1, \quad B^{-1}SB = S,$$
$$A_1^{-1}SA_1 = S^4, \quad \text{and} \quad A_1^{-1}BA_1 = B \quad \text{or} \quad B^2.$$

If the sub-groups of order 4 are not cyclical, G must contain an operation A' of order 2, which is permutable with A; and A' transforms S into its inverse and $\{B\}$ into itself. In this case, if the self-conjugate sub-group is of type (α), there are two types given by

$$A'^2 = 1, \quad A^2 = 1, \quad B^3 = 1, \quad S^5 = 1, \quad ASA = S, \quad B^{-1}SB = S,$$
$$A'AA' = A, \quad A'SA' = S^4, \quad ABA = B, \quad A'BA' = B \quad \text{or} \quad B^2.$$

If the self-conjugate sub-group is of type (β), there is a single type in which the last two of the preceding equations are replaced by

$$ABA = B^2, \quad A'BA' = B.$$

Secondly, let the operations of order 5 form a single conjugate set. The sub-groups of order 4 must then be cyclical since G contains an operation T, such that T^4 is the lowest power of T which is permutable with S. Also S is permutable in a sub-group of order 15. This sub-group must be self-conjugate; and therefore G contains a self-conjugate sub-group of order 3. Let

$$B^3 = 1, \quad S^5 = 1, \quad B^{-1}SB = S$$

define the self-conjugate sub-group of order 15; and let A_1 be an operation of order 4, none of whose powers is permutable with S. We may then take

$$A_1^{-1}SA_1 = S^2;$$

since $\{B\}$ is self-conjugate, A_1 must transform this sub-group into itself. There are therefore two types given by

$$A_1^4 = 1, \quad B^3 = 1, \quad S^5 = 1, \quad B^{-1}SB = S, \quad A_1^{-1}SA_1 = S^2,$$

and

$$A_1^{-1}BA_1 = B \quad \text{or} \quad B^2.$$

Hence there are in all twelve distinct groups of order 60, each of which has a self-conjugate sub-group of order 5.

(iii) Next, suppose that G contains 6 conjugate sub-groups of order 5. No operation of order 3 can be permutable with an operation of order 5, and therefore by Sylow's theorem there must be 10 conjugate sub-groups of order 3. Hence G contains 24 operations of order 5 and 20 operations of order 3. If any one operation of order 5 were permutable with an operation of order 2, all its powers would be permutable with the same operation, and therefore, since the sub-groups of order 5 form a single conjugate set, every operation of order 5 would be permutable with an operation of order 2. The group would then contain at least 24 operations of order 10. This is clearly impossible, since the sum of the numbers of operations of orders 3, 5 and 10 would be greater than the order of the group. Hence the sub-group of order 10, which contains self-conjugately a sub-group of order 5, must be of the type

$$A^2 = 1, \quad S^5 = 1, \quad ASA = S^4.$$

In a similar way, we shew that a sub-group of order 6, which contains self-conjugately a sub-group of order 3, is of the type

$$A^2 = 1, \quad B^3 = 1, \quad ABA = B^{-1}.$$

Since no operation of order 3 or 5 is permutable with an operation of order 2, it follows (Theorem III, § 80) that no two sub-groups of order 4 can have a common operation other than identity. Hence there must be 5 sub-groups of order 4; for if there were 3 or 15, some of them would necessarily have common operations. Each sub-group of order 4 is therefore contained self-conjugately in a sub-group of order 12. Such a sub-group of order 12 can contain no self-conjugate operation of order 2, since G contains no operation of order 6. Hence the sub-groups of order 4 are non-cyclical, and the 3 operations of order 2 in any sub-group of order 4 are conjugate operations in the sub-group of order 12 containing it. This sub-group must therefore be of the type

$$B^3 = 1, \quad B^{-1}A_1B = A_2, \quad B^{-1}A_2B = A_1A_2,$$

where A_1 and A_2 are two permutable operations of order 2.

The 5 sub-groups of order 4 contain therefore 15 distinct operations of order 2; and these form a conjugate set. We have already seen that the 20 operations of order 3 form a conjugate set, and that the 24 operations of order 5 form two conjugate sets of 12 each. Hence the 60 operations of the group are distributed in 5 conjugate sets, containing respectively 1, 12, 12, 15 and 20 operations. It follows at once (§ 27, p. 39) that the group, when it exists, is simple.

A sub-group of order 12, the existence of which has been proved, must be one of 5 conjugate sub-groups; and, since the group is simple, no operation can transform each of these into itself. Hence if the 5 conjugate sub-groups

$$H_1,\ H_2,\ H_3,\ H_4,\ H_5$$

are transformed by any operation of the group into

$$H_1',\ H_2',\ H_3',\ H_4',\ H_5',$$

and if we regard

$$\begin{pmatrix} H_1, H_2, H_3, H_4, H_5 \\ H_1', H_2', H_3', H_4', H_5' \end{pmatrix}$$

as a substitution performed on 5 symbols, the group is simply isomorphic with a substitution group of 5 symbols. In other words, the group can be represented as a group of substitutions of 5 symbols. Now there are just 60 even substitutions of 5 symbols; and it is easy to verify that the group they form satisfy all the conditions above determined. Moreover it will be formally proved in Chapter VIII, and it is indeed almost obvious, that no group of substitutions can be simple if it contains odd substitutions. Hence finally, there is one and only one type of group of order 60 which contains 6 sub-groups of order 5.

Ex. 1. If p, q, r are distinct primes, shew that a group of order p^2qr, which contains a self-conjugate operation of order r, must be the direct product of two groups of orders p^2q and r respectively.

Ex. 2. Shew that there is a single type of group of order 84 which contains 28 sub-groups of order 3; and determine its defining relations.

Ex. 3. If p^α $(\alpha > 1)$ is the highest power of p which divides the order of G, and if $1 + kp$ be the number of sub-groups of G of order p^α, shew that (i) if $1 + kp < p^2$, (ii) if a group of order p^x is cyclical and $1 + kp < p^\alpha$, G is composite. (Maillet, *Comptes Rendus*, CXVIII, (1894), p. 1188.)

86. A remarkable and important extension of Sylow's theorem has recently been given by Herr Frobenius*. In the theorem, as stated above,

*Frobenius: *Berliner Sitzungsberichte* (1895), p. 988.

p^α is the highest power of a prime p that divides the order of a group. Herr Frobenius shews that, if p^k is any power of a prime p that divides the order of a group, the number of sub-groups of order p^k is congruent to unity, (mod. p). These groups do not however, in general, form a single conjugate set.

For a group whose order is a power of p higher than p^k, this result has been already proved in § 61. Suppose now that

$$G_k,\ G'_k,\ G''_k,\ \dots,$$

are the sub-groups of order p^k contained in a group G, and that p^α is the highest power of p dividing the order of G. It has been seen, in the proof of Sylow's theorem, that G contains at least one sub-group G_α of order p^α. The above set of sub-groups of order p^k may then be divided into two classes, those namely which are contained in G_α, and those which are not. The result of § 61 shews that the number of the sub-groups contained in the first of these classes is congruent to unity, (mod. p); and if G_α is a self-conjugate sub-group of G, all the sub-groups must be contained in this class. If G_α is not self-conjugate in G, and if G_k, any one of the sub-groups belonging to the second class, be transformed by all the operations of G_α, a set of p^x sub-groups of order p^k will result.

Now it is easy to see that x is not less than unity, and that every one of these p^x sub-groups belongs to the second class. For suppose first that x is zero, so that G_k is transformed into itself by every operation of G_α. Then $\{G_\alpha, G_k\}$ is a sub-group of G whose order is a power of p; and since G_k is not contained in G_α, the order of this sub-group is not less than $p^{\alpha+1}$. This is impossible, since $p^{\alpha+1}$ is not a factor of the order of G. Secondly, if $P^{-1}G_kP$ were contained in G_α, P being some operation of G_α, then G_k would be contained in $PG_\alpha P^{-1}$ or in G_α, contrary to supposition. If the sub-groups of the second class are not thus exhausted, and if G'_k is a new one, we may on transforming G'_k by the operations of G_α form a fresh set of $p^{x'}$ sub-groups of the second class which are distinct from each other and from the previous set; and this process may be continued till the second class is exhausted. The number of sub-groups in the second class is therefore a multiple of p. Hence:—

THEOREM VI. *If p^k, where p is prime, divides the order of a group G, the number of sub-groups of G of order p^k is congruent to unity,* (mod. p).

87. If p is a prime which divides the order of a group G, it immediately follows from the foregoing theorem that the number of operations of G, which satisfy the relation

$$S^p = 1,$$

is a multiple of p. For the number of sub-groups of G of order p is k_1p+1, and no two of these sub-groups can contain a common operation except identity. There are therefore

$$(k_1p+1)(p-1)$$

distinct operations of order p in G; these, together with the identical operation, which also satisfies

$$S^p = 1,$$

give in all $(k_1p - k_1 + 1)p$ operations.

This result has been generalized by Herr Frobenius* in the following form:—

THEOREM VII. *If n is a factor of the order N of a group G, the number of operations of G, including identity, whose orders are factors of n, is a multiple of n.*

It is easy to verify the truth of this theorem directly for small values of N; and we may therefore assume it true for every group whose order is less than that of the given group G. Again, when n is equal to N, the theorem is obviously true. If then, on the assumption that the theorem is true for all factors of N which are greater than n, we shew that it is true for n, the general truth of the theorem will follow by induction.

If p is any factor of $\frac{N}{n}$, we assume that the number of operations of G whose orders divide np is a multiple of np, and therefore also of n; and we

*"Verallgemeinerung des Sylow'schen Satzes," *Berliner Sitzungsberichte* (1895), pp. 984, 985.

have to shew that the number of operations of G, whose orders divide np and do not divide n, is also a multiple of n. Let this set of operations be denoted by A. If $p^{\lambda-1}$ is the highest power of p that divides n, the order of every operation of the set A must be equal to or be a multiple of p^{λ}. Let P be one of these operations and m its order. Then, if $\phi(m)$ is the number of integers less than and prime to m, the cyclical sub-group $\{P\}$ contains $\phi(m)$ operations of order m, and each of these belongs to the set A. If these do not exhaust the set, let P' of order m' be another operation belonging to it. Then no one of the $\phi(m')$ operations of order m', contained in the cyclical sub-group $\{P'\}$, can be identical with any of the preceding set of $\phi(m)$ operations, since P' itself is not contained in that set; at the same time, the new set of $\phi(m')$ operations all belong to A. This process may be continued till the set A is exhausted. Now $m, m', \ldots$ are all divisible by p^{λ}, and therefore $\phi(m), \phi(m'), \ldots$ are all divisible by $p^{\lambda-1}(p-1)$. Hence the number of operations in the set A is a multiple of $p^{\lambda-1}$.

If now $n = p^{\lambda-1}s$, where by supposition s is relatively prime to p, it remains to shew that the number of operations in the set A is a multiple of s. For this purpose, let P be any operation of G of order p^{λ}; and let those operations of G, which are permutable with P, form a sub-group H of order $p^{\lambda}r$. The number of operations of H whose orders divide s is the same as the number whose orders divide t, where t is the greatest common measure of r and s. Now the order of $\dfrac{H}{\{P\}}$ is less than N, and therefore we may assume that the number of operations of $\dfrac{H}{\{P\}}$ whose orders divide t is a multiple of t, say kt. H therefore contains kt operations of the form PT, where P and T are permutable and the order of T divides s. Now P is one of a set of $\dfrac{N}{p^{\lambda}r}$ conjugate operations in G; and corresponding to each of these, there is a similar set of kt operations. Moreover no two of these operations can be identical; for we have seen (§ 16) that, if m and n are relatively prime, an operation of order mn of a group G, can be expressed in only one way as the product of two permutable operations of G of orders m and n.

The complete set of $\dfrac{N}{p^{\lambda}r}kt$ operations of the form PT belongs to A;

if A is not thus exhausted, its remaining operations can be divided into similar sets. Now N is divisible by both r and s, and therefore by their least common multiple $\frac{rs}{t}$. Hence $\frac{Nkt}{p^\lambda r}$ is divisible by s, and therefore also the number of operations in the set A is divisible by s. The number of operations in A, being divisible both by $p^{\lambda-1}$ and by s, is therefore divisible by n. Hence finally, the number of operations of G, whose orders divide n, is a multiple of n; and the theorem is proved.

Corollary I. Let N_n be the number of operations of G whose orders divide n, and suppose that

$$N_n = n.$$

Let p be the smallest prime factor of n, and p^α the highest power of p that divides n, so that

$$n = p^\alpha n_1,$$

where every prime factor of n_1 is greater than p. If the order of G is divisible by a higher power of p than p^α, then

$$N_{pn} = kpn,$$

where k is an integer. Now if the order m of any operation divides pn and does not divide n, then m must be a multiple of $p^{\alpha+1}$; and therefore among the operations, whose orders divide n, there must be operations whose orders are equal to or are multiples of p^α. Hence

$$N_{p^{\alpha-1}n_1} = \lambda p^{\alpha-1} n_1,$$

where

$$\lambda < p.$$

Now $N_{p^\alpha n_1} - N_{p^{\alpha-1}n_1}$ is the number of operations whose orders are factors of n and multiples of p^α; and it has been shewn, in the proof of the theorem, that this number is a multiple of $p^{\alpha-1}(p-1)$. Hence

$$(p-\lambda)p^{\alpha-1}n_1 = \mu p^{\alpha-1}(p-1).$$

Since every prime factor of n_1 is greater than p, this equation requires that λ should be unity; therefore

$$N_{p^{\alpha-1}n_1} = p^{\alpha-1}n_1.$$

This process may be repeated to shew that, for each value of β which is not greater than α, we have

$$N_{p^{\alpha-\beta}n_1} = p^{\alpha-\beta}n_1,$$

so that, finally,

$$N_{n_1} = n_1.$$

Moreover it is easy to see that the reasoning holds when p^α is the highest power of p that divides the order of G, provided that then the sub-groups of G, of order p^α, are cyclical.

Suppose now that

$$n = p_1^{\alpha_1} p_2^{\alpha_2} \dots p_n^{\alpha_n},$$

where $p_1, p_2, \dots, p_n$ are primes in ascending order; and either that, for each value of r from 1 to $n-1$, $p_r^{\alpha_r}$ is not the highest power of p_r which divides the order of G: or that, if $p_r^{\alpha_r}$ is the highest power of p_r that divides the order of G, the sub-groups of order $p_r^{\alpha_r}$ are cyclical.

Then we may prove as above, first, that G contains operations of each of the orders $p_1^{\alpha_1}, p_2^{\alpha_2}, \dots, p_{n-1}^{\alpha_{n-1}}$; and secondly that, for each value of r from 2 to n,

$$N_{p_r^{\alpha_r} \dots p_n^{\alpha_n}} = p_r^{\alpha_r} \dots p_n^{\alpha_n}.$$

The equation

$$N_{p_n^{\alpha_n}} = p_n^{\alpha_n}$$

implies that G has a self-conjugate sub-group of order $p_n^{\alpha_n}$; for G has a sub-group of this order, and if G had more than one, it would necessarily contain more than $p_n^{\alpha_n}$ operations whose orders divide $p_n^{\alpha_n}$.

Again, since G contains a self-conjugate sub-group of order $p_n^{\alpha_n}$ and also operations of order $p_{n-1}^{\alpha_{n-1}}$, it must contain a sub-group of order $p_{n-1}^{\alpha_{n-1}} p_n^{\alpha_n}$.

If it had more than one sub-group of this order, it would contain more than $p_{n-1}^{\alpha_{n-1}}p_n^{\alpha_n}$ operations whose orders divide $p_{n-1}^{\alpha_{n-1}}p_n^{\alpha_n}$. But since

$$N_{p_{n-1}^{\alpha_{n-1}}p_n^{\alpha_n}} = p_{n-1}^{\alpha_{n-1}}p_n^{\alpha_n},$$

this is impossible. Hence G contains a single sub-group of order $p_{n-1}^{\alpha_{n-1}}p_n^{\alpha_n}$, which is necessarily self-conjugate. In the same way we shew that, for each order

$$p_r^{\alpha_r}\dots p_n^{\alpha_n} \quad (r = n-1, n-2, \dots, 1),$$

G contains a single sub-group.

Finally then, under the conditions stated above, the equation

$$N_n = n$$

involves the property that G has a self-conjugate sub-group of order n and no other sub-group of the same order.

Corollary II*. If m and n are relatively prime factors of the order of G, and if the numbers of operations of G whose orders divide m and n respectively are equal to m and n; then every operation whose order is a factor of m is permutable with every operation whose order is a factor of n, and the number of operations of G whose orders divide mn is equal to mn.

Every operation of G whose order divides mn can be expressed as the product of two permutable operations whose orders divide m and n respectively; and therefore the number of operations whose orders divide mn cannot be greater than mn. On the other hand, it follows from the theorem that the number of such operations cannot be less than mn; and therefore every operation whose order divides m must be permutable with every operation whose order divides n.

Corollary III†. If a group of order mn, where m and n are relatively prime, contains a self-conjugate sub-group of order n, the group contains exactly n operations whose orders divide n.

*Frobenius, *Berliner Sitzungsberichte*, (1895), p. 987.

†Frobenius: "Ueber endliche Gruppen," *Berliner Sitzungsberichte* (1895), p. 170.

If the group G has an operation S whose order divides n, and if S is not contained in the self-conjugate sub-group H of order n, $\{S, H\}$ would be a sub-group of G, whose order is greater than n and at the same time contains no factor in common with m. This is impossible, and therefore H must contain all the operations of G whose orders divide n.

Corollary IV*. If G has a self-conjugate sub-group H of order mn, where m and n are relatively prime, and if H has a self-conjugate sub-group K of order n, then K is a self-conjugate sub-group of G.

For by the preceding Corollary, H contains exactly n operations whose orders divide n, namely the operations forming the sub-group K; and every operation that transforms H into itself must interchange these n operations among themselves. Hence every operation of G is permutable with K.

Ex. 1. Shew that if, the conditions of Corollary II being satisfied, the order of G is mn, then either G is the direct product of two groups of orders m and n, or G contains an Abelian self-conjugate sub-group.

Ex. 2. If p and q are distinct primes, there cannot be more than one type of group of order $p^\alpha q^\beta$ which contains no operation of order pq.

88. We have seen in § 53 that a group G, whose order is the power of a prime, contains a series of self-conjugate sub-groups

$$H_1,\ H_2,\ \ldots,\ H_{n-1},\ H_n,\ G,$$

such that in $\dfrac{G}{H_r}$ every operation of $\dfrac{H_{r+1}}{H_r}$ is self-conjugate. We shall conclude the present chapter by shewing that any group which has such a series of self-conjugate sub-groups is the direct product of two or more groups whose orders are powers of primes.

THEOREM VIII. *If a group G, of order $p^\alpha q^\beta \ldots r^\gamma$, where p, q, $\ldots$, r are distinct primes, has a series of self-conjugate sub-groups*

$$H_1,\ H_2,\ \ldots,\ H_{n-1},\ H_n,\ G,$$

*Frobenius, *loc. cit.* p. 170.

such that in $\frac{G}{H_r}$ *every operation of* $\frac{H_{r+1}}{H_r}$ *is self-conjugate, then* G *is the direct product of groups of order* p^α, q^β, ..., r^γ.

Suppose, if possible, that p divides the order of H_2 and does not divide the order of H_1. If P is an operation of order p contained in H_2, $\{P, H_1\}$ is a self-conjugate sub-group of G; and every operation conjugate to P is contained in the set PH_1. But the only operation of this set, whose order is p, is P. Hence P must be a self-conjugate operation, contrary to the supposition that has been made. Hence if the order of H_1 is not divisible by p, neither is the order of H_2. Suppose, next, that p^x is the highest power of p that divides the orders of both H_r and H_{r-1}. Then the order of the sub-group $\frac{H_r}{H_{r-1}}$ of $\frac{G}{H_{r-1}}$, formed of the self-conjugate operations of the latter, is not divisible by p; and therefore the order of $\frac{H_{r+1}}{H_{r-1}}$ is not divisible by p. Hence p^x is the highest power of p that divides the order of H_{r+1}. This reasoning may be repeated to shew that p^x is the highest power of p that divides the order of each of the groups H_{r+2}, H_{r+3}, ... Hence x must be equal to α; and therefore the order of H_1 must be divisible by each of the primes p, q, ..., r.

Suppose now that, for each prime p which divides the order of G, every operation of H_r, whose order is a power of p, is permutable with every operation of G whose order is relatively prime to p. Let P be any operation, whose order is a power of p, belonging to H_{r+1} and not to H_r; and let Q be any operation of G whose order is relatively prime to p. If Q is not permutable with P, then

$$Q^{-1}PQ = Ph_r,$$

where h_r is some operation of H_r. The order of h_r must be a power of p. For let $h_r = h'_r h''_r$, where the order of h'_r is a power of p and the order of h''_r is relatively prime to p. Then, from the supposition made with regard to the sub-group H_r, the operation Ph_r is the product of the permutable operations Ph'_r and h''_r. But, since the order of Ph_r is a power of p, this is impossible unless h''_r is identity. If the order of h_r is p^β, then

$$Q^{-p^\beta} P Q^{p^\beta} = Ph_r^{p^\beta} = P;$$

and this equation implies that Q is permutable with P, since p^β and the order of Q are relatively prime. Hence if the supposition that has been made holds for H_r, it also holds for H_{r+1}. But it certainly holds for H_1, and therefore it is true for G. Hence every operation of G whose order is a power of p is permutable with every operation of G whose order is relatively prime to p. The group therefore contains self-conjugate sub-groups of each of the orders $p^\alpha, q^\beta, \ldots, r^\gamma$; and it follows, from the definition of § 31, that G is the direct product of these groups.

We add here two examples in further illustration of the applications of Sylow's theorem.

Ex. 1. If p is a prime, greater than 3, shew that the number of distinct types of group of order $6p$ is 6 or 4, according as p is congruent to 1 or 5, (mod. 6).

Ex. 2. If p is a prime, greater than 5, shew that the number of distinct types of group of order $12p$ is 18, 12, 15 or 10, according as p is congruent to 1, 5, 7 or 11, (mod. 12).

CHAPTER VII.

ON THE COMPOSITION-SERIES OF GROUP.

89. LET G_1 be a maximum self-conjugate sub-group (§ 27) of a given group G, G_2 a maximum self-conjugate sub-group of G_1, and so on. Since G is a group of finite order, we must, after a finite number of sub-groups, arrive in this way at a sub-group G_{n-1}, whose only self-conjugate sub-group is that formed of the identical operation alone, so that G_{n-1} is a simple group.

Definitions. The series of groups

$$G,\ G_1,\ G_2,\ \ldots,\ G_{n-1},\ 1,$$

obtained in the manner just described is called a *composition-series* of G.

The set of groups

$$\frac{G}{G_1},\ \frac{G_1}{G_2},\ \ldots,\ \frac{G_{n-2}}{G_{n-1}},\ G_{n-1},$$

is called a set of *factor-groups* of G, and the orders of these groups are said to form a set of *composition-factors* of G.

Each of the set of factor-groups is necessarily (§ 30) a simple group.

The set of groups forming a composition-series of G is not, in general, unique. Thus G may have more than one maximum self-conjugate sub-group, in which case the second term in the series may be taken different from G_1. Moreover the groups succeeding G_1 are not all necessarily self-conjugate in G; and when some of them are not so, we obtain a new composition-series on transforming the whole set by a suitably chosen operation of G. That the new set thus obtained is again a composition-series is obvious; for if G_{r+1} is a maximum self-conjugate sub-group of G_r, so also is $S^{-1}G_{r+1}S$ of $S^{-1}G_rS$. We proceed to prove that, if a group has two different composition-series, the number of terms in them is the same and the factor-groups derived from them are identical except as regards the sequence in which they occur.

This result, which is of great importance in the subsequent theory, is due to Herr Hölder*; and the proof we here give does not differ materially from his.

The less general result, that, however the composition-series may be chosen, the composition-factors are always the same except as regards their sequence, had been proved by M. Jordan† some years before the date of Herr Hölder's memoir.

90. THEOREM I. *If H is any self-conjugate sub-group of a group G; and if K, K' are two self-conjugate sub-groups of G contained in H, such that there is no self-conjugate sub-group of G contained in H and containing either K or K' except H, K and K' themselves; and if L is the greatest common sub-group of K and K', so that L is necessarily self conjugate in G; then the groups $\frac{H}{K}$ and $\frac{K'}{L}$ are simply isomorphic, as also are the groups $\frac{H}{K'}$ and $\frac{K}{L}$.*

Since K and K' are self-conjugate sub-groups of G contained in H, $\{K, K'\}$ must also be a self-conjugate sub-group of G contained in H; and since, by supposition, there is in H no self-conjugate sub-group of G other than H itself, which contains either K or K', $\{K, K'\}$ must coincide with H. Hence (§ 33) the product of the orders of K and K' is equal to the product of the orders of H and L.

If the order of $\frac{K}{L}$ is m, the operations of K may be divided into the m sets

$$L,\ S_1L,\ S_2L,\ \ldots,\ S_{m-1}L,$$

such that any operation of one set multiplied by any operation of a second gives some operation of a definite third set, and the group $\frac{K}{L}$ is defined by the laws according to which the sets combine.

Consider now the m sets of operations

$$K',\ S_1K',\ S_2K',\ \ldots,\ S_{m-1}K'.$$

*"Zurückführung einer beliebigen algebraischen Gleichung auf eine Kette von Gleichungen," *Math. Ann.* XXXIV, (1889), p. 33.

†"Traité des substitutions," (1870), p. 42.

No two operations of any one set can be identical. If operations from two different sets are the same, say

$$S_p k'_1 = S_q k'_2,$$

where k'_1 and k'_2 are operations of K', then

$$S_q^{-1} S_p = k'_2 k'^{-1}_1,$$

some operation of K'. But $S_q^{-1} S_p$ is an operation of K; hence, as it belongs both to K and K', it must belong to L, so that

$$S_p = S_q l,$$

where l is some operation of L. This however contradicts the supposition that the operations $S_p L$ and $S_q L$ are all distinct. It follows that the operations of the above m sets are all distinct.

Now they all belong to the group H; and their number, being the order of K' multiplied by the order of $\frac{K}{L}$, is equal to the order of H. Hence in respect of the self-conjugate sub-group K', which H contains, the operations of the group H can be divided into the sets

$$K',\ S_1K',\ S_2K',\ \ldots,\ S_{m-1}K',$$

and the group $\frac{H}{K'}$ is defined by the laws according to which these sets combine. But if

$$S_pL \cdot S_qL = S_rL,$$

then necessarily

$$S_pK' \cdot S_qK' = S_rK'.$$

Hence the groups $\frac{H}{K'}$ and $\frac{K}{L}$ are simply isomorphic. In precisely the same way it is shewn that $\frac{H}{K}$ and $\frac{K'}{L}$ are simply isomorphic.

Corollary. If H coincides with G, K and K' are maximum self-conjugate sub-groups of G. Hence if K and K' are maximum self-conjugate sub-groups of G, and if L is the greatest group common to K and K', then $\frac{G}{K}$ and $\frac{K'}{L}$ are simply isomorphic; as also are $\frac{G}{K'}$ and $\frac{K}{L}$.

Now $\frac{G}{K}$ and $\frac{G}{K'}$ are simple groups; and therefore, $\frac{K}{L}$ and $\frac{K'}{L}$ being simple groups, L must be a maximum self-conjugate sub-group of both K and K'.

91. We may now at once proceed to prove by a process of induction the properties of the composition-series of a group stated at the end of § 89. Let us suppose that, for groups whose orders do not exceed a given number n, it is already known that any two composition-series contain the same number of groups and that the factor-groups defined by them are the same except as regards their sequence. If G, a group whose order does not exceed $2n$, has more than one composition-series, let two such series be

$$G,\ G_1,\ G_2, \ldots,\ 1;$$
$$\text{and}\quad G,\ G_1',\ G_2', \ldots,\ 1.$$

If H is the greatest common sub-group of G_1 and G_1', and if

$$H,\ I,\ J,\ \ldots,\ 1$$

is a composition-series of H, then, by the Corollary in the preceding paragraph,

$$G,\ G_1,\ H,\ I,\ J, \ldots,\ 1,$$
$$\text{and}\quad G,\ G_1',\ H,\ I,\ J, \ldots,\ 1$$

are two composition-series of G which contain the same number of terms and give the same factor-groups. For it has there been shewn that $\frac{G}{G_1}$ and $\frac{G_1'}{H}$ are simply isomorphic; as also are $\frac{G}{G_1'}$ and $\frac{G_1}{H}$. Now the order of G_1, being a factor of the order of G, cannot exceed n. Hence the two composition-series

$$G_1, G_2,\ \ldots,\ 1,$$
$$\text{and}\quad G_1, H,\ I, \ldots,\ 1,$$

by supposition contain the same number of groups and give the same factor-groups; and the same is true of the two composition-series

$$G_1', \; G_2', \quad \ldots, \; 1,$$
$$\text{and} \quad G_1', \; H, \; I, \ldots, \; 1.$$

Hence finally, the two original series are seen, by comparing them with the two new series that have been formed, to have the same number of groups and to lead to the same factor-groups. The property therefore, if true for groups whose order does not exceed n, is true also for groups whose order does not exceed $2n$. Now the simplest group, which has more than one composition-series, is that defined by

$$A^2 = 1, \quad B^2 = 1, \quad AB = BA.$$

For this group there are three distinct composition-series, viz.

$$\{A, \; B\}, \; \{A\}, \quad 1;$$
$$\{A, \; B\}, \; \{B\}, \quad 1;$$
$$\text{and} \quad \{A, \; B\}, \; \{AB\}, \; 1:$$

and for these the theorem is obviously true. It is therefore true generally. Hence:—

THEOREM II. *Any two composition-series of a group consist of the same number of sub-groups, and lead to two sets of factor-groups which, except as regards the sequence in which they occur, are identical with each other.*

The definite set of simple groups, which we thus arrive at from whatever composition-series we may start, are essential constituents of the group: the group is said to be *compounded* from them. The reader must not, however, conclude either that the group is defined by its set of factor-groups, or that it necessarily contains a sub-group simply isomorphic with any given one of them.

92. It has been already pointed out that the groups in a composition-series of G are not necessarily, all of them, self-conjugate groups of G.

Suppose now that a series of groups

$$G,\ H_1,\ H_2,\ \ldots,\ H_{m-1},\ 1$$

are chosen so that each one is a self-conjugate sub-group of G, while there is no self-conjugate sub-group of G contained in any one group of the series and containing the next group.

Definition. The series of groups, obtained in the manner just described, is called a *chief composition-series*, or a *chief-series* of G.

It should be noticed that such a series is not necessarily obtained by dropping out from a composition-series those of its groups which are not self-conjugate in the original group. Thus it follows immediately, from the results of § 54, that the composition-series of a group whose order is the power of a prime can be chosen, either (i) so that every group of the series is a self-conjugate sub-group, or (ii) so as to contain any given sub-group, self-conjugate or not.

A chief composition-series of a group is not necessarily unique; and when a group has more than one, the following theorem, exactly analogous to Theorem II, holds:—

THEOREM III. *Any two chief-composition-series of a group consist of the same number of terms and lead to two sets of factor-groups, which, except as regards the sequence in which they occur, are identical with each other.*

The formal proof of this theorem would be a mere repetition of the proof of § 91, Theorem I itself being used to start from instead of its Corollary; it is therefore omitted.

Although it is not always possible to pass from a composition-series to a chief-series, the process of forming a composition-series on the basis of a given chief-series can always be carried out. Thus if, in a chief-series, H_{r+1} is not a maximum self-conjugate sub-group of H_r, the latter group must have a maximum self-conjugate sub-group $G_{r,1}$ which contains H_{r+1}. If H_{r+1} is not a maximum self-conjugate sub-group of $G_{r,1}$, then such a group, $G_{r,2}$, may be found still containing H_{r+1}; and this process may be continued till we arrive at a group $G_{r,s-1}$, of which H_{r+1} is a maximum self-conjugate sub-group. A similar process may be carried out for each pair of consecutive terms in the chief-series; the resulting series so obtained is a composition-series of the original group.

93. The factor-groups $\frac{H_r}{H_{r+1}}$ arising from a chief-series are not necessarily simple groups. If between H_r and H_{r+1} no groups of a corresponding composition-series occur, the group $\frac{H_r}{H_{r+1}}$ is simple; but when there are such intermediate groups, $\frac{H_r}{H_{r+1}}$ cannot be simple. We proceed to discuss the nature of this group in the latter case.

Let G be multiply isomorphic with G', so that the self-conjugate subgroup H_{r+1} of G corresponds to the identical operation of G'. Also let

$$H'_1,\ H'_2,\ \ldots,\ H'_p,\ H'_{p+1},\ \ldots,\ H'_r,\ 1$$

be the sub-groups of G' which correspond to the sub-groups

$$H_1,\ H_2,\ \ldots,\ H_p,\ H_{p+1},\ \ldots,\ H_r,\ H_{r+1}$$

of G. Since H_p contains H_{p+1}, H'_p must contain H'_{p+1}; and since H_{p+1} is self-conjugate in G, H'_{p+1} is self-conjugate in G'. Also if G' had a self-conjugate sub-group contained in H'_p and containing H'_{p+1}, G would have a self-conjugate sub-group contained in H_p and containing H_{p+1}. This is not the case, and therefore

$$G',\ H'_1,\ H'_2,\ \ldots,\ H'_r,\ 1$$

is a chief-series of G'. Hence $\frac{H_r}{H_{r+1}}$ is simply isomorphic with H'_r, the last group but one in the chief-series of G'.

Definition. If Γ is a self-conjugate sub-group of G, and if G has no self-conjugate sub-group, contained in Γ, whose order is less than that of Γ, then Γ is called a *minimum* self-conjugate sub-group of G.

Making use of the phrase thus defined, the discussion of the factor-groups $\frac{H_r}{H_{r+1}}$ of a chief-series is the same as that of the minimum self-conjugate sub-groups of a given group.

94. To simplify the notation as much as possible, let I be a minimum self-conjugate sub-group of G; and, if I is not a simple group, suppose that in a composition-series of G the term following I is g_1. Since g_1 cannot be self-conjugate in G, it must be one of a set of conjugate sub-groups

$$(\alpha) \quad g_1,\ g_2,\ \ldots,\ g_n.$$

Now, if $g_r = S^{-1}g_1S$, then as g_1 is a maximum self-conjugate sub-group of I, $S^{-1}g_1S$ or g_r is a maximum self-conjugate sub-group of $S^{-1}IS$ or I; and hence every one of the above set of conjugate sub-groups is a maximum self-conjugate sub-group of I. If g_{rs} represents the greatest sub-group common to g_r and g_s, then, by Theorem I of the present chapter (§ 90),

$$\ldots,\ I,\ g_1,\ g_{1r},\ \ldots$$
$$\ldots,\ I,\ g_r,\ g_{1r},\ \ldots$$

are composition-series of G. Hence $\frac{g_1}{g_{1r}}$ is simply isomorphic with $\frac{I}{g_r}$. Now $\frac{I}{g_r}$ and $\frac{SIS^{-1}}{Sg_rS^{-1}}$, or $\frac{I}{g_1}$ are simply isomorphic; therefore $\frac{g_1}{g_{1r}}$ and $\frac{I}{g_1}$ are simply isomorphic. Similarly we may shew that, whatever r and s may be, $\frac{g_r}{g_{rs}}$ and $\frac{I}{g_1}$ are simply isomorphic.

If g_{1r} and g_{1s} are the same group, whatever r and s may be, this group is common to the whole set of conjugate groups (α). If these groups have any common sub-group, except identity, it is (Theorem V, § 27) a self-conjugate sub-group of G, and this self-conjugate sub-group would be contained in I, contrary to supposition. Hence if g_{1r} and g_{1s} are the same group, for all values of r and s, this group consists of the identical operation alone, and a composition-series is

$$\ldots,\ I,\ g_1,\ 1.$$

If g_{1r} and g_{1s} are distinct and if g_{1rs} is their greatest common sub-group, then

$$\ldots,\ I,\ g_1,\ g_{1r},\ g_{1rs},\ \ldots$$
$$\ldots,\ I,\ g_1,\ g_{1s},\ g_{1rs},\ \ldots$$

are two composition-series, and $\frac{g_{1r}}{g_{1rs}}$ is simply isomorphic with with $\frac{I}{g_1}$ for all values of r and s.

This reasoning may be repeated. If g_{1rs} and g_{1rt} are the same group whatever s and t may be, they must each be the identical operation. If not, g_{1rst} is another term of the composition-series, and $\frac{g_{1rs}}{g_{1rst}}$ is simply isomorphic with $\frac{I}{g_1}$.

Hence however the composition-series from I onwards be constructed, the corresponding factor-groups are all simply isomorphic with each other. Moreover every group after I in the composition-series is self-conjugate in I. For g_1 and g_r being self-conjugate sub-groups of I, so also is their greatest common sub-group g_{1r}; and g_{1r}, g_{1s} being self-conjugate sub-groups of I, g_{1rs} is also self-conjugate; and so on. Let the composition-series thus arrived at be now written

$$\ldots,\ I,\ \gamma_1,\ \gamma_2,\ \ldots,\ \gamma_{s-1},\ 1;$$

where, as has been proved, $\frac{I}{\gamma_1}$, $\frac{\gamma_1}{\gamma_2}$, ..., γ_{s-1}, are simply isomorphic.

The final group γ_{s-1} must be one of a conjugate set of, say, ν groups in G, no one of which has any operation except identity in common with any other. Since each of these ν groups is self-conjugate in I, and since no two of them have a common operation except identity, it follows by Theorem IX, § 34, that every operation of any one of them is permutable with every operation of the remaining $\nu - 1$. The group generated by the ν groups conjugate to γ_{s-1}, being self-conjugate in G and contained in I, must coincide with I. Now, if γ_{s-1}^1 and γ_{s-1}^2 are any two of this set of ν groups, $\{\gamma_{s-1}^1, \gamma_{s-1}^2\}$ is their direct product, and it is a self-conjugate sub-group of I. If $s > 2$, $\{\gamma_{s-1}^1, \gamma_{s-1}^2\}$ does not coincide with I; and therefore there must be another sub-group γ_{s-1}^3, of the set to which γ_{s-1} belongs, which is not contained in $\{\gamma_{s-1}^1, \gamma_{s-1}^2\}$. Since both the latter group and γ_{s-1}^3 are self-conjugate in I, while γ_{s-1}^3 is a simple group, no operation of γ_{s-1}^3 except identity can be contained in $\{\gamma_{s-1}^1, \gamma_{s-1}^2\}$. Hence γ_{s-1}^1, γ_{s-1}^2 and γ_{s-1}^3 are independent, i.e. no operation of one of these groups can be expressed in terms of operations of the other two. The group $\{\gamma_{s-1}^1, \gamma_{s-1}^2, \gamma_{s-1}^3\}$ is the direct product of γ_{s-1}^1, γ_{s-1}^2 and γ_{s-1}^3, and it is self-conjugate in I. If $s > 3$, the same reasoning may be repeated. Finally, from the set of ν groups conjugate to γ_{s-1}, it must be possible to choose s independent groups

$$\gamma_{s-1}^1,\ \gamma_{s-1}^2,\ \ldots,\ \gamma_{s-1}^s,$$

such that no operation of any one of them can be expressed in terms of the operations of the remaining $s-1$; and I will then be the direct product of these s groups. Hence:—

THEOREM IV. *If between two consecutive terms H_r and H_{r+1} in the chief-composition-series of a group there occur the groups $G_{r,1}$, $G_{r,2}$, ..., $G_{r,s-1}$ of a composition-series; then* (i) *the factor groups*

$$\frac{H_r}{G_{r,1}},\ \frac{G_{r,1}}{G_{r,2}},\ \ldots,\ \frac{G_{r,s-1}}{H_{r+1}}$$

are all simply isomorphic, and (ii) $\frac{H_r}{H_{r+1}}$ *is the direct product of s groups of the type* $\frac{H_r}{G_{r,1}}$.

Corollary. If the order of $\frac{H_r}{H_{r+1}}$ is a power, p^s, of a prime, $\frac{H_r}{H_{r+1}}$ must be an Abelian group of type $(1, 1, \ldots, \text{to } s \text{ units})$.

95. A chief-series of a group G can always be constructed which shall contain among its terms any given self-conjugate sub-group of G. For if Γ is a self-conjugate sub-group of G, and if $\frac{G}{\Gamma}$ is simple, we may take Γ for the group which follows G in the chief-series. If on the other hand $\frac{G}{\Gamma}$ is not simple, it must contain a minimum self-conjugate sub-group. Then Γ_1, the corresponding self-conjugate sub-group of G, contains Γ; and if there were a self-conjugate sub-group of G contained in Γ_1, and containing Γ, the self-conjugate sub-group of $\frac{G}{\Gamma}$, which corresponds to Γ_1, would not be a minimum self-conjugate sub-group. We may now repeat the same process with Γ_1, and so on; the sub-groups thus introduced will, with G and Γ, clearly form the part of a chief-series extending from G to Γ. The series may be continued from Γ, till we arrive at the identical operation, in the usual way.

96. It will perhaps assist the reader if we illustrate the foregoing theory by one or two simple examples. We take first a group of order 12, defined by the relations

$$A^2 = 1, \quad B^2 = 1, \quad AB = BA,$$
$$R^3 = 1, \quad R^{-1}AR = B, \quad R^{-1}BR = AB^*.$$

From the last two equations, it follows that

$$R^{-1}ABR = A,$$

and therefore R transforms the sub-group $\{A, B\}$ of order 4 into itself; so that this sub-group is self-conjugate, and the order of the group is 12 as stated. The self-conjugate sub-group $\{A, B\}$ thus determined is clearly a maximum self-conjugate sub-group. Also it is the only one. For if there were another its order would be 6, and it would contain all the operations of order 3 in the group. Now since $\{R\}$ is only permutable with its own operations, the group contains 4 sub-groups of order 3, and therefore there can be no self-conjugate sub-group of order 6. The three cyclical sub-groups $\{A\}$, $\{B\}$ and $\{AB\}$ of order 2 are transformed into each other by R, and therefore no one of them is self-conjugate.

Hence the only chief-series is

$$\{R, A, B\}, \quad \{A, B\}, \quad 1,$$

and there are three composition-series, viz.

$$\{R, A, B\}, \quad \{A, B\}, \quad \{A\}, \quad 1;$$
$$\{R, A, B\}, \quad \{A, B\}, \quad \{B\}, \quad 1;$$
$$\text{and} \quad \{R, A, B\}, \quad \{A, B\}, \quad \{AB\}, \quad 1.$$

*The reader will notice that B can be eliminated from these relations, and that the group can be defined by $A^2 = 1$, $R^3 = 1$, $(AR)^3 = 1$. The structure of the group however is given, at a glance, by the equations in the text.

The orders of the factor-groups in the chief-series are 3 and 2^2, and the group of order 2^2 is, as it should be, an Abelian group whose operations are all of order 2. The composition-factors are 3, 2, 2 in the order written.

97. As a rather less simple instance, we will now take a group generated by four permutable independent operations A, B, P, Q, of orders 2, 2, 3, 3 respectively and an operation R of order 3, for which

$$R^{-1}AR = B, \quad R^{-1}BR = AB, \quad R^{-1}PR = P, \quad R^{-1}QR = QP^{*}.$$

The sub-group $\{A, B, P, Q\}$, of order 36, is clearly a maximum self-conjugate sub-group, and therefore the order of the group is 108. Since A, B and AB are conjugate operations, every self-conjugate sub-group that contains A must contain B; and since Q and QP are conjugate, every self-conjugate sub-group that contains Q must contain P. Hence the only other possible maximum self-conjugate sub-groups are those of the form $\{A, B, P, RQ^\alpha\}$; and since

$$Q^{-1}RQ^\alpha Q = RQ^\alpha P^{-1},$$

these groups actually are self-conjugate. The same reasoning shews that the only maximum self-conjugate sub-group of $\{A, B, P, Q\}$ or of $\{A, B, P, RQ^\alpha\}$, which is self-conjugate in the original group, is $\{A, B, P\}$; and the only maximum self-conjugate sub-groups of the latter, which are self-conjugate in the original group, are $\{A, B\}$ and $\{P\}$. Hence all the chief-series of the group are given by

$$\{R, A, B, P, Q\}, \quad \begin{matrix}\{A, B, P, Q\}, \\ \text{or} \\ \{A, B, P, RQ^\alpha\},\end{matrix} \quad \{A, B, P\}, \quad \begin{matrix}\{A, B\}, \\ \text{or} \\ \{P\},\end{matrix} \quad 1.$$

Since $\{A, B, P, Q\}$ is an Abelian group, all of its sub-groups are self-conjugate. Hence if G_1, G_2, and G_3 are *any* maximum sub-groups of $\{A, B, P, Q\}$, G_1 and G_2 respectively, then

$$\{R, A, B, P, Q\}, \quad \{A, B, P, Q\}, \quad G_1, \quad G_2, \quad G_3, \quad 1$$

is a composition-series.

Again, since A and B are conjugate in $\{A, B, P, RQ^\alpha\}$, the only maximum self-conjugate sub-groups of this group are those of the form $\{A, B, P^x(RQ^\alpha)^y\}$.

*Here again the group can clearly be defined in terms of A, Q and R.

If y is zero, this sub-group is Abelian; and we may take for the next term in the composition series any maximum sub-group g_2 of this Abelian sub-group, and for the last term but one any sub-group g_3 of g_2. But if y is not zero, $\{A, B, P^x(RQ^\alpha)^y\}$ can only be followed by $\{A, B\}$. Hence the remaining composition-series are of the forms:—

$$\{R, A, B, P, Q\}, \quad \{A, B, P, RQ^\alpha\}, \quad \{A, B, P\}, \quad g_2, \quad g_3, \quad 1,$$

and

$$\{R, A, B, P, Q\}, \quad \{A, B, P, RQ^\alpha\}, \quad \{A, B, P^x(RQ^\alpha)^y\},$$
$$\{A, B\}, \quad \{A\} \text{ or } \{B\} \text{ or } \{AB\}, \quad 1.$$

It should be noticed that if, in the last of these series, we drop out the terms which are not self-conjugate in the original group, here the third and fifth terms, we do not arrive at a chief-series. This illustrates a remark made in § 92.

98. THEOREM V. *If H is a sub-group of G, each composition-factor of H must be equal to or be a factor of some composition-factor of G.*

If G is simple, its only composition-factor is equal to its order: the theorem in this case is obvious.

If G is not simple, let G_{r+1} be the first term in a composition-series of G which does not contain H; and let G_r be the term preceding G_{r+1}. If H_1 is the greatest common sub-group of G_{r+1} and H, then H_1 is a self-conjugate sub-group of H. For every operation of H transforms both H and G_{r+1} into themselves; and therefore every operation of H transforms H_1, the greatest common sub-group of H and G_{r+1}, into itself. Now the order of $\{H, G_{r+1}\}$ is equal to the product of the orders of H and G_{r+1} divided by the order of H_1; and $\{H, G_{r+1}\}$ is contained in G_r. Hence the order of $\frac{H}{H_1}$ is equal to or is a factor of the order of $\frac{G_r}{G_{r+1}}$. If then a composition-series of H be taken, containing the term H_1, the orders of the factor-groups, formed by those terms of the series terminating with H_1, are equal to or are factors of the order of $\frac{G_r}{G_{r+1}}$. The same reasoning may now be used for H_1 that has been applied to H; and the theorem is therefore true.

Corollary. If all the composition-factors of G are primes, so also are the composition-factors of every sub-group of G.

99. Definition. A group, all whose composition-factors are primes, is called a *soluble* group.

The Corollary of the last theorem may be stated in the form:—if a group is soluble, so also are all its sub-groups.

A soluble group of order $p^\alpha q^\beta \dots r^\gamma$, where p, q, $\dots$, r are distinct primes, has $\alpha + \beta + \dots + \gamma$ composition-factors; these are capable of $\dfrac{(\alpha+\beta+\dots+\gamma)!}{\alpha!\,\beta!\,\dots\,\gamma!}$ distinct arrangements.

For a specified group the composition-factors may, as we have already seen, occur in two or more distinct arrangements; but it is immediately obvious that two groups of the same order cannot be of the same type, i.e. simply isomorphic, unless the distinct arrangements, of which the composition-factors are capable, are the same for both. A first step therefore towards the enumeration of all distinct types of soluble groups of a given order, will be to classify them according to the distinct arrangements of which the composition-factors are capable; for no two groups belonging to different classes can be of the same type.

The case, in which the composition-factors are capable of all possible arrangements, is one which will always occur. Taking in this case β q's followed by α p's for the last $\alpha+\beta$ composition-factors, the group contains a sub-group G' of order $p^\alpha q^\beta$. In the composition-series of this group, with the composition-factors taken as proposed, there is a sub-group H of order p^α contained self-conjugately in a sub-group H_1 of order $p^\alpha q$. This sub-group H_1 is contained self-conjugately in a group H_2 of order $p^\alpha q^2$. Hence (Theorem VII, Cor. IV, § 87) H is contained self-conjugately in H_2. Again, H_2 is contained self-conjugately in a group H_3 of order $p^\alpha q^3$, and therefore again H is self-conjugate in H_3. Proceeding thus, we shew that H is self-conjugate in G'. It follows that n, the number of conjugate sub-groups of order p^α contained in the group, is not a multiple of q. Now q may be any one of the distinct primes other than p that divide the order of the group. Hence finally the group contains a self-conjugate sub-group of order p^α. In the same way we shew that it contains self-conjugate sub-groups of orders $q^\beta, \ldots, r^\gamma$. The group must therefore be the direct product of groups whose orders are p^α, q^β, $\ldots$, r^γ.

Hence:—

THEOREM VI. *A soluble group, the composition-factors of which may be taken in any order, is the direct product of groups whose orders are powers of primes.*

100. THEOREM VII. *If G is a soluble group of order $p^\alpha m$, where p is prime and does not divide m, and if every operation of G whose order is a power of p is permutable with every operation whose order is relatively prime to p; then G is the direct product of two groups of orders p^α and m.*

Let H be a sub-group of G of order p^α. This sub-group, from the conditions of the theorem, is necessarily self-conjugate. Let a chief-series of G be formed which contains H, say

$$G, \ldots, H', H, \ldots, 1.$$

The order of H' must be of the form $p^\alpha q^\beta$; and H' must contain a single sub-group h of order q^β. This sub-group h is self-conjugate in G, since H' is self-conjugate in G; and h is the only sub-group of order q^β contained in H'.

Consider now the group $\frac{G}{h}$. Every operation of this group whose order is a power of p is permutable with every operation whose order is relatively prime to p. Hence we may repeat the same reasoning to shew that $\frac{G}{h}$ contains a self-conjugate sub-group of order r^γ where r is a prime distinct from p, but possibly the same as q. It follows that G has a self-conjugate sub-group h_1 of order $q^\beta r^\gamma$. We may now repeat the same reasoning with $\frac{G}{h_1}$; and in this way we must at last reach a self-conjugate sub-group of G of order m. Hence, since G contains self-conjugate sub-groups of orders p^α and m, which are relatively prime, G must be the direct product of these sub-groups.

It should be noticed that the above reasoning does not necessarily hold if G is not soluble; for then the order of H' may be of the form $p^\alpha \mu^\beta$, where μ, the order of a simple group, contains more than one prime factor. In that case it would not be necessarily true that H' contains a group of order μ^β.

101. Though the actual determination of all types of soluble groups of a given order more properly forms part of the subject of Chapter XI, we will here, as a further illustration of the subject of the present Chapter, deal with the problem for groups whose orders are of the form p^2q, p and q being distinct primes.

Every such group must be soluble. In fact, if $p > q$, the group must contain a self-conjugate sub-group of order p^2; and if $p < q$, there must be a self-conjugate sub-group of order q unless $p = 2$, $q = 3$; while in this case if there are 4 sub-groups of order 3, there must be a self-conjugate sub-group of order 4. These statements all follow immediately from Sylow's theorem.

The Abelian groups of order p^2q may be specified immediately; and therefore in what follows we will assume that the group is not Abelian. There are 3 possible arrangements for the composition-factors, viz.

$$p,\ p,\ q; \quad p,\ q,\ p; \quad q,\ p,\ p.$$

If the factors are capable of all three arrangements, the group is the direct product of groups of orders p^2 and q; it is therefore an Abelian group.

If the two arrangements p, p, q and q, p, p are possible, there are self-conjugate sub-groups of orders p^2 and q; the group again is Abelian, and all three arrangements are possible.

There are now five other possibilities.

I. p, p, q and p, q, p, the only possible arrangements.

There must be here a sub-group of order pq, containing self-conjugate sub-groups of orders p and q and therefore Abelian. Let this be generated by operations P and Q, of orders p and q. Since the group has sub-groups of order p^2, there must be operations of orders p or p^2, not contained in the sub-group of order pq, and permutable with P. Let R be such an operation, so that R^p belongs to the sub-group of order pq. R cannot be permutable with Q, as the group would be then Abelian; hence

$$R^{-1}QR = Q^\alpha,$$

so that

$$R^{-p}QR^p = Q^{\alpha^p},$$

and

$$\alpha^p \equiv 1 \text{ (mod. } q).$$

This case can therefore only occur if p is a factor of $q-1$. There are two distinct types, according as R is of order p or of order p^2; i.e. according as the sub-groups of order p^2 are non-cyclical or cyclical. If α and β are any two distinct primitive roots of the congruence

$$\alpha^p \equiv 1 \text{ (mod. } q),$$

the relations

$$R^{-1}QR = Q^\alpha,$$

and

$$R^{-1}QR = Q^\beta,$$

do not lead to distinct types, since the latter may be reduced to the form by replacing R by R^x, where

$$\beta^x \equiv \alpha \text{ (mod. } q).$$

The two types are respectively defined by the relations

$$Q^q = 1, \quad P^p = 1, \quad R^p = 1, \quad P^{-1}QP = Q,$$
$$R^{-1}PR = P, \quad R^{-1}QR = Q^\alpha;$$

and

$$Q^q = 1, \quad R^{p^2} = 1, \quad R^{-1}QR = Q^\alpha.$$

In each case, α is a primitive root of the congruence $\alpha^p \equiv 1$ (mod. q).

II. p, q, p and q, p, p, the only possible arrangements.

There must be a self-conjugate sub-group of order pq, in which the sub-group of order q is not self-conjugate, and a self-conjugate sub-group of order p^2. The sub-group of order pq must be given by

$$P^p = 1, \quad Q^q = 1, \quad Q^{-1}PQ = P^\alpha,$$
$$\alpha^q \equiv 1 \text{ (mod. } p);$$

so that in this case q must be a factor of $p-1$. If the sub-group of order p^2 is not cyclical, there must be an operation R of order p, not contained in the sub-group of order pq. Any such operation must be permutable with P. Moreover since the sub-group of order pq is self-conjugate and contains only p sub-groups of order q, the sub-group $\{Q\}$ must be permutable with some operation of order p. Hence we may assume that R is permutable with $\{Q\}$, and, since $p > q$, with Q.

We thus obtain a single type defined by

$$P^p = 1, \quad R^p = 1, \qquad PR = RP,$$
$$Q^q = 1, \quad QR = RQ, \quad Q^{-1}PQ = P^{\alpha}.$$

If the sub-group of order p^2 is cyclical, all the operations, which have powers of p for their orders and are not contained in the sub-group of order pq, must be of order p^2. There can therefore be no operation of order p, which is permutable with $\{Q\}$; therefore there is no corresponding type.

III. p, p, q, the only possible arrangement.

There must be a self-conjugate sub-group of order pq, which has no self-conjugate sub-group of order p; it is therefore defined by

$$P^p = 1, \quad Q^q = 1, \quad P^{-1}QP = Q^{\alpha},$$
$$\alpha^p \equiv 1 \text{ (mod. } q);$$

so that here p must be a factor of $q - 1$.

If the sub-groups of order p^2 are not cyclical, there must be an operation R' of order p, not contained in this sub-group and permutable with P. Hence

$$R'^{-1}QR' = Q^{\beta};$$

and if

$$\beta \equiv \alpha^x \text{ (mod. } q),$$

then $R'P^{-x}$ is an operation of order p, which is not contained in the sub-group of order pq and is permutable with Q. It is therefore a self-conjugate operation of order p. Hence p, q, p is a possible arrangement of the composition-factors, and there is in this case no type.

If the sub-groups of order p^2 are cyclical, there must be an operation R of order p^2, such that

$$R^p = P.$$

Hence

$$R^{-1}QR = Q^{\beta},$$

where β is a primitive root of the congruence

$$\beta^{p^2} = 1 \text{ (mod. } q).$$

This case then can only occur when p^2 is a factor of $q-1$; and we again have a single type defined by

$$R^{p^2}=1, \quad Q^q=1, \quad R^{-1}QR=Q^\beta.$$

IV. p, q, p, the only possible arrangement.

Here the self-conjugate sub-group of order pq must be given by

$$P^p=1, \quad Q^q=1, \quad Q^{-1}PQ=P^\alpha,$$
$$\alpha^q \equiv 1 \text{ (mod. } p),$$

and q must be a factor of $p-1$. As in II, there must be an operation R of order p, permutable with $\{Q\}$ and therefore with Q; and since R transforms $\{P,Q\}$ into itself, it must be permutable with P. This however makes the sub-group $\{P,R\}$ self-conjugate, which requires q, p, p to be a possible arrangement of the composition-factors. Hence there is no type corresponding to this case.

V. q, p, p, the only possible arrangement.

If the sub-group of order p^2 is cyclical, and is generated by P, while Q is an operation of order q, we must have

$$Q^{-1}PQ=P^\alpha,$$

where

$$\alpha^q \equiv 1 \text{ (mod. } p^2).$$

Here q must be a factor of $p-1$; since the congruence has just $q-1$ primitive roots, there is a single type of group.

If the sub-group of order p^2 is not cyclical, it can be generated by two permutable operations P_1 and P_2 of order p. Since a sub-group of order q is not self-conjugate, either p or p^2 must be congruent to unity (mod. q); and therefore q must be a factor of either $p-1$ or $p+1$.

Suppose first, that q is a factor of $p-1$.

There are $p+1$ sub-groups of order p. When these are transformed by any operation Q of order q, those which are not permutable with Q must be interchanged in sets of q. Hence at least two of these sub-groups must be permutable with Q, and we may take the generating operations of two such sub-groups for P_1 and P_2. Therefore

$$Q^{-1}P_1Q=P_1^\alpha, \quad Q^{-1}P_2Q=P_2^\beta.$$

Now if either α or β, say β, were unity, then $\{Q, P_1\}$ would be a self-conjugate sub-group and p, q, p would be a possible arrangement of the composition-factors. Hence neither α nor β can be unity, and we may take

$$Q^{-1}P_1Q = P_1^{\alpha}, \quad Q^{-1}P_2Q = P_2^{\alpha^x},$$

where α is a primitive root of

$$\alpha^q \equiv 1 \text{ (mod. } p),$$

and x is not zero.

It remains to determine how many distinct types these equations contain. When $q = 2$, the only possible value of x is unity; and there is a single type. When q is an odd prime, and we take

$$Q^y = Q', \quad P_1 = P_2', \quad P_2 = P_1', \quad xy \equiv 1 \text{ (mod. } q),$$

the equations become

$$Q'^{-1}P_1'Q' = P_1'^{\alpha}, \quad Q'^{-1}P_2'Q' = P_2'^{\alpha^y},$$

and therefore the values x and y of the index of α, where

$$xy \equiv 1 \text{ (mod. } q),$$

give the same type. Now the only way, in which the two equations can be altered into two equations of the same form, is by replacing Q by some other operation of the group whose order is q and by either interchanging P_1 and P_2 or leaving each of them unchanged. Moreover the other operations of the group whose orders are q are those of the form $Q^l P_1^m P_2^n$, where l is not zero, and this operation transforms P_1 and P_2 in the same way as Q^l. Hence finally, the values x and y of the index will only give groups of the same type when

$$xy \equiv 1 \text{ (mod. } q).$$

There are therefore $\frac{1}{2}(q+1)$ distinct types, when q is an odd prime; they are given by the above equations.

Suppose next, that q is a factor of $p+1$.

Any operation Q, of order q, will transform the sub-groups of order p, with which it is not permutable, so as to interchange them in sets of q; and hence if it is permutable with any sub-group, it must be permutable with q at least. This, by the last case, is clearly impossible, and hence Q is permutable with no sub-group of order p. We may therefore, by suitably choosing the generating operations of the group of order p^2, assume that

$$Q^{-1}P_1Q = P_2, \quad Q^{-1}P_2Q = P_1^{\alpha}P_2^{\beta}.$$

If now

$$Q^{-x-1}P_1Q^{x+1} = P_1^{\alpha_x}P_2^{\beta_x},$$

then

$$\alpha_{x+1} \equiv \alpha\beta_x, \quad \beta_{x+1} \equiv \alpha_x + \beta\beta_x \text{ (mod. } p),$$

and therefore

$$\beta_{x+1} - \beta\beta_x - \alpha\beta_{x-1} \equiv 0 \text{ (mod. } p).$$

Hence if ι_1 and ι_2 are the roots of the congruence

$$\iota^2 - \beta\iota - \alpha \equiv 0 \text{ (mod. } p),$$

then

$$\beta_x \equiv \frac{\iota_2^{x+1} - \iota_1^{x+1}}{\iota_2 - \iota_1}.$$

Now since Q^q is the lowest power of Q that is permutable with P_1, β_{q-1} must be the first term of the series $\beta_1, \beta_2, \ldots$ which vanishes. Hence q is the least value of z for which

$$\iota_2^z \equiv \iota_1^z,$$

and therefore the congruence

$$\iota^2 - \beta\iota - \alpha \equiv 0$$

is irreducible. Moreover α_{q-1} must be congruent to unity, and therefore

$$1 \equiv -\iota_1\iota_2\frac{\iota_2^{q-1} - \iota_1^{q-1}}{\iota_2 - \iota_1} \equiv \iota_1^q.$$

From the quadratic congruence satisfied by ι, it follows that

$$\alpha \equiv -\iota^{p+1} \equiv -1, \quad \beta \equiv \iota^p + \iota \text{ (mod. } p);$$

and thence

$$\alpha_x \equiv -\frac{\iota^{px} - \iota^x}{\iota^p - \iota}, \quad \beta_x \equiv \frac{\iota^{p(x+1)} - \iota^{x+1}}{\iota^p - \iota}.$$

Finally, we may shew that, when q is a factor of $p+1$, the equations

$$P_1^p = 1, \quad P_2^p = 1, \quad Q^q = 1, \quad P_1 P_2 = P_2 P_1,$$
$$Q^{-1} P_1 Q = P_2, \quad Q^{-1} P_2 Q = P_1^{-1} P_2^{\iota^p + \iota},$$

where ι is a primitive root of the congruence

$$\iota^q \equiv 1 \text{ (mod. } p),$$

define a single type of group, whatever primitive root of the congruence is taken for ι.

Thus from the given equations it follows that

$$Q^{-x} P_1 Q^x = P_1^{\alpha_{x-1}} P_2^{\beta_{x-1}} = P_3, \text{ say},$$

and

$$\begin{aligned} Q^{-x} P_2' Q^x &= (P_1^{\alpha_{x-1}} P_2^{\beta_{x-1}})^{\alpha_{x-1}} (P_1^{\alpha_x} P_2^{\beta_x})^{\beta_{x-1}}, \\ &= P_1^{\alpha_x \beta_{x-1} - \alpha_{x-1} \beta_x} P_3^{\alpha_{x-1} + \beta_x} \\ &= P_1^{-1} P_3^{\iota^{px} + \iota^x}. \end{aligned}$$

If then we take P_1, P_3 and Q^x as generating operations in the place of P_1, P_2 and Q, the defining relations are reproduced with ι^x in the place of ι. The relations therefore define a single type of group*.

We have, for the sake of brevity, in each case omitted the verification that the defining relations actually give a group of order $p^2 q$. This presents no difficulty, even for the last type; for the previous types it is immediately obvious.

*On groups whose order is of the form $p^2 q$ the reader may consult; Hölder, "Die Gruppen der Ordnungen p^3, pq^2, pqr, p^4," *Math. Ann.* XLIII (1893), in particular pp. 335–360; and Cole and Glover, "On groups whose orders are products of three prime factors," *Amer. Journal*, XV (1893), pp. 202–214.

CHAPTER VIII.

ON SUBSTITUTION-GROUPS: TRANSITIVE AND INTRANSITIVE GROUPS.

102. It has been proved, in the theorem in § 20, that every group is capable of being represented as a group of substitutions performed on a number of symbols equal to the order of the group. For applications to Algebra, and in particular to the Theory of Equations, the presentation of a group as a group of substitutions is of special importance; and we shall now proceed to consider the more important properties of this special mode of representing groups*.

Definition. When a group is represented by means of substitutions performed on a finite set of n distinct symbols, the integer n is called the *degree* of the group.

It is obvious, by a consideration of simple cases, that a group can always be represented in different forms as a group of substitutions, the number of symbols which are permuted in two forms not being necessarily the same; examples have already been given in Chapter II. The "degree of a group" is therefore only an abbreviation of "the degree of a special representation of the group as a substitution-group."

The $n!$ substitutions, including the identical substitution, that can be performed upon n distinct symbols, clearly form a group; for they satisfy the conditions of the definition (§ 12). Moreover they form the greatest group of substitutions that can be performed on the n symbols, because every possible substitution occurs among them. When a group then is spoken of as of degree n, it is implicitly being regarded as a sub-group of this most general group of order $n!$ which can be represented by substitutions of the n symbols; and therefore (Theorem I, § 22) the order of a substitution-group of degree n must be a factor of $n!$.

103. It has been seen in § 11 that any substitution performed on

*When it is necessary to call attention directly to the fact that the group we are dealing with is supposed to be presented as a group of substitutions, the group will be spoken of as a *substitution-group*.

n symbols can be represented in various ways as the product of transpositions; but that the number of transpositions entering in any such representation of the substitution is either always even or always odd. In particular, the identical substitution can only be represented by an even number of transpositions. Hence if S and S' are any two even (§ 11) substitutions of n symbols, and T any substitution at all of n symbols, then SS' and $T^{-1}ST$ are even substitutions. The even substitutions therefore form a self-conjugate sub-group H of the group G of all substitutions.

If now T is any odd substitution, the set of substitutions TH are all odd and all distinct. Moreover they give all the odd substitutions; for if T' is any odd substitution distinct from T, then $T^{-1}T'$ is an even substitution and must be contained in H. Hence the number of even substitutions is equal to the number of odd substitutions: and the order of G is twice that of H.

Definitions. The group of order $n!$ which consists of all the substitutions that can be performed on n symbols is called the *symmetric* group of degree n.

The group of order $\frac{1}{2}n!$ which consists of all the even substitutions of n symbols is called the *alternating** group of degree n.

If the substitutions of a group of degree n are not all even, the preceding reasoning may be repeated to shew that its even substitutions form a self-conjugate sub-group whose order is half the order of the group; and this sub-group is a sub-group of the alternating group of the n symbols.

104. Definition. A substitution-group is called *transitive* when, by means of its substitutions, a given symbol a_1 can be changed into every other symbol $a_2, a_3, \dots, a_n$ operated on by the group. When it has not this property, the group is called *intransitive.*

A transitive group contains substitutions changing any one symbol into any other. For if S and T respectively change a_1 into a_s and a_t, then $S^{-1}T$ changes a_s into a_t.

*The symmetric group has been so called because the only functions of the n symbols which are unaltered by all the substitutions of the group are the symmetric functions.

All the substitutions of the alternating group leave the square root of the discriminant unaltered (§ 11).

THEOREM I. *The substitutions of a transitive group G, which leave a given symbol a_1 unchanged, form a sub-group; and the number of substitutions, which change a_1 into any other symbol a_r, is equal to the order of this sub-group.*

The substitutions which leave a_1 unchanged must form a sub-group H of G; for if S and S' both leave a_1 unchanged, so also does SS'.

Let the operations of G be divided into the sets

$$H,\ HS_1,\ HS_2,\ \ldots,\ HS_{m-1}.$$

No operation of the set HS_p leaves a_1 unchanged; and each operation of the set HS_p changes a_1 into a_p, if S_p does so. If the operations of any other set HS_q also changed a_1 into a_p, then $S_pS_q^{-1}$ would leave a_1 unchanged and would belong to H, which it does not. Hence each set changes a_1 into a distinct symbol. The number of sets must therefore be equal to the number of symbols, while from their formation each set contains the same number of substitutions. If N is the order and n the degree of the transitive group G, then $\frac{N}{n}$ is the order of the sub-group that leaves any symbol a_1 unchanged; and there are $\frac{N}{n}$ substitutions changing a_1 into any other given symbol a_p.

Corollary. The order of a transitive group must be divisible by its degree.

Every group conjugate to H leaves one symbol unchanged. For if S changes a_1 into a_p, then $S^{-1}HS$ leaves a_p unchanged. The sub-groups which leave the different symbols unchanged form therefore a conjugate set.

A transitive group of degree n and order mn has, as we have just seen, $m-1$ substitutions other than identity which leave a given symbol a_1 unchanged. Hence there must be at least $mn-1-n(m-1)$, i.e. $n-1$, substitutions in the group that displace every symbol. If the $m-1$ substitutions, other than identity, that leave a_1 unchanged are all distinct from the $m-1$ that leave a_p unchanged, whatever other symbol a_p may be, $n-1$ will be the actual number of substitutions that displace all the symbols; and no operation other than identity will displace less than $n-1$ symbols. If however the sub-groups that leave a_1 and a_p unchanged have other substitutions besides identity in common, these substitutions must displace less than $n-1$ symbols; and there will be more than $n-1$ substitutions which displace all the symbols.

Ex. If the substitutions of two transitive groups of degree n which displace all the symbols are the same, the groups can only differ in the substitutions that keep just one symbol unchanged. (Netto.)

105. We have seen that every group can be represented as a substitution-group whose degree is equal to its order. A reference to the proof of this theorem (§ 20) will shew that such a substitution-group is transitive, and that the identical substitution is the only one which leaves any symbol unchanged.

We will now consider some of the properties of a transitive group of degree n, whose operations, except identity, displace all or all but one of the n symbols. It has just been seen that such a group has exactly $n-1$ operations which displace all the n symbols. If these $n-1$ operations, with identity, form a sub-group, the sub-group must clearly be self-conjugate.

Suppose now that nm is the order of the group. Then the order of the sub-group, that leaves one symbol a_1 unchanged, is m; and if a_2 is any other of the n symbols, no two operations of this sub-group can change a_2 into the same symbol. For if S, S' were two operations of the sub-group both changing a_2 into a_r, then SS'^{-1} would be an operation, distinct from identity, which would keep both a_1 and a_2 unchanged. Let now P be any operation that displaces all the symbols. Then the set of m operations $S^{-1}PS$, where for S is put in turn each operation of the sub-group that keeps a_1 unchanged, are all distinct; for each of them changes a_1 into a different symbol. If this set does not exhaust the operation conjugate to P, and if P_1 is another such operation, then the set of m operations $S^{-1}P_1S$ are all distinct from each other and from those of the previous set. This process may be continued till the operations conjugate to P are exhausted. The number of operations conjugate to P is therefore a multiple of m. Since P itself does not belong to any one of the n conjugate sub-groups that each keep one symbol fixed, no operation conjugate to P can belong to any of them. Hence each of the km operations of the conjugate set to which P belongs displaces all the symbols. The $n-1$ operations that displace all the symbols can therefore be divided into sets, so that the number in each set is a multiple of m; and hence m must be a factor of $n-1$.

Suppose now that p is any prime factor of n, and that p^α is the highest power of p which divides n. If P is an operation whose order is a power of p, and if $p^\beta\mu$ is the order of the greatest sub-group h that contains P self-conjugately, then P is one of $\frac{nm}{p^\beta\mu}$ conjugate operations. Now (§ 87) the sub-group h contains $k\mu$ ($k \nless 1$) operations whose orders are relatively prime to p; and therefore there are $k\mu$ operations of the form PQ, where Q is permutable with P and the order of Q is relatively prime to the order of P. If P' is any operation conjugate to P, there are similarly $k\mu$ operations of the form $P'Q'$; and (§ 16) no one of these operations can be identical with any one of those of the previous set. The group therefore will contain $\frac{nmk}{p^\beta}$ distinct operations, which are conjugate to the various operations of the set PQ. Moreover since P displaces all the symbols, each one of these operations must displace all the symbols.

If then the group has r distinct sets of conjugate operations whose orders are powers of p, the number of operations, whose orders are divisible by p, is equal to

$$\sum_{s=1}^{s=r} \frac{nmk_s}{p^{\beta_s}}.$$

Also the number of operations, which displace all the symbols and the orders of which are not divisible by p, is of the form $\frac{nk_0}{p^\alpha} - 1$ (§ 87).

Hence finally

$$\begin{aligned} m\sum_{s=1}^{s=r} \frac{k_s}{p^{\beta_s}} &= \frac{1}{n}\left\{n-1-\left(\frac{nk_0}{p^\alpha}-1\right)\right\} \\ &= \frac{p^\alpha - k_0}{p^\alpha}. \end{aligned}$$

The greatest possible value of m will correspond to the suppositions

$$r = 1, \quad k_1 = 1, \quad k_0 = 1;$$

and these give

$$m = p^\alpha - 1.$$

Hence if $n = p^\alpha q^\beta \dots r^\gamma$, where $p, q, \dots, r$ are distinct primes, m cannot be greater than the least of the numbers

$$p^\alpha - 1, \quad q^\beta - 1, \quad \dots, \quad r^\gamma - 1.$$

Certain particular cases may be specially noticed. First, a group of degree n and order $n(n-1)$, whose operations other than identity displace all or all but one of the symbols, can exist only when n is the power of a prime*. Groups which satisfy these conditions will be discussed in § 112.

Similarly, a group of degree n and order nm, where m is not less than $\sqrt{n}$, whose operations other than identity displace all or all but one of the symbols, can only exist when n is the power of a prime.

If n is equal to twice an odd number, a transitive group of degree n, none of whose operations except identity leave two symbols unchanged, must be of order n.

Lastly we may shew that, if m is even, a group of degree n and order nm, none of whose operations except identity leave two symbols unchanged, must contain a self-conjugate Abelian sub-group of order and degree n.

A sub-group that keeps one symbol fixed must, if m is even, contain an operation of order 2. If it contained r such operations, the group would contain nr; and each of these could be expressed as the product of $\frac{1}{2}(n-1)$ independent transpositions. Now from n symbols $\frac{1}{2}n(n-1)$ transpositions can be formed. If then r were greater than 1, among the operations of order 2 that keep one symbol fixed there would be pairs of operations with a common transposition; and the product of two such operations would be an operation, distinct from identity, which would keep two symbols at least fixed. This is impossible; therefore r must be unity. Now let

$$A_1,\ A_2,\ \ldots,\ A_n$$

be the n operations of order 2 belonging to the group. Since no two of these operations contain a common transposition,

$$A_1A_r,\ A_2A_r,\ \ldots,\ A_{r-1}A_r,\ A_{r+1}A_r,\ \ldots, A_nA_r$$

are the $n-1$ operations which displace all the symbols. These operations may also be expressed in the form

$$A_rA_1,\ A_rA_2,\ \ldots,\ A_rA_{r-1},\ A_rA_{r+1},\ \ldots, A_rA_n;$$

and since

$$A_pA_r \cdot A_rA_q = A_pA_q,$$

*Jordan, "Récherches sur les substitutions," *Liouville's Journal*, 2me sér. Vol. XVII (1872), p. 355.

the product of any two of these operations is either identity or another operation which displaces all the symbols. Hence the $n-1$ operations which displace all the symbols, with identity, form a self-conjugate sub-group. Now

$$A_r \cdot A_p A_r \cdot A_r = A_r A_p,$$

so that A_r transforms every operation of this sub-group into its inverse. Hence

$$A_r A_p \cdot A_q A_r = A_q A_p = A_q A_r \cdot A_r A_q;$$

i.e. every two operations of this sub-group are permutable, and the sub-group is therefore Abelian.

106. If

$$S = (a_1 a_2 \ldots a_i)(a_{i+1} a_{i+2} \ldots a_j) \ldots,$$

and

$$T = \begin{pmatrix} a_1, a_2, \ldots, a_n \\ b_1, b_2, \ldots, b_n \end{pmatrix},$$

are any two substitutions of a group, then (§ 10)

$$T^{-1} S T = (b_1 b_2 \ldots b_i)(b_{i+1} b_{i+2} \ldots b_j) \ldots.$$

Hence every substitution of the group, which is conjugate to S, is also similar to S. It does not necessarily or generally follow that two similar substitutions of a group are conjugate. That this is true however of the symmetric group is obvious, for then the substitution T may be chosen so as to replace the n symbols by any permutation of them whatever.

A self-conjugate substitution of a transitive group of degree n must be a regular substitution (§ 9) changing all the n symbols. For if it did not change all the n symbols, it would belong to one of the sub-groups that keep a symbol unchanged. Hence, since it is a self-conjugate substitution, it would belong to each sub-group that keeps a symbol unchanged, which is impossible unless it is the identical substitution. Again if it were not regular, one of its powers would keep two or more symbols unchanged, and this cannot be the case since every power of a self-conjugate substitution must be self-conjugate. On the other hand, a self-conjugate sub-group of a transitive group need not contain any substitution which displaces all the symbols. Thus if

$$S = (12)(34),$$
$$T = (135)(246),$$

then $\{S, T\}$ is a transitive group of degree 6. The only substitutions conjugate to S are

$$T^{-1}ST = (34)(56) \quad \text{and} \quad TST^{-1} = (12)(56);$$

and these, with S and identity, form a self-conjugate sub-group of order 4, none of whose substitutions displace more than 4 symbols. The form of a self-conjugate sub-group of a transitive group will be considered in greater detail in the next Chapter.

107. Since a self-conjugate substitution of a transitive substitution-group G of degree n must be a regular substitution which displaces all the symbols, the self-conjugate sub-group H of G which consists of all its self-conjugate operations must have n or some submultiple of n for its order. For if S and S' are two self-conjugate substitutions of G, so also is $S^{-1}S'$; and therefore S and S' cannot both change a into b. The order of H therefore cannot exceed n; and if the order is n', the substitutions of H must interchange the n symbols in sets of n', so that n' is a factor of n. Let now S, some substitution performed on the n symbols of the transitive substitution-group G of degree n, be permutable with every substitution of G. Then S is a self-conjugate operation of the transitive substitution-group $\{S, G\}$ of degree n, and it is therefore a regular substitution in all the n symbols. The totality of the substitutions S, which are permutable with every substitution of G, form a group (not necessarily Abelian); and the order of this group is n or a factor of n.

The special case, in which G is a group whose order n is equal to its degree, is of sufficient importance to merit particular attention. If

$$S_1 (= 1),\ S_2,\ \ldots,\ S_n$$

are the operations of G, it has been seen (§ 20) that the substitution-group can be expressed as consisting of the n substitutions

$$\begin{pmatrix} S_1 \;, & S_2 \;, \ldots, & S_n \\ S_1S_x, & S_2S_x, \ldots, & S_nS_x \end{pmatrix}, \quad (x = 1, 2, \ldots, n),$$

performed upon the symbols of the operations of the group.

Now if pre-multiplication be used in the place of post-multiplication, it may be verified exactly as in § 20 that the n substitutions

$$\begin{pmatrix} S_1 \;, & S_2 \;, \ldots, & S_n \\ S_xS_1, & S_xS_2, \ldots, & S_xS_n \end{pmatrix}, \quad (x = 1, 2, \ldots, n)$$

form a group G' simply isomorphic with G; and the substitution of G' which is given corresponds to the operation S_x^{-1} of G. Moreover

$$\begin{aligned} \begin{pmatrix} S_1 \;, \ldots, & S_n \\ S_xS_1, \ldots, & S_xS_n \end{pmatrix}^{-1} \begin{pmatrix} S_1 \;, \ldots, & S_n \\ S_1S_y, \ldots, & S_nS_y \end{pmatrix} \begin{pmatrix} S_1 \;, \ldots, & S_n \\ S_xS_1, \ldots, & S_xS_n \end{pmatrix} \\ = \begin{pmatrix} S_xS_1, \ldots, & S_xS_n \\ S_1S_y, \ldots, & S_1S_n \end{pmatrix} \begin{pmatrix} S_1S_y \;, \ldots, & S_nS_y \\ S_xS_1S_y, \ldots, & S_xS_nS_y \end{pmatrix} \\ = \begin{pmatrix} S_xS_1 \;, \ldots, & S_xS_n \\ S_xS_1S_y, \ldots, & S_xS_nS_y \end{pmatrix} = \begin{pmatrix} S_1 \;, \ldots, & S_n \\ S_1S_y, \ldots, & S_nS_y \end{pmatrix}, \end{aligned}$$

so that every operation of G' is permutable with every operation of G, while G and G' can be expressed as transitive substitution-groups in the same n symbols. Hence:—

THEOREM II. *Those substitutions of n symbols which are permutable with every substitution of a substitution-group G of order n, transitive in the n symbols, form a group G' of order and degree n, simply isomorphic with G**.

It has, in fact, been seen that there is a substitution-group G' of order and degree n, every one of whose substitutions is permutable with every substitution of G; and also that the order of any such group cannot exceed n.

*Jordan, *Traité des Substitutions* (1870), p. 60.

If S_x is a self-conjugate substitution of G, the substitutions

$$\begin{pmatrix} S_1, \ldots, S_n \\ S_1S_x, \ldots, S_nS_x \end{pmatrix} \quad \text{and} \quad \begin{pmatrix} S_1, \ldots, S_n \\ S_xS_1, \ldots, S_xS_n \end{pmatrix}$$

are the same. Hence G and G' have for their greatest common sub-group, that which is constituted by the self-conjugate substitutions of either; and if n' is the order of this sub-group, the order of $\{G, G'\}$ is $\frac{n^2}{n'}$. In particular, if G is Abelian, G and G' coincide; and if G has no self-conjugate operation except identity, $\{G, G'\}$ is the direct product of G and G'.

The sub-group of $\{G, G'\}$ which leaves one symbol, say S_1, unchanged, is formed of the distinct substitutions of the set

$$\begin{pmatrix} S_1, & S_2, & \ldots, & S_n \\ S_x^{-1}S_1S_x, & S_x^{-1}S_2S_x, & \ldots, & S_x^{-1}S_nS_x \end{pmatrix}, \quad (x = 1, 2, \ldots, n).$$

When G has no self-conjugate operation except identity, the order of this sub-group is n, and it is simply isomorphic with G. In fact, in this case the order of $\{G, G'\}$, a transitive group of degree n, is n^2, and therefore the order of a sub-group that keeps one symbol unchanged is n. Again

$$\begin{aligned}
&\begin{pmatrix} S_1, & S_2, & \ldots, & S_n \\ S_x^{-1}S_1S_x, & S_x^{-1}S_2S_x, & \ldots, & S_x^{-1}S_nS_x \end{pmatrix} \begin{pmatrix} S_1, & S_2, & \ldots, & S_n \\ S_y^{-1}S_1S_y, & S_y^{-1}S_2S_y, & \ldots, & S_y^{-1}S_nS_y \end{pmatrix} \\
&\quad = \begin{pmatrix} S_1, & S_2, & \ldots, & S_n \\ S_x^{-1}S_1S_x, & S_x^{-1}S_2S_x, & \ldots, & S_x^{-1}S_nS_x \end{pmatrix} \\
&\qquad\qquad \begin{pmatrix} S_x^{-1}S_1S_x, & S_x^{-1}S_2S_x, & \ldots, & S_x^{-1}S_nS_x \\ S_y^{-1}S_x^{-1}S_1S_xS_y, & S_y^{-1}S_x^{-1}S_2S_xS_y, & \ldots, & S_y^{-1}S_x^{-1}S_nS_xS_y \end{pmatrix} \\
&\quad = \begin{pmatrix} S_1, & S_2, & \ldots, & S_n \\ S_y^{-1}S_x^{-1}S_1S_xS_y, & S_y^{-1}S_x^{-1}S_2S_xS_y, & \ldots, & S_y^{-1}S_x^{-1}S_nS_xS_y \end{pmatrix};
\end{aligned}$$

thus giving a direct verification that the sub-group is isomorphic with the group whose operations are

$$S_1,\ S_2,\ \ldots,\ S_n.$$

When G contains self-conjugate operations, it will be multiply isomorphic with the sub-group K_1 of $\{G, G'\}$ which keeps the symbol S_1 fixed;

and if g is the group constituted by the self-conjugate operations of G (or G'), then K_1 is simply isomorphic with $\frac{G}{g}$.

If K_1 is not a maximum sub-group of $\{G, G'\}$, let I be a greater sub-group containing K_1. Then I and G' (or G) must contain common substitutions. For every substitution of $\{G, G'\}$ is of the form

$$\begin{pmatrix} S_1 & , & S_2 & ,\ldots, & S_n \\ S_yS_1S_x, & & S_yS_2S_x, & \ldots, & S_yS_nS_x \end{pmatrix};$$

and if this substitution belongs to I, then

$$\begin{pmatrix} S_1 & , & S_2 & ,\ldots, & S_n \\ S_yS_1S_x, & & S_yS_2S_x, & \ldots, & S_yS_nS_x \end{pmatrix}\begin{pmatrix} S_1 & , & S_2 & ,\ldots, & S_n \\ S_xS_1S_x^{-1}, & & S_xS_2S_x^{-1}, & \ldots, & S_xS_nS_x^{-1} \end{pmatrix},$$

or

$$\begin{pmatrix} S_1 & , & S_2 & ,\ldots, & S_n \\ S_xS_yS_1, & & S_xS_yS_2, & \ldots, & S_xS_yS_n \end{pmatrix},$$

is a substitution of G' which belongs to I. Moreover, since G' is a self-conjugate sub-group of $\{G, G'\}$, the substitutions of G' which belong to I form a self-conjugate sub-group of I: this sub-group we will call H'.

Now every substitution of the group can be represented as the product of a substitution of K_1 by a substitution of G: and therefore all the sub-groups conjugate to I will be obtained on transforming I by the operations of G. Hence, because every substitution of G transforms H' into itself, H' is common to the complete set of conjugate sub-groups to which I belongs; and H' is therefore a self-conjugate sub-group of $\{G, G'\}$. Finally then, K_1 is a maximum sub-group of $\{G, G'\}$, if and only if G is a simple group.

108. Definition. A substitution-group, that contains one or more substitutions changing k given symbols $a_1, a_2, \ldots, a_k$ into any other k symbols, is called *k-ply transitive.*

Such a group clearly contains substitutions changing any set of k symbols into any other set of k; and the order of the sub-group keeping any j $(\ngtr k)$ symbols unchanged is independent of the particular j symbols chosen.

THEOREM III. *The order of a k-ply transitive group of degree n is $n(n-1)\ldots(n-k+1)m$, where m is the order of the sub-group that leaves* *any k symbols unchanged. This sub-group is contained self-conjugately in a sub-group of order $k!m$.*

If N is the order of the group, the order of the sub-group which keeps one symbol fixed is $\frac{N}{n}$, by Theorem I (§ 104). Now this sub-group is a transitive group of degree $n-1$; and therefore the order of the sub-group that keeps two symbols unchanged is $\frac{N}{n(n-1)}$. If $k > 2$, this sub-group again is a transitive group of degree $n-2$; and so on. Proceeding thus, the order of the sub-group which keeps k symbols unchanged is seen to be

$$\frac{N}{n(n-1)\dots(n-k+1)},$$

which proves the first part of the theorem.

Let $a_1, a_2, \dots, a_k$ be the k symbols which are left unchanged by a sub-group H of order m. Since the group is k-ply transitive, it must contain substitutions of the form

$$\begin{pmatrix} a_1, a_2, \dots, a_k, b\ , c\ , \dots \\ a'_1, a'_2, \dots, a'_k, b', c', \dots \end{pmatrix},$$

where $a'_1, a'_2, \dots, a'_k$ are the same k symbols as $a_1, a_2, \dots, a_k$ arranged in any other sequence. Also every substitution of this form is permutable with H, since it interchanges among themselves the symbols left unchanged by H. Further, if S_1 and S_2 are any two substitutions of this form, $S_1^{-1}S_2$ will belong to H if, and only if, S_1 and S_2 give the same permutation of the symbols $a_1, a_2, \dots, a_k$. Hence finally, since $k!$ distinct substitutions can be performed on the k symbols, the order of the sub-group that contains H self-conjugately is $k!m$.

If m is unity, the identical substitution is the only one that keeps any k symbols fixed, and there is just one substitution that changes k symbols into any other k. In the same case, the group contains substitutions which displace $n-k+1$ symbols only, and there are none, except the identical substitution, which displace less.

If $m > 1$, the group will contain $m-1$ substitutions besides identity, which leave unchanged any k given symbols, and therefore displace $n-k$ symbols at most.

It follows from § 105 that a k-ply transitive group of degree n and order $n(n-1)\dots(n-k+1)$ can exist only if $n-k+2$ is the power of a prime. For such a group must contain sub-groups of order $(n-k+2)(n-k+1)$, which keep $k-2$ symbols unchanged and are doubly transitive in the remaining $n-k+2$. When k is n, the group is the symmetric group; and when k is $n-2$, we shall see (in § 110) that the group is the alternating group. If k is less than $n-2$, M. Jordan* has shewn that, with two exceptions for $n=11$ and $n=12$, the value of k cannot exceed 3. The actual existence of triply transitive groups of order $(p^n+1)p^n(p^n-1)$, for all prime values of p, will be established in § 113.

109. A k-ply transitive group, of degree n and order N, is not generally contained in some $(k+1)$-ply transitive group of degree $n+1$ and order $N(n+1)$. To determine whether this is the case for any given group, M. Jordan† has suggested the following tentative process, which for moderate values of n is always practicable.

Let G be a transitive group of order N in the n symbols

$$a_1,\ a_2,\ \dots,\ a_n;$$

and suppose that G is that sub-group of the transitive group Γ in the $n+1$ symbols

$$a_1,\ a_2,\ \dots,\ a_n,\ a_{n+1},$$

which leaves the symbol a_{n+1} unchanged. Then Γ must be at least doubly transitive, and it therefore contains a substitution of order 2 which interchanges the two symbols a_1 and a_{n+1}. Let A be such a substitution. Then

$$\Gamma = \{G, A\};$$

for $\{G, A\}$ is contained in Γ, and its order cannot be less than the order of Γ. Also if S is any substitution of G which displaces a_1, two other substitutions S' and S'' of G can always be found such that

$$ASA = S'AS''.$$

*"Récherches sur les substitutions," *Liouville's Journal*, 2me sér., Vol. XVII (1872), pp. 357–363.

† *Traité des Substitutions* (1870), pp. 31, 32.

In fact, if ASA changes a_r into a_{n+1}, and if S' changes a_r into a_1, then $AS'^{-1}ASA$ leaves a_{n+1} unchanged, and it therefore belongs to G.

Conversely, if A is any operation of order 2 which changes a_1 into a_{n+1} and permutes the remaining a's among themselves; and if, whatever substitution of G is represented by S,two other substitutions of G can be found such that

$$ASA = S'AS'';$$

then $\{G, A\}$ is a group with the required properties. In fact, when these conditions are satisfied, every substitution of the group $\{G, A\}$ can be expressed in one of the two forms

$$S_1, \ S_2AS_3,$$

where S_1, S_2, S_3 are substitutions of G. For instance, the substitution

$$\begin{aligned} AS_pAS_qAS_r &= S'_pAS''_p \cdot S_qAS_r, \\ &= S'_pAS_tAS_r, \quad \text{if} \quad S''_pS_q = S_t, \\ &= S'_pS'_tAS''_tS_r, \end{aligned}$$

which is of the second form. The reduction is here carried out on supposition that S_p and S_t displace a_1. The modification, when this is not the case, is obvious.

Moreover every operation of the form S_2AS_3 displaces a_{n+1}, and therefore the sub-group of $\{G, A\}$ which leaves a_{n+1} unchanged is G.

It is clearly sufficient that the conditions

$$ASA = S'AS''$$

should be satisfied, when each of a set of independent generating operations of G is taken for S.

Ex. Construct a doubly transitive group of degree 12 of which the sub-group that keeps one symbol unchanged is

$$\{(123456789ab), \ (256a4)(39b87)\}.$$

110. Let

$$S = (a_1 a_2 \dots a_i) \dots (\dots a_{j-1} a_j)(a_{j+1} \dots a_{k-1} a_k \dots) \dots$$

be a substitution of a k-ply transitive group displacing s $(> k)$ symbols. If $j < k - 1$, take

$$T = \begin{pmatrix} a_1, a_2, \dots, a_{k-1}, a_k, \dots \\ a_1, a_2, \dots, a_{k-1}, b_k, \dots \end{pmatrix},$$

where b_k is some other symbol occurring in S. Since the group is k-ply transitive, it must contain a substitution such as T. Now

$$T^{-1}ST = (a_1 a_2 \dots a_i) \dots (\dots a_{j-1} a_j)(a_{j+1} \dots a_{k-1} b_k \dots) \dots;$$

and this is certainly not identical with S, so that $T^{-1}STS^{-1}$ cannot be the identical substitution. Moreover a_1, a_2, $\dots$, a_{k-2} are not affected by $T^{-1}STS^{-1}$; and therefore this substitution will displace at most $2s-2k+2$ symbols.

If $j = k - 1$, take

$$T = \begin{pmatrix} a_1, a_2, \dots, a_{k-1}, a_k, \dots \\ a_1, a_2, \dots, a_{k-1}, c_k, \dots \end{pmatrix},$$

where c_k is a symbol that does not occur in S. Then

$$T^{-1}ST = (a_1 a_2 \dots a_i) \dots (\dots a_{k-2} a_{k-1})(c_k \dots),$$

and this cannot coincide with S. Now in this case, a_1, a_2, $\dots$, a_{k-1} are not affected by $T^{-1}STS^{-1}$; and therefore this operation will again displace at most $2s - 2k + 2$ symbols.

If then

$$2s - 2k + 2 < s,$$

or

$$s < 2k - 2,$$

the group must contain a substitution affecting fewer symbols than S. This process may be repeated till we arrive at a substitution

$$\Sigma = (\alpha_1 \alpha_2 \dots \alpha_i)(\alpha_{i+1} \dots \alpha_j) \dots (\alpha_{j+1} \dots \alpha_k),$$

which affects exactly k symbols; and if this substitution be transformed by

$$T = \begin{pmatrix} \alpha_1, \alpha_2, \dots, \alpha_{k-1}, \alpha_k, \dots \\ \alpha_1, \alpha_2, \dots, \alpha_{k-1}, \beta_k, \dots \end{pmatrix},$$

then

$$T^{-1}\Sigma T = (\alpha_1\alpha_2 \dots \alpha_i)(\alpha_{i+1} \dots \alpha_j) \dots (\alpha_{j+1} \dots a_{k-1}\beta_k),$$

and

$$\Sigma^{-1}T^{-1}\Sigma T = (\alpha_k\beta_k\alpha_{j+1}).$$

Thus in the case under consideration the group contains one, and therefore every, circular substitution of three symbols; and hence (§ 11) it must contain every even substitution. It is therefore either the alternating or the symmetric group. If then a k-ply transitive group of degree n does not contain the alternating group of n symbols, no one of its substitutions, except identity, must displace fewer than $2k - 2$ symbols. It has been shewn that such a group contains substitutions displacing not more than $n-k+1$ symbols; and therefore, for a k-ply transitive group of degree n, other than the alternating or the symmetric group, the inequality

$$n - k + 1 \nless 2k - 2,$$

or

$$k \ngtr \tfrac{1}{3}n + 1,$$

must hold. Hence:—

THEOREM IV. *A group of degree n, which does not contain the alternating group of n symbols, cannot be more than $(\frac{1}{3}n + 1)$-ply transitive.*

The symmetric group is n-ply transitive; and, since of the two substitutions

$$\begin{pmatrix} a_1, a_2, \dots, a_{n-2}, a_{n-1}, a_n \\ b_1, b_2, \dots, b_{n-2}, b_{n-1}, b_n \end{pmatrix} \quad \text{and} \quad \begin{pmatrix} a_1, a_2, \dots, a_{n-2}, a_{n-1}, a_n \\ b_1, b_2, \dots, b_{n-2}, b_n, b_{n-1} \end{pmatrix},$$

one is evidently even and the other odd, the alternating group is $(n-2)$-ply transitive. The discussion just given shews that no other group of degree n can be more than $(\frac{1}{3}n+1)$-ply transitive*.

111. The process used in the preceding paragraph may be applied to shew that, unless $n = 4$, the alternating group of n symbols is simple. It has just been shewn that the alternating group is $(n-2)$-ply transitive. Therefore, if S is a substitution of the alternating group displacing fewer than $n-1$ symbols, a substitution $T^{-1}ST$ can certainly be found such that $S^{-1}T^{-1}ST$ is a circular substitution of three symbols. In this case, the self-conjugate group generated by S and its conjugate substitutions contains all the circular substitutions of three symbols, and therefore it coincides with the alternating group itself. If S displaces $n-1$ symbols, then $T^{-1}ST$ can be taken so that $S^{-1}T^{-1}ST$ displaces not more than $2(n-1)-2(n-2)+2$, or 4 symbols; and if S displaces n symbols, $S^{-1}T^{-1}ST$ can be found to displace not more than $2n-2(n-2)+2$, or 6 symbols.

It is therefore only necessary to consider the case $n = 5$, when S displaces $n-1$ symbols; and the cases $n = 4, 5, 6$, when S displaces n symbols; in all other cases, the group generated by S and its conjugate substitutions must contain circular substitutions of 3 symbols.

When $n = 5$, and S is an even substitution displacing 4 symbols, we may take

$$S = (12)(34).$$

If

$$T = (12)(35),$$

then

$$T^{-1}ST = (12)(45),$$

and

$$S^{-1}T^{-1}ST = (345).$$

Hence, in this case again, we are led to the alternating group itself.

---

*For a further discussion of the limits of transitivity of a substitution-group, compare Jordan, *Traité des Substitutions*, pp. 76–87; and Bochert, *Math. Ann.*, XXIX, (1886) pp. 27–49; XXXIII, (1888) pp. 573–583.

When $n = 6$, and S is an even substitution displacing all the symbols, we may take

$$S = (12)(3456),$$

or

$$S' = (123)(456).$$

If now

$$T = (12)(3645),$$

then

$$S^{-1}T^{-1}ST = (356),$$

and

$$S'^{-1}T^{-1}S'T = (14263);$$

and, in either case, we are led to the alternating group.

When $n = 5$, and S is an even substitution displacing all the symbols, we may put

$$S = (12345).$$

If

$$T = (345),$$

then

$$S^{-1}T^{-1}ST = (134);$$

and again the alternating group is generated.

When $n = 4$, and S is an even substitution displacing all the symbols, we may take

$$S = (12)(34).$$

Here the only two substitutions conjugate to S are clearly (13)(24) and (14)(23), which are permutable with each other and with S. Hence the alternating group of 4 symbols, which is of order 12, has a self-conjugate sub-group of order 4.

Finally when $n = 3$, the alternating group, being the group $\{(123)\}$, is a simple cyclical group of order 3. Hence:—

THEOREM V. *The alternating group of n symbols is a simple group except when $n = 4$.*

112. It has been seen in § 108 that the order of a doubly transitive group of degree n is equal to or is a multiple of $n(n-1)$. If it is equal to this number, every substitution of the group, except identity, must displace either all or all but one of the symbols; for a sub-group of order $n-1$ which keeps one symbol fixed is transitive in the remaining $n-1$ symbols, and therefore all its substitutions, except identity, displace all the $n-1$ symbols.

Now it has been shewn in § 105 that a transitive group of degree n and order $n(n-1)$, whose operations displace all or all but one of the symbols can exist only if n is the power of a prime p. The $n-1$ operations displacing all the symbols are the only operations of the group whose orders are powers of p; and therefore with identity they form a self-conjugate sub-group of order n. Moreover it also follows from § 105 that the $n-1$ operations of this sub-group other than identity form a single conjugate set. Hence this sub-group must be Abelian, and all its operations are of order p.

Suppose first that n is a prime p, and that P is any operation of the group of order p. If α is a primitive root of p, there must be an operation S in the group such that

$$S^{-1}PS = P^{\alpha};$$

then S^{p-1} is the lowest power of S which is permutable with P. Now S must belong to a sub-group of order $p-1$ that keeps one symbol fixed, and we have just seen that the order of $\{S\}$ is not less than $p-1$. The sub-group of order $p-1$ is therefore cyclical, and

$$S^{p-1} = 1.$$

Hence the group, if it exists, must be defined by

$$P^p = 1, \quad S^{p-1} = 1, \quad S^{-1}PS = P^{\alpha}.$$

It is an immediate result of a theorem, which will be proved in the next chapter (§ 123), that this group can be actually represented as a transitive substitution-group of degree p; this may be also verified directly as follows.

Let

$$P = (a_1 a_2 \dots a_p),$$

so that

$$P^{\alpha} = (a_1 a_{\alpha+1} a_{2\alpha+1} \dots a_{(p-1)\alpha+1}),$$

where the suffixes are to be reduced (mod. p); and suppose that S is a substitution that keeps unchanged. Then since

$$S^{-1}PS = P^{\alpha},$$

S must change a_2 into $a_{\alpha+1}$, a_3 into $a_{2\alpha+1}$, and generally, a_r into $a_{(r-1)\alpha+1}$. Hence

$$S = (a_2 a_{\alpha+1} a_{\alpha^2+1} \ldots) \ldots;$$

and since α is a primitive root of p, there is only a single cycle; so that

$$S = (a_2 a_{\alpha+1} a_{\alpha^2+1} \ldots a_{\alpha^{p-2}+1}).$$

The substitutions P and S thus constructed actually generate a doubly transitive substitution-group of degree p and order $p(p-1)$.

Without making a complete investigation of the case in which n is the power of a prime, we go on to shew that, p being any prime, there is always a doubly transitive group of degree p^m and order $p^m(p^m - 1)$, in which a sub-group of order $p^m - 1$ is cyclical*.

Let i be a primitive root of the congruence

$$i^{p^m-1} \equiv 1 \text{ (mod. } p),$$

so that the distinct roots of the congruence are

$$i,\ i^2,\ i^3,\ \ldots,\ i^{p^m-1}.$$

Every rational function of i with real integral coefficients satisfies the same congruence; and therefore every such function is congruent (mod. p) to some power of i not exceeding the $(p^m - 1)$th.

Consider now a set of transformations of the form

$$x' \equiv \alpha x + \beta \text{ (mod. } p),$$

where α is a power of i, and β is either a power of i or zero. Two such transformations, performed successively, give another transformation of the same form; and since α cannot be zero, the inverse of each transformation is another definite

*On the subject of this and the following paragraph, the reader should consult the memoirs by Mathieu, *Liouville's Journal*, 2me Sér., t. V (1860), pp. 9–42; *ib.* t. VI (1861), pp. 241–323; where the groups here considered were first shewn to exist.

transformation; so that the totality of transformations of this form constitute a group. Moreover

$$\left.\begin{array}{r} x' \equiv \alpha x + \beta, \\ \text{and} \quad x' \equiv \alpha' x + \beta' \end{array}\right. (\text{mod. } p),$$

are not the same transformation unless

$$\alpha \equiv \alpha' \quad \text{and} \quad \beta \equiv \beta' \ (\text{mod. } p).$$

Hence, since α can take $p^m - 1$ distinct values and β can take p^m distinct values, the order of the group, formed of the totality of these transformations, is $p^m(p^m - 1)$.

The transformations for which α is unity clearly form a sub-group. If S and T represent

$$x' \equiv \alpha x + \beta \quad \text{and} \quad x' \equiv x + \gamma$$

respectively, $S^{-1}TS$ represents

$$x' \equiv x + \alpha\gamma.$$

Hence the transformations for which α is unity form a self-conjugate sub-group whose order is p^m. Every two transformations of this sub-group are clearly permutable; and the order of each of them except identity is p.

Again, the transformations for which β is zero form a sub-group. Since every one of them is a power of the transformation

$$x' \equiv ix,$$

this sub-group is a cyclical sub-group of order $p^m - 1$. If the transformation just written be denoted by I, then $S^{-1}IS$ is

$$x' \equiv ix + \beta(1 - i).$$

Hence the only operations permutable with $\{I\}$ are its own operations, and therefore $\{I\}$ is one of p^m conjugate sub-groups.

The set of transformations

$$x' \equiv \alpha x + \beta$$

therefore forms a group of order $p^m(p^m - 1)$. This group contains a self-conjugate Abelian sub-group of order p^m and type $(1, 1, \ldots, 1)$, and p^m conjugate cyclical sub-groups of order $p^m - 1$, none of whose operations are permutable with any of the operations of the self-conjugate sub-group.

Now if the operation

$$x' \equiv \alpha x + \beta$$

be performed on each term of the series

$$0,\ i,\ i^2,\ \ldots,\ i^{p^m-1},$$

it will, since every rational integral function of i with real integral coefficients is congruent (mod. p) to some power of i, change the term into another of the same series; and since the congruence

$$\alpha i^x + \beta \equiv \alpha i^y + \beta$$

gives

$$x \equiv y \text{ (mod. } p^{m-1}),$$

no two terms of the series can thus be transformed into the same term. Moreover the only operation that leaves every term of the series unchanged is clearly the identical operation.

To each operation of the form

$$x' \equiv \alpha x + \beta$$

therefore will correspond a single substitution performed on the p^m symbols just written, so that to the product of two operations will correspond the product of the two homologous substitutions. The group is therefore simply isomorphic with a substitution group of degree p^m. Moreover since the linear congruence

$$x \equiv \alpha x + \beta \text{ (mod. } p)$$

has only a single solution when α is different from unity, and none when α is unity, every substitution except identity must displace all or all but one of the symbols. The substitution group is therefore doubly transitive*.

*The author has shewn (*Messenger of Mathematics*, Vol. XXV (1896), pp. 147–153) that the type of group considered in the text is the only type of doubly transitive group of degree p^m and order $p^m(p^m - 1)$ when $m = 3$; and that, when $m = 2$ and $p > 3$, the same is true. When $m = 2$ and $p = 3$, there is one other type.

Ex. 1. Apply the method just explained to the actual construction of a doubly transitive group of degree 8 and order 56.

Ex. 2. Shew that the equations

$$A^2 = 1, \quad S^{2^m-1} = 1, \quad AS^{-1}AS = S^{-n}AS^n,$$

where n is such that a primitive root of the congruence,

$$i^{2^m-1} - 1 \equiv 0 \text{ (mod. 2)},$$

satisfies the congruence

$$i^n + i + 1 \equiv 0 \text{ (mod. 2)},$$

suffice to define a group which can be expressed as a doubly transitive group of degree 2^m and order $2^m(2^m - 1)$.

(*Messenger of Mathematics*, Vol. XXV, p. 189.)

113. A slight extension of the method of the preceding paragraph will enable us to shew that, for every prime p, it is possible to construct a triply transitive group of degree $p^m + 1$ and order $(p^m + 1)p^m(p^m - 1)$. The analysis of this group will form the subject of investigation in Chapter XIV; here we shall only demonstrate its existence.

In the place of the operations of the last paragraph, we now consider those of the form

$$x' \equiv \frac{\alpha x + \beta}{\gamma x + \delta} \text{ (mod. } p),$$

where again α, β, γ, δ are powers of i, limited now by the condition that $\alpha\delta - \beta\gamma$ is not congruent to zero (mod. p). When this relation is satisfied, the set of operations again clearly form a group. Moreover if we represent $\frac{i^x}{0}$ by ∞ for all values of x, any operation of this group, when carried out on the set of quantities

$$\infty, \ 0, \ i, \ i^2, \ \ldots, \ i^{p^m-1},$$

will change each of them into another of the set; while no operation except identity will leave each symbol of the set unchanged. Hence the group can be represented as a substitution group of degree $p^m + 1$.

Now

$$\frac{x' - i^{a'}}{x' - i^{b'}} \frac{i^{c'} - i^{b'}}{i^{c'} - i^{a'}} \equiv \frac{x - i^a}{x - i^b} \frac{i^c - i^b}{i^c - i^a}$$

is an operation of the above form, which changes the three symbols i^a, i^b, i^c into $i^{a'}$, $i^{b'}$, $i^{c'}$ respectively; and it is easy to modify this form so that it holds when 0 or ∞ occurs in the place of i^a, etc. Hence the substitution group is triply transitive, since it contains an operation transforming any three of the $p^m + 1$ symbols into any other three.

On the other hand, if the typical operation keeps the symbol x unchanged, then

$$\gamma x^2 + (\delta - \alpha)x - \beta \equiv 0 \text{ (mod. } p),$$

and this congruence cannot have more than two roots among the set of $p^m + 1$ symbols. Hence no substitution of the group, except identity, keeps more than two symbols fixed.

Finally then, since the group is triply transitive and since it contains no operation, except identity, that keeps more than two symbols fixed, its order must (§ 108) be $(p^m + 1)p^m(p^m - 1)$.

114. An intransitive substitution group, as defined in § 104, is one which does not contain substitutions changing a_1 into each of the other symbols $a_2, a_3, \ldots, a_n$ operated on by the group. Let us suppose that the substitutions of such a group change a_1 into $a_2, a_3, \ldots, a_k$ only. Then all the substitutions of the group must interchange these k symbols among themselves; for if the group contained a substitution changing a_2 into a_{k+1}, then the product of any substitution changing a_1 into a_2 by this latter substitution would change a_1 into a_{k+1}. Hence the n symbols operated on by the group can be divided into a number of sets, such that the substitutions of the group change the symbols of any one set transitively among themselves, but do not interchange the symbols of two distinct sets. It follows immediately that the order of the group must be a common multiple of the numbers of symbols in the different sets.

Suppose now that $a_1, a_2, \ldots, a_k$ is a set of symbols which are interchanged transitively by all the substitutions of a group G of degree n. If for a time we neglect the effect of the substitutions of G on the remaining $n-k$ symbols, the group G will reduce to a transitive group H of degree k. The group G is isomorphic with the group H; for if we take as the substitutions of G, that correspond to a given substitution of H, those which produce the same permutation of the symbols $a_1, a_2, \ldots, a_k$ as that produced by the substitution of H, then to the product of any two substitutions of G will correspond the product of the corresponding two substitutions of H. The isomorphism thus shewn to exist may be simple or multiple. In the former case, the order of H is the same as that of G; in the latter case, the substitutions of G which correspond to the identical substitution of H, i.e. those substitutions of G which change none of the symbols $a_1, a_2, \ldots, a_k$ form a self-conjugate sub-group.

We will consider in particular an intransitive group G which interchanges the symbols in *two* transitive sets; these we will refer to as the α's and the β's. Let G_α and G_β be the two groups transitive in the α's and β's respectively, to which G reduces when we alternately leave out of account the effect of the substitutions on the β's and the α's. Also let g_α and g_β be the self-conjugate sub-groups of G, which keep respectively all the β's and all the α's unchanged; and denote the group $\{g_\alpha, g_\beta\}$ by g. This last group g, which is the direct product of g_α and g_β, is self-conjugate in G, since it is generated by the two self-conjugate groups g_α and g_β. Now g_α is self-conjugate not only in G but also in G_α; for G_α permutes the α's in the same way that G does, while any substitution of g_α, not affecting the β's, is necessarily permutable with every substitution performed on the β's. The group G_α is simply isomorphic with the group $\dfrac{G}{g_\beta}$, and G_β with $\dfrac{G}{g_\alpha}$; hence, using n_H to denote the order of a group H,

$$n_G = n_{G_\alpha} n_{g_\beta} = n_{G_\beta} n_{g_\alpha}.$$

Let the substitutions of G be now divided into sets in respect of the self-conjugate sub-group g, so that

$$G = g,\ Sg,\ Tg,\ \ldots.$$

The group $\dfrac{G}{g}$ is defined by the laws according to which these sets combine among themselves, the sets being such that, if any substitution of one set be multiplied by any substitution of a second set, the resulting substitution will belong to a definite third set.

If we now neglect the effect of the substitutions on the symbols β, the group G reduces to G_α and g reduces to g_α, and hence

$$G_\alpha = g_\alpha,\ S_\alpha g_\alpha,\ T_\alpha g_\alpha,\ \ldots,$$

where S_α, T_α, ... represent the substitutions S, T, ..., so far as they affect the α's. Moreover the substitutions in the different sets into which G_α is thus divided must be all distinct since, by the preceding relations between the orders of the groups, their number is just equal to the order of G_α. Hence $\dfrac{G_\alpha}{g_\alpha}$ is defined by the laws according to which these sets of substitutions combine. But if

$$Sg \cdot Tg = Ug,$$

then necessarily

$$S_\alpha g_\alpha \cdot T_\alpha g_\alpha = U_\alpha g_\alpha,$$

and therefore, finally, the three groups $\dfrac{G}{g}$, $\dfrac{G_\alpha}{g_\alpha}$, and $\dfrac{G_\beta}{g_\beta}$ are simply isomorphic.

115. The relation of simple isomorphism between $\dfrac{G_\alpha}{g_\alpha}$ and $\dfrac{G_\beta}{g_\beta}$ thus arrived at establishes between the groups G_α and G_β an isomorphism of the most general kind (§ 32).

To every operation of G_α correspond n_{g_β} operations of G_β, and to every operation of G_β correspond n_{g_α} operations of G_α; so that to the product of any two operations of G_α (or G_β) there corresponds a definite set of n_{g_β} operations of G_β (or n_{g_α} operations of G_α).

Returning now to the intransitive group G, its genesis from the two transitive groups G_α and G_β, with which it is isomorphic, may be represented as follows. The n_{g_α} to n_{g_β} correspondence, such as has just been described, having been established between the groups G_α and G_β, each substitution of G_α is multiplied by the n_{g_β} substitutions that correspond to it in G_β. The set of $n_{G_\alpha} n_{g_\beta}$ substitutions so obtained form a group, for

$$S_\alpha S_\beta \cdot S'_\alpha S'_\beta = S_\alpha S'_\alpha \cdot S_\beta S'_\beta = S''_\alpha S''_\beta,$$

where, if S_β, S'_β are substitutions corresponding to S_α, S'_α, then S''_β is a substitution corresponding to S''_α. Moreover, this group may be equally well generated by multiplying every one of the substitutions of G_β by the n_{g_α} corresponding substitutions of G_α; and by a reference to the representations of G, G_α and G_β, as divided into sets of substitutions given above, it is immediately obvious that all these substitutions occur in G. Hence, as their number is equal to the order of G, the group thus formed coincides with G.

116. The general result for any intransitive group, the simplest case of which has been considered in the two last paragraphs, may be stated in the following form:—

THEOREM VI. *If G is an intransitive group of degree n which permutes the n symbols in s transitive sets, and if* (i) *G_r is what G becomes when the substitutions of G are performed on the rth set of symbols only,* (ii) *Γ_r is what G becomes when the substitutions of G are performed on all the sets except the rth,* (iii) *g_r is that sub-group of G which changes the symbols of the rth set only,* (iv) *γ_r is that sub-group of G which keeps all the symbols of the rth set unchanged: then the groups $\frac{G_r}{g_r}$ and $\frac{\Gamma_r}{\gamma_r}$ are simply isomorphic, and n_r, ν_r being the orders of g_r, γ_r, an n_r to ν_r correspondence is thus established between the substitutions of the groups G_r and Γ_r. Moreover, the substitutions of G are given, each once and once only, by multiplying each substitution of G_r by the ν_r substitutions of Γ_r that correspond to it**.

It is not necessary to give an independent proof of this theorem, since if, in the discussion of the two preceding paragraphs, G_α, G_β, g_a, g_b, g be replaced by G_r, Γ_r, g_r, γ_r, $\{g_r, \gamma_r\}$, it will be found each step of the process there carried out may be repeated without alteration.

If we regard G_α and G_β as two given transitive groups in distinct sets of

*On intransitive groups, reference may be made to Bolza, "On the construction of intransitive groups," *Amer. Journal*, Vol. XI, (1889), pp. 195–214. The general isomorphism which underlies the construction of these groups is considered by Klein and Fricke, *Vorlesungen über die Theorie der elliptischen Modulfunctionen*, Vol. I, (1890), pp. 402–406.

symbols, the determination of all the intransitive groups in the combined symbols, which reduce to G_α or G_β when the symbols of the second or first set are neglected, involves a knowledge of the composition of the two groups. To each distinct m to n isomorphism, that can be established between the two groups, there will correspond a distinct intransitive group. If G_α is a simple group, containing therefore only itself and identity self-conjugately, then to each substitution of G_β there must correspond either one or all of the substitutions of G_α; and the former can be the case only when G_β contains a self-conjugate sub-group H, such that $\frac{G_\beta}{H}$ is simply isomorphic with G_α. Hence, if the order of G_β is less than twice the order of G_α, the only possible intransitive group is the direct product of G_α and G_β, unless G_β is simply isomorphic with G_α.

117. In illustration of the preceding paragraphs, we will determine the number of distinct intransitive groups of degree 7, when the symbols are interchanged in transitive sets of 4 and 3 respectively. The four symbols will be referred to as the α's, and the three symbols as the β's.

If a transitive group of degree 4 contains operations of order 3, it must be either the symmetric or the alternating group. If it contains no operation of order 3, its order must be either 8 or 4. By Sylow's theorem, the symmetric group of degree 4, being of order 24, contains a single set of conjugate sub-groups of order 8, so that there is only one type of transitive group of degree 4 and order 8. This is given by

$$S = (1234), \quad T = (13),$$

so that

$$S^4 = 1, \quad T^2 = 1, \quad TST = S^{-1}.$$

This group contains three self-conjugate sub-groups of order 4, namely,

$$\begin{array}{lllll} \text{(i)} & 1, & (1234), & (13)(24), & (1432), \\ \text{(ii)} & 1, & (12)(34), & (13)(24), & (14)(23), \\ \text{(iii)} & 1, & (13), & (13)(24), & (24). \end{array}$$

The two latter are simply isomorphic; but as substitution groups, they are of distinct form. Hence for G_α in the construction of the intransitive group, we may take either G_{24}, the symmetric group of the α's, or G_{12}, the alternating group of the α's, or G_8, the above group of order 8, or finally G_4, G_4', G_4'', the above three groups of order 4. The only transitive groups of 3 symbols are G_6, the symmetric group, and G_3, the alternating group: one of these must be taken for G_β.

I. $G_\alpha = G_{24}$.

The only self-conjugate sub-groups of G_{24} are G_{12} and G_4'. Also $\dfrac{G_{24}}{G_{12}}$ is a group of order 2; and it may easily be verified that $\dfrac{G_{24}}{G_4'}$ is simply isomorphic with the symmetric group of degree 3.

Hence (i) if G_β is G_6, we may take

$$\begin{aligned} g_\alpha &= G_{24}, & g_\beta &= G_6, \text{ so that } n_G = 144,\\ g_\alpha &= G_{12}, & g_\beta &= G_3, \quad\text{"}\quad n_G = 72,\\ g_\alpha &= G_4', & g_\beta &= 1, \quad\text{"}\quad n_G = 24; \end{aligned}$$

and (ii) if G_β is G_3, we must take

$$g_\alpha = G_{24}, \quad g_\beta = G_3, \text{ giving } n_G = 72.$$

II. $G_\alpha = G_{12}$.

The only self-conjugate sub-group of G_{12} is G_4'; and $\dfrac{G_{12}}{G_4'}$ is a cyclical group of order 3.

Hence (i) if G_β is G_6, we must take

$$g_\alpha = G_{12}, \quad g_\beta = G_6, \text{ giving } n_G = 72;$$

and (ii) if G_β is G_3, we may take

$$\begin{aligned} & g_\alpha = G_{12}, & g_\beta &= G_3, \text{ giving } n_G = 36,\\ \text{or } & g_\alpha = G_4', & g_\beta &= 1, \quad\text{"}\quad n_G = 12. \end{aligned}$$

III. $G_\alpha = G_8$.

The self-conjugate sub-groups of G_8, of order 4, are determined above.

If (i) G_β is G_6, we may take

$$\begin{aligned} g_\alpha &= G_8, & g_\beta &= G_6, \text{ giving } n_G = 48,\\ g_\alpha &= G_4, & g_\beta &= G_3, \quad\text{"}\quad n_G = 24,\\ g_\alpha &= G'_4, & g_\beta &= G_3, \quad\text{"}\quad n_G = 24,\\ g_\alpha &= G''_4, & g_\beta &= G_3, \quad\text{"}\quad n_G = 24. \end{aligned}$$

The two latter groups are simply isomorphic; but regarded as substitution groups, they are of distinct forms.

If (ii) G_β is G_3, we must take

$$g_\alpha = G_8, \quad g_\beta = G_3, \text{ giving } n_G = 24.$$

IV. $G_\alpha = G_4$.

If (i) G_β is G_6, we may take

$$g_\alpha = G_4, \quad g_\beta = G_6, \text{ giving } n_G = 24,$$

or

$$g_\alpha = G_2, \quad g_\beta = G_3, \quad\text{"}\quad n_G = 12;$$

G_2 representing the group $\{1, (13)(24)\}$.

If (ii) G_β is G_3, we must take

$$g_\alpha = G_4, \quad g_\beta = G_3, \text{ giving } n_G = 12.$$

V. $G_\alpha = G'_4$.

There are, exactly as in the last case, three possibilities: G'_4 taking the place of G_4. These groups are not however simply isomorphic with the preceding three.

VI. $G_\alpha = G''_4$.

There are in this case four possibilities. Of these three correspond to those of IV, G''_4 taking the place of G_4. The remaining one is given by $G_\beta = G_6$, $g_\beta = G_3$, $g_\alpha = G'_2$, where G'_2 represents the group $\{1, (13)\}$. Regarded as substitution groups all these are of distinct form from the groups of V.

There are thus 22 distinct intransitive substitution groups of degree 7, in which the symbols are interchanged transitively in two sets of 4 and 3 respectively.

118. Let ν_r $(r = 0, 1, 2, \ldots, n)$ be the number of substitutions of a group of degree n and order N which leave exactly r symbols unchanged, so that

$$N = \sum_{r=0}^{r=n} \nu_r.$$

Suppose first that the group is transitive; and in a sub-group, which keeps one symbol unchanged, let ν'_r $(r = 1, 2, \ldots, n)$ be the number of substitutions that leave exactly r symbols unchanged, so that

$$\frac{N}{n} = \sum_{r=1}^{r=n} \nu'_r.$$

Each of the n sub-groups, which leave a single symbol unchanged, have ν'_r substitutions which leave exactly r symbols unchanged; and each of these substitutions belong to r sub-groups which leave one symbol unchanged. Hence

$$n\nu'_r = r\nu_r,$$

and therefore

$$N = \sum_{r=1}^{r=n} r\nu_r;$$

or the number of unchanged symbols in all the substitutions of a transitive group is equal to the order of the group.

Suppose, next, that the group is intransitive; and consider a set of s symbols among the n, which are permuted transitively among themselves by the operations of the group. Let N_1 be the order of the self-conjugate sub-group H_1, which leaves unchanged each of this set of s symbols. Then if we consider the effect of the substitutions on this set of s symbols only, the group reduces to a transitive group of order $\frac{N}{N_1}$ with which the original group is multiply isomorphic. If S' is any substitution of this group of order $\frac{N}{N_1}$, and if SH_1 denote the corresponding N_1 operations of the original group, then every substitution of the set SH_1 produces the same effect on the s symbols that S' produces. Now the number of unchanged symbols in all the substitutions of the transitive group of degree s and order $\frac{N}{N_1}$

is $\frac{N}{N_1}$; therefore, in all the substitutions of the original group, the number of symbols of the set of s that remain unchanged is N. The same reasoning applies to each separate set of the n symbols, which are permuted transitively among themselves by the operations of the group. Hence if there are t_1 such transitive sets, the total number of symbols which remain unchanged in all the substitutions of the group is Nt_1; or

$$Nt_1 = \sum_{r=1}^{r=n} r\nu_r.$$

119. The formula just obtained is the first of a series of similar formulæ, due to Herr Frobenius,* which are capable of many useful applications.

Consider the symbol $(a_1, a_2, \ldots, a_p)$, where the letters are p distinct letters chosen from the n which are operated on by the group, the sequence in which the p letters are arranged being regarded as essential. The number of such symbols that can be formed from the n letters is $n(n-1)\ldots(n-p+1)$. Every substitution of the group interchanges this set of symbols among themselves; and no substitution can leave one of the symbols unchanged unless it leaves each of the letters forming it unchanged. Moreover no substitution of the group, other than identity, can leave each of the set of symbols unchanged. Hence the group can be expressed as a substitution group in this set of $n(n-1)\ldots(n-p+1)$ symbols. Every substitution of the group which leaves exactly r letters unchanged will, if $r < p$, leave none of the set of symbols unchanged; while if $r \geqslant p$, it will leave exactly

$$r(r-1)\ldots(r-p+1)$$

unchanged. Hence, if the set of symbols are interchanged by the substitutions of the group in t_p transitive sets, then

$$Nt_p = \sum_{r=p}^{r=n} \frac{r!}{r-p!}\nu_r.$$

*"Ueber die Congruenz nach einem aus zwei endlichen Gruppen gebildeten Doppelmodul," *Crelle*, t. CI, (1887), p. 288.

From the symbol $(a_1, a_2, \ldots, a_p)$, containing p letters, may be formed $n-p$ symbols containing $p+1$ letters, by adding any one of the remaining $n-p$ letters in the last place. If the symbol $(a_1, a_2, \ldots, a_p)$ is one of a transitive set of s symbols, to these there will correspond $s(n-p)$ symbols of $p+1$ letters. No symbol of $p+1$ letters, which is not included among these $s(n-p)$ symbols, can enter into a transitive set with any one of the $s(n-p)$; since if it did, the s symbols of p letters would not form a transitive set. Hence the $s(n-p)$ symbols must form, by themselves, a number of transitive sets of the symbols of $p+1$ letters; and this number clearly cannot be less than 1 nor greater than $n-p$. Accordingly, to every transitive set of the symbols of p letters, there correspond x $(1 \leqslant x \leqslant n-p)$ transitive sets of the symbols of $p+1$ letters; and therefore

$$t_p \leqslant t_{p+1} \leqslant (n-p)t_p.$$

Ex. 1. If t_p and t_{p+1} are equal and if $p < n-1$, shew that t_{p+1} is unity and that the group is $(p+1)$-ply transitive.

Ex. 2. Apply the method of § 119 to shew that no substitution, except identity, of a k-ply transitive group, which does not contain the alternating group, can displace less than $2k-2$ letters.

Note to § 108.

The results of M. Jordan, stated on p. 168, may be established as follows. Let G be a quadruply transitive group of degree n and order $n(n-1)(n-2)(n-3)$, so that $n-2$ is the power of a prime. There is a single operation of G which changes any four symbols into any other four symbols. Let a, b, c, d be four of the symbols operated on by G. The operations of G which permute these symbols among themselves form a sub-group H, simply isomorphic with the symmetric group of degree 4. Suppose that the remaining $n-4$ symbols are permuted by H in transitive sets of n_1, n_2, ... symbols each.

The only groups with which H is multiply isomorphic are (i) the symmetric group of degree 3, (ii) a group of order 2. If then, when we consider the effect of H on a set of n_1 symbols which are permuted transitively by it, the group of degree n_1 so obtained is one with which H is multiply isomorphic, this group must be either the symmetric group of degree 3 or a group of order 2. Hence n_1 must be either 6, 3 or 2. Since in this case H will contain operations leaving all the n_1 symbols unchanged, the value 6 for n_1 is inadmissible. If n_1 is 3, the operations of H, which give the substitutions

$$1, \quad (ab)(cd), \quad (ad)(bc), \quad (ac)(bd),$$

of a, b, c, d, leave the 3 symbols unchanged; and if n_1 is 2, the operations of H which give all the even substitutions of a, b, c, d, leave the 2 symbols unchanged.

If, on the other hand, when we consider the effect of H on the set of n_1 symbols only, the transitive group of degree n_1 so obtained is simply isomorphic with H, then n_1 must be either 4, 6, 8, 12 or 24. In this case, the value 4 for n_1 is inadmissible. In fact, if n_1 were 4 the operation of H which leaves c and d unchanged would also leave two of the n_1 symbols unchanged; and this is impossible since no operation of H, except identity, can leave more than 3 symbols unchanged. For a similar reason, the value 12 for n_1 is inadmissible; while, if n_1 is 6, the sub-group that keeps one of the n_1 symbols unchanged must be cyclical.

The only other possible value for n_1 is unity.

Suppose, first, that n_1 is 2. If any of the remaining numbers $n_2, n_3, \dots$ differ from 24, it may be shewn immediately that H contains operations which keep more than 3 symbols fixed. Hence in this case n is congruent to 6 (mod. 24); and since $n-2$ cannot then be a power of a prime unless $n = 6$, this case gives the alternating group of degree 6.

Next, suppose that n_1 is 3. The only admissible values for $n_2, n_3, \dots$ are 8 and 24. In this case, the sub-group of H which keeps all the n_1 $(= 3)$ symbols unchanged is a non-cyclical sub-group of order 4. But we have seen in § 105 that, if $n-3$ is even, a sub-group of order $n-3$ which keeps 3 symbols unchanged can only have a single operation of order 2. Hence this case cannot occur.

Next, suppose that n_1 is 8. The only admissible values for $n_2, n_3, \dots$ are again 24. This case cannot occur since no number congruent to 10, (mod. 24), can be a power of a prime.

The only remaining possibilities are:

$$\text{(i)} \quad n_1 = n_2 = \dots\dots\dots\dots\dots = 24,$$
$$\text{(ii)} \quad n_1 = 1, \quad n_2 = n_3 = \dots\dots = 24,$$
$$\text{(iii)} \quad n_1 = 6, \quad n_2 = n_3 = \dots\dots = 24,$$

$$\begin{aligned}
&\text{(iv)} \quad n_1 = 1, \quad n_2 = 6, \quad n_3 = \ldots = 24,\\
&\text{(v)} \quad n_1 = 1, \quad n_2 = 6, \quad n_3 = \ldots = 0,\\
&\text{(vi)} \quad n_1 = 1, \quad n_2 = n_3 = \;\ldots\ldots = 0.
\end{aligned}$$

The first of them cannot occur, since then $n - 2$ is not the power of a prime. In the second case, an operation of H which leaves c and d unchanged would be of the form

$$(ab)(\alpha\beta)(\gamma\delta)\ldots,$$

where there are $\frac{1}{2}(n-3)$ independent transpositions; while an operation of H, of order 2, which leaves none of the symbols a, b, c, d unchanged, will consist of the product of $\frac{1}{2}(n-1)$ independent transpositions. Now the operation

$$(ab)(\alpha\beta)(\gamma\delta)\ldots$$

occurs in the group, conjugate to H, which permutes a, b, α, β, among themselves; and as an operation of this group, it must be the product of $\frac{1}{2}(n-1)$ independent transpositions. This is a contradiction: hence this case cannot occur.

In case (iii), let 1, 2, 3, 4, 5, 6 be the six symbols that are permuted transitively. The self-conjugate sub-group of H of order 4 will consist of identity and the three substitutions

$$\begin{aligned}
&(ab)(cd)(34)(56)(\alpha\beta)(\gamma\delta)\ldots,\\
&(ac)(bd)(56)(12)(\alpha\gamma)(\beta\delta)\ldots,\\
&(ad)(bc)(12)(34)(\alpha\delta)(\beta\gamma)\ldots.
\end{aligned}$$

This sub-group will also occur as the self-conjugate sub-group of order 4 of the group, conjugate to H, which permutes α, β, γ, δ among themselves. If S, S' are two operations, of order 3, of these sub-groups which give the same permutation of 1, 2, 3, 4, 5, 6, then SS'^{-1} is an operation of G, distinct from identity, which keeps the six symbols unchanged. Hence this case cannot occur. Precisely the same reasoning applies to case (iv).

Finally then, cases (v) and (vi), in which n is respectively 11 and 5 are the only remaining possibilities.

When n is 5, G is the symmetric group of degree 5.

When n is 11, the group G, if it exists, is of degree 11 and order $11 \cdot 10 \cdot 9 \cdot 8$. A sub-group of order 8 which keeps three symbols fixed must contain a single operation of order 2; hence it must either be cyclical or of the type given in Theorem V, § 63. When expressed in 8 symbols, this is generated by

$$(1254)(3867) \quad \text{and} \quad (1758)(2643),$$

while we may take a cyclical group of order 8 to be generated by

$$(12835476).$$

It is easy to verify that each of these substitutions transforms the group of order 9, generated by

$$(123)(456)(789) \quad \text{and} \quad (147)(258)(369),$$

into itself.

If we now apply the method of § 109, we find that each of these doubly transitive groups of degree 9 is contained in a triply transitive group of degree 10; a repetition of the same process of trial shews that, while the group

$$\{(123)(456)(789),\ (12835476)\}$$

is not contained in a quadruply transitive group of degree 11 and order $11 \cdot 10 \cdot 9 \cdot 8$, the group

$$\{(123)(456)(789),\ (1254)(3867),\ (1758)(2643)\}$$

and the two substitutions

$$(a2)(58)(64)(79),\ (ab)(57)(68)(49)$$

actually generate such a group.

A further trial will shew that this group and the substitution

$$(bc)(47)(58)(69)$$

generate a group of degree 12 and order $12 \cdot 11 \cdot 10 \cdot 9 \cdot 8$; but that there is no group of degree 13 which contains the last group as the sub-group that keeps one symbol fixed.

The three substitutions, of order two, just given generate that sub-group of order 24, of the transitive group of degree 11 and order $11 \cdot 10 \cdot 9 \cdot 8$ in the symbols

$$a,\ b,\ c,\ 2,\ 3,\ 4,\ 5,\ 6,\ 7,\ 8,\ 9,$$

which permutes a, b, c, 2 among themselves.

If $k \nless 4$, a k-ply transitive group, of degree n and order $n(n-1)\dots(n-k+1)$, must contain a quadruply transitive group of degree $n-k+4$ and order $(n-k+4)(n-k+3)(n-k+2)(n-k+1)$. Hence, if the group is neither the alternating group nor the symmetric group of degree n, it must be either the group of degree 11 or the group of degree 12 that have been determined above.

Note to § 110.

It may be shewn that, with a single exception when $n = 12$, the inequality (p. 171)

$$k \ngtr \tfrac{1}{3}n + 1$$

may be replaced by

$$k < \tfrac{1}{3}n + 1.$$

Since k is an integer, it is only necessary to consider the case in which n is a multiple of 3, so that we may write $3m$ for n. If, in this case, $k = m + 1$, then $2k - 2 = 2m$; and a $(m+1)$-ply transitive group of degree $3m$, which does not contain the alternating group, can therefore have no substitution which displaces fewer than $2m$ symbols. Its order must therefore be $3m(3m-1)\dots(2m+1)2m$. From M. Jordan's results, which have just been proved, such a group exists only when $n = 12$; and therefore when n is not 12, a k-ply transitive group of degree n, which does not contain the alternating group, can only exist if $k < \frac{1}{3}n + 1$.

CHAPTER IX.

ON SUBSTITUTION GROUPS: PRIMITIVE AND IMPRIMITIVE GROUPS.

120. WE have seen that the symbols permuted by the operations of an intransitive substitution group may be divided into sets, such that every substitution of the group permutes the symbols of each set among themselves. For a transitive group the symbols must, from this point of view, be regarded as forming a single set. It may however in particular cases be possible to divide the symbols permuted by a transitive group into sets in such a way, that every substitution of the group either interchanges the symbols of any set among themselves or else changes them all into the symbols of some other set. That this may be possible, it is clearly necessary that each set shall contain the same number of symbols.

In the present Chapter we shall discuss those properties of transitive groups which depend on their possessing or not possessing the property indicated.

121. Definition. When the symbols operated on by a transitive substitution group can be divided into sets, each set containing the same number of distinct symbols and no symbol occurring in two different sets, and when the sets are such that all the symbols of any set are either interchanged among themselves or changed into the symbols of another set by every substitution of the group, the group is called *imprimitive*. When no such division into sets is possible, the group is called *primitive*. The sets of symbols which are interchanged by an imprimitive group are called *imprimitive systems*.

A simple example of an imprimitive group is given by group VII of § 17. An examination of the substitutions of this group will shew that they all either transform the systems of symbols xyz and abc into themselves or else interchange them, and that the same is true of the systems xa, yb, zc; so that, in this case, the symbols may be divided into two distinct sets of imprimitive systems.

It follows at once, from the definition, that an imprimitive group cannot be more than simply transitive. For if it were doubly transitive, it would contain substitutions changing any two symbols into any other two, and of these the first pair might be chosen from the same imprimitive system and the second pair from distinct systems.

The question as to whether a given group can be expressed as a transitive group of given degree, and the further question as to whether such a representation of the group, when possible, is imprimitive or primitive, finds its complete solution in the following investigation due to Herr Dyck*.

122. In § 20 it was shewn how any group G of order N could be represented as a substitution group of N symbols. This form of the group, defined as the regular form, is simply transitive; for all its substitutions except identity displace all the symbols, and therefore there must be just one substitution changing a given symbol into any other. Let us now suppose that $N = \mu\nu$, and that G has a sub-group H of order μ, consisting of the operations

$$S_1(=1),\ S_2,\ S_3,\ \ldots,\ S_\mu,$$

so that every operation of the group can be represented uniquely in the form

$$S_mT_n \quad (m = 1, 2, 3, \ldots, \mu;\ n = 1, 2, 3, \ldots, \nu);\ T_1 = 1.$$

The tableau representing the group as a substitution group of N symbols will, in terms of these symbols, take the form given on the following page. Every symbol in this tableau is of the form

$$S_mT_nS_{m'}T_{n'};$$

and such a symbol will belong to the column headed by S_mT_n and to the line beginning with $S_{m'}T_{n'}$. The symbols in any line (or column) differ in arrangement only from those in the leading line (or column); hence T_k must occur in the line beginning with $S_{m'}T_{n'}$. We may therefore suppose that

$$S_mT_nS_{m'}T_{n'} = T_k;$$

*Dyck, "Gruppentheoretische Studien, II," *Math. Ann.* XXII, (1883), pp. 86–95.

and then

$$\begin{aligned} S_2 S_m T_n S_{m'} T_{n'} &= S_2 T_k; \\ S_3 S_m T_n S_{m'} T_{n'} &= S_3 T_k, \\ &\cdots\cdots \\ S_\mu S_m T_n S_{m'} T_{n'} &= S_\mu T_k. \end{aligned}$$

Since

$$1,\ S_2,\ S_3,\ \ldots,\ S\mu$$

form a group, these symbols differ from

$$S_m,\ S_2 S_m,\ S_3 S_m,\ \ldots,\ S_\mu S_m$$

only in the sequence in which they are written; and therefore the set of symbols

$$S_p T_n S_{m'} T_{n'} \quad (p = 1, 2, \ldots, \mu)$$

is identical, except as regards arrangement, with the set

$$S_p T_k \quad (p = 1, 2, \ldots, \mu).$$

Hence the substitution of G, represented by the line beginning with $S_{m'} T_{n'}$, changes the set of symbols $S_p T_n$ into the set $S_p T_k$ in some sequence or other.

Every substitution of the group therefore changes the symbols of each of the ν sets, into which the first line of the tableau is divided, either into themselves or into the symbols of some other of the ν sets. Hence:—

THEOREM I. *If a group of order $\mu\nu$ contains a sub-group of order μ, the regular form of the group will be imprimitive in such a way that the $\mu\nu$ symbols may be divided into ν imprimitive systems of μ symbols each.*

The converse of this theorem is also true. For if the $\mu\nu$ symbols

$$1,\ S_2,\ S_3,\ \ldots,\ S_{\mu\nu},$$

by whose permutations the group can be expressed in regular form, are

1	S_2	S_3	$\ldots$	S_μ	$\ldots$	T_a	S_2T_a	S_3T_a	$\ldots$	$S_\mu T_a$	$\ldots$	T_ν	S_2T_ν	S_3T_ν	$\ldots$	$S_\mu T_\nu$
S_2	S_2S_2	S_3S_2	$\ldots$	$S_\mu S_2$	$\ldots$	T_aS_2	$S_2T_aS_2$	$S_3T_aS_2$	$\ldots$	$S_\mu T_aS_2$	$\ldots$	$T_\nu S_2$	$S_2T_\nu S_2$	$S_3T_\nu S_2$	$\ldots$	$S_\mu T_\nu S_2$
$\ldots$	$\ldots$	$\ldots$	$\ldots$	$\ldots$	$\ldots$	$\ldots$	$\ldots$	$\ldots$	$\ldots$	$\ldots$	$\ldots$	$\ldots$	$\ldots$	$\ldots$	$\ldots$	$\ldots$
S_μ	S_2S_μ	S_3S_μ	$\ldots$	$S_\mu S_\mu$	$\ldots$	T_aS_μ	$S_2T_aS_\mu$	$S_3T_aS_\mu$	$\ldots$	$S_\mu T_aS_\mu$	$\ldots$	$T_\nu S_\mu$	$S_2T_\nu S_\mu$	$S_3T_\nu S_\mu$	$\ldots$	$S_\mu T_\nu S_\mu$
$\ldots$	$\ldots$	$\ldots$	$\ldots$	$\ldots$	$\ldots$	$\ldots$	$\ldots$	$\ldots$	$\ldots$	$\ldots$	$\ldots$	$\ldots$	$\ldots$	$\ldots$	$\ldots$	$\ldots$
T_b	S_2T_b	S_3T_b	$\ldots$	$S_\mu T_b$	$\ldots$	T_aT_b	$S_2T_aT_b$	$S_3T_aT_b$	$\ldots$	$S_\mu T_aT_b$	$\ldots$	$T_\nu T_b$	$S_2T_\nu T_b$	$S_3T_\nu T_b$	$\ldots$	$S_\mu T_\nu T_b$
S_2T_b	$S_2S_2T_b$	$S_3S_2T_b$	$\ldots$	$S_\mu S_2T_b$	$\ldots$	$T_aS_2T_b$	$S_2T_aS_2T_b$	$S_3T_aS_2T_b$	$\ldots$	$S_\mu T_aS_2T_b$	$\ldots$	$T_\nu S_2T_b$	$S_2T_\nu S_2T_b$	$S_3T_\nu S_2T_b$	$\ldots$	$S_\mu T_\nu S_2T_b$
$\ldots$	$\ldots$	$\ldots$	$\ldots$	$\ldots$	$\ldots$	$\ldots$	$\ldots$	$\ldots$	$\ldots$	$\ldots$	$\ldots$	$\ldots$	$\ldots$	$\ldots$	$\ldots$	$\ldots$
$S_\mu T_b$	$S_2S_\mu T_b$	$S_3S_\mu T_b$	$\ldots$	$S_\mu S_\mu T_b$	$\ldots$	$T_aS_\mu T_b$	$S_2T_aS_\mu T_b$	$S_3T_aS_\mu T_b$	$\ldots$	$S_\mu T_aS_\mu T_b$	$\ldots$	$T_\nu S_\mu T_b$	$S_2T_\nu S_\mu T_b$	$S_3T_\nu S_\mu T_b$	$\ldots$	$S_\mu T_\nu S_\mu T_b$
$\ldots$	$\ldots$	$\ldots$	$\ldots$	$\ldots$	$\ldots$	$\ldots$	$\ldots$	$\ldots$	$\ldots$	$\ldots$	$\ldots$	$\ldots$	$\ldots$	$\ldots$	$\ldots$	$\ldots$
T_ν	S_2T_ν	S_3T_ν	$\ldots$	$S_\mu T_\nu$	$\ldots$	T_aT_ν	$S_2T_aT_\nu$	$S_3T_aT_\nu$	$\ldots$	$S_\mu T_aT_\nu$	$\ldots$	$T_\nu T_\nu$	$S_2T_\nu T_\nu$	$S_3T_\nu T_\nu$	$\ldots$	$S_\mu T_\nu T_\nu$
S_2T_ν	$S_2S_2T_\nu$	$S_3S_2T_\nu$	$\ldots$	$S_\mu S_2T_\nu$	$\ldots$	$T_aS_2T_\nu$	$S_2T_aS_2T_\nu$	$S_3T_aS_2T_\nu$	$\ldots$	$S_\mu T_aS_2T_\nu$	$\ldots$	$T_\nu S_2T_\nu$	$S_2T_\nu S_2T_\nu$	$S_3T_\nu S_2T_\nu$	$\ldots$	$S_\mu T_\nu S_2T_\nu$
$\ldots$	$\ldots$	$\ldots$	$\ldots$	$\ldots$	$\ldots$	$\ldots$	$\ldots$	$\ldots$	$\ldots$	$\ldots$	$\ldots$	$\ldots$	$\ldots$	$\ldots$	$\ldots$	$\ldots$
$S_\mu T_\nu$	$S_2S_\mu T_\nu$	$S_3S_\mu T_\nu$	$\ldots$	$S_\mu S_\mu T_\nu$	$\ldots$	$T_aS_\mu T_\nu$	$S_2T_aS_\mu T_\nu$	$S_3T_aS_\mu T_\nu$	$\ldots$	$S_\mu T_aS_\mu T_\nu$	$\ldots$	$T_\nu S_\mu T_\nu$	$S_2T_\nu S_\mu T_\nu$	$S_3T_\nu S_\mu T_\nu$	$\ldots$	$S_\mu T_\nu S_\mu T_\nu$

divisible into ν imprimitive systems of μ each, let

$$1,\ S_2,\ S_3,\ \ldots,\ S_\mu$$

be that system which contains identity. Then this system and the systems

$$S_m,\ S_2S_m,\ S_3S_m,\ \ldots,\ S_\mu S_m,\quad m = 2,\ 3,\ \ldots,\ \mu,$$

having the symbol S_m in common, must have all their symbols in common; therefore the product of any two of the operations of this set of μ operations is another operation of the set. The set therefore forms a group.

123. Let us now represent the imprimitive system

$$T_m,\ S_2T_m,\ S_3T_m,\ \ldots,\ S_\mu T_m$$

by the single symbol U_m, for all values of m. If we then, in the preceding tableau representing the group, pay attention only to the way in which the systems are permuted among themselves, without regarding the permutations of the symbols within the individual systems, we obtain a substitution group of the ν symbols

$$U_1,\ U_2,\ U_3,\ \ldots,\ U_\nu.$$

This group is isomorphic with the original group G; and if no substitution of the original group interchanges among themselves the symbols of each imprimitive system, the isomorphism must be simple. Now a substitution of G, which does not change any imprimitive system into another, must, if it exists, be a substitution S' of the sub-group H, which is constituted by

$$1,\ S_2,\ S_3,\ \ldots,\ S_\mu,$$

such that T_mS' belongs to the set U_m for each suffix m; and therefore, for each suffix m, we must have an equation

$$T_mS' = S''T_m,$$

where S'' is another substitution of H. The sub-group H must therefore contain every substitution of G which is conjugate to S': in other words, it must contain a self-conjugate sub-group of G. Hence:—

THEOREM II. *If H, of order μ, is a sub-group of G of order $\mu\nu$, and if no self-conjugate sub-group of G is contained in H, then G can be expressed as a transitive group of degree ν.*

That the converse of this theorem is true is immediately obvious.

124. If the sub-group H of G is contained in a greater sub-group K of order $\mu\nu'$, where $\nu = \nu'\nu''$, the operations of K consist of the sets

$$H, \; HT_2, \; HT_3, \; \ldots, \; HT_{\nu'},$$

and the operations of G of the sets

$$K, \; KR_2, \; KR_3, \; \ldots, \; KR_{\nu''};$$

while the set KR_m is made up of the sets

$$HR_m, \; HT_2R_m, \; HT_3R_m, \; \ldots, \; HT_{\nu'}R_m.$$

The regular form of the group has $\nu'\nu''$ imprimitive systems corresponding to the sub-group H, and ν'' imprimitive systems corresponding to the sub-group K; the above method of representing K shews that each of the latter systems contain ν' complete systems of the former set.

Now it has just been proved that, if H contains no self-conjugate sub-group of G, the group can be represented as a transitive substitution group of the symbols

$$HT_nR_m \quad (n = 1, 2, \ldots, \nu', \; m = 1, 2, \ldots, \nu'').$$

But from the division of the symbols in the regular form of the group into ν'' imprimitive systems, it follows that the set of symbols

$$HT_nR_m \quad (n = 1, 2, \ldots, \nu')$$

must either be permuted among themselves or be changed into another set

$$HT_nR_{m'} \quad (n = 1, 2, \ldots, \nu')$$

by every substitution of the group. The representation of the group as a transitive group of $\nu\nu''$ symbols is therefore imprimitive. Hence:—

THEOREM III. *If a group G of order $\mu\nu'\nu''$ has a sub-group H of order μ, which contains no self-conjugate sub-group of G; and if H is contained in a sub-group K of G of order $\mu\nu'$; then the representation of G as a transitive group of degree $\nu'\nu''$ (in respect of H) is imprimitive, the $\nu'\nu''$ symbols being divisible into ν'' systems of ν' symbols each.*

Corollary I. A transitive group of order $\mu\nu$ and degree ν will be primitive if, and only if, a sub-group of order μ that keeps one symbol fixed is a maximum sub-group.

Corollary II. A group, which contains other self-conjugate operations besides identity, cannot be represented in primitive form.

For if a sub-group H of order μ contains a self-conjugate operation, the group (of order $\mu\nu$) cannot be represented as a transitive group of degree ν in respect of H; and if H contains none of the self-conjugate operations, and is not a self-conjugate sub-group, it cannot be a maximum sub-group.

In particular, a group whose order is the power of a prime cannot be represented as a primitive group.

Corollary III. An Abelian group when represented as a transitive substitution group, must be in regular form.

Corollary IV. A simple group can always be represented in primitive form.

125. Every possible representation of a group as a transitive substitution group is given by the method of the preceding paragraphs. There is another method of dealing with the same problem which we may shortly consider here in view of its utility in many special cases, though it does not in general lead to all possible modes of representation. Let

$$H_1,\ H_2,\ \ldots,\ H_\nu$$

be a conjugate set of sub-groups (or operations) of a given group G, and let

$$I_1,\ I_2,\ \ldots,\ I_\nu$$

be a set of sub-groups of G, such that I_r is the greatest sub-group containing H_r self-conjugately. The latter set of groups are not necessarily all distinct;

in fact, we have seen in § 55 that, when the order of G is the power of a prime, they cannot be all distinct.

If S is any operation of G, then

$$\begin{pmatrix} H_1 & , & H_2 & ,\ldots, & H_\nu \\ S^{-1}H_1S, & & S^{-1}H_2S, & \ldots, & S^{-1}H_\nu S \end{pmatrix}$$

is a substitution performed on the set of symbols $H_1, H_2, \ldots, H_\nu$; and if for S each substitution of the group is written in turn, these substitutions form a transitive substitution group of degree ν. The substitution corresponding to S, followed by the substitution

$$\begin{pmatrix} H_1 & , & H_2 & ,\ldots, & H_\nu \\ T^{-1}H_1T, & & T^{-1}H_2T, & \ldots, & T^{-1}H_\nu T \end{pmatrix}$$

corresponding to T, gives the substitution

$$\begin{pmatrix} H_1 & , & H_2 & ,\ldots, & H_\nu \\ T^{-1}S^{-1}H_1ST, & & T^{-1}S^{-1}H_2ST, & \ldots, & T^{-1}S^{-1}H_\nu ST \end{pmatrix};$$

and therefore the substitutions form a group isomorphic with G, since the product of the substitutions corresponding to S and T is the substitution corresponding to ST.

Moreover, since there are operations of G which transform H_1 into each of the other sub-groups (or operations) of the conjugate set, the substitution group is transitive in the ν symbols. The substitution group will be simply isomorphic with G if, and only if, there is no operation of G which transforms each of the ν sub-groups into itself. Now the only operations of G which transform H_1 into itself are the operations of I_1; and hence the substitution group will be simply isomorphic with G if, and only if, the conjugate set of sub-groups

$$I_1,\ I_2,\ \ldots,\ I_\nu,$$

has no common sub-group except identity. This will be the case only when I_1 contains no self-conjugate sub-group of G.

It has been seen (§ 123) that, when this condition is satisfied, G can be represented as a transitive substitution group whose degree is ν, the ratio of the orders of G and I_1. That the form there obtained is identical with the form obtained in the present paragraph may be easily verified. Thus in the earlier form, the substitution corresponding to S is

$$\begin{pmatrix} I_1T_1\ , & I_1T_2\ , \ldots, & I_1T_\nu \\ I_1T_1S, & I_1T_2S, \ldots, & I_1T_\nu S \end{pmatrix}, \quad (T_1 = 1),$$

or in abbreviated form

$$\begin{pmatrix} I_1T_x \\ I_1T_{x'} \end{pmatrix},$$

if

$$T_xS = iT_{x'},$$

i being some operation of I_1.

In the present mode of representation, the substitution corresponding to S can be written in the form

$$\begin{pmatrix} T_1^{-1}H_1T_1 & , & T_2^{-1}H_1T_2 & ,\ldots, & T_\nu^{-1}H_1T_\nu \\ S^{-1}T_1^{-1}H_1T_1S, & & S^{-1}T_2^{-1}H_1T_2S, & \ldots, & S^{-1}T_\nu^{-1}H_1T_\nu S \end{pmatrix},$$

since each operation of the set I_1T_n will transform H_1 into the same sub-group. Now, if a corresponding abbreviated form be used, this substitution may be written

$$\begin{pmatrix} T_x^{-1}H_1T_x \\ T_{x'}^{-1}H_1T_{x'} \end{pmatrix},$$

and therefore the symbols I_1T_x in the one form are permuted by the substitutions in identically the same manner as the corresponding symbols $T_x^{-1}H_1T_x$ in the other form.

It should be noticed that, if G contains self-conjugate operations other than identity, these operations necessarily occur in I_1; and therefore in such a case the present method cannot lead to a substitution group which is simply isomorphic with G. In any case, if K is the greatest self-conjugate sub-group of G contained in I_1, the substitution group is simply isomorphic with $\frac{G}{K}$.

126. As an illustration of the preceding paragraphs, we will determine the different modes in which the alternating group of degree 5 can be represented as a transitive group.

The only cyclical sub-groups contained in G_{60}, the alternating group of degree 5, are groups of orders 2, 3 and 5; and of each of these cyclical sub-groups there is a single conjugate set.

The non-cyclical sub-groups may be determined as follows. The lowest possible order for such a sub-group is 4; since this is the highest power of 2 that divides 60, there is a single conjugate set of sub-groups of order 4. The next lowest possible order is 6. Now no operation of order 3 is permutable with an operation of order 2, as the group contains no operations of order 6; on the other hand, every sub-group of order 3 is permutable with an operation of order 2; thus

$$(12)(45)(123)(12)(45) = (132).$$

There is therefore a single set of conjugate sub-groups of order 6. The next lowest possible order is 10. The group contains no operation of order 10; but every sub-group of order 5 is permutable with an operation of order 2; thus

$$(14)(23)(12345)(14)(23) = (15432).$$

There is therefore a single conjugate set of sub-groups of order 10. The next lowest possible order is 12. If the group contains a sub-group of this order, it must be transitive in 4 symbols. Now the alternating group of 4 symbols is of order 12. Hence G_{60} must contain a single conjugate set of sub-groups of order 12. The only other possible orders are 15, 20 and 30. The reader will readily verify directly that there are no sub-groups of these orders. This can also be seen indirectly, since G_{60} is a simple group and therefore, if there were a sub-group of order 15, the group could be expressed transitively in 4 symbols. Since the group contains operations of order 5, this is clearly impossible.

Hence finally, since each of the sub-groups leads to a transitive representation of the group, G_{60} can be represented as a transitive substitution group in 30, 20, 15, 12, 10, 6 and 5 symbols, and in one distinct form in each case. The second method, as given in § 125, does not lead to all these modes of representation. The group will be found to contain 15 conjugate operations (or sub-groups) of order 2: 10 conjugate sub-groups and 20 conjugate operations of order 3: 5 conjugate sub-groups of order 4: 6 conjugate sub-groups and two sets of 12 conjugate operations, each of order 5: 10 conjugate sub-groups of order 6: 6 conjugate sub-groups of order 10: and 5 conjugate sub-groups of order 12. Hence, by using the second method, the representation of the group as transitive in 30 symbols would be missed.

Since a sub-group of order 2 is contained in sub-groups of orders 4, 6, 10 and 12, the 30 symbols permuted by G_{60}, when it is expressed as a transitive group of degree 30, can be divided into sets of imprimitive systems, containing respectively 2, 3, 5 and 6 symbols each. Similarly, when G_{60} is represented as a transitive group of degree 20, 15 or 12, it is imprimitive. When expressed as a group of order 10, 6 or 5, it is primitive.

127. As a further illustration, we shall determine all the distinct forms of imprimitive groups of degree 6. Let G be such a group, and H that sub-group of G which interchanges among themselves the symbols of each imprimitive system.

We will first suppose that there are two systems of three symbols each, viz.

$$1,\ 2,\ 3 \quad \text{and} \quad 4,\ 5,\ 6.$$

In this case, H cannot consist of the identical operation only; for there must be a substitution changing 1 into 2, and this must permute 1, 2 and 3 among themselves, and therefore also 4, 5 and 6 among themselves.

Let H contain substitutions which leave 4, 5 and 6 unchanged. These must (§ 114) form a self-conjugate sub-group of H, which will be either

$$\{(123),\ (12)\} \quad \text{or} \quad \{(123)\}.$$

Since G has substitutions interchanging the systems, H must similarly contain

$$\{(456),\ (45)\} \quad \text{or} \quad \{(456)\}.$$

In the first alternative, H is the group

$$\text{(i)} \quad \{(123),\ (12),\ (456),\ (45)\};$$

for H contains this group, and on the other hand, this is the most general group that interchanges the six symbols in two intransitive systems of three each. The order of this group is $2^2 \cdot 3^2$.

In the second alternative, H contains the self-conjugate sub-group

$$\{(123),\ (456)\}.$$

Now if H is of order $2^2 \cdot 3^2$, it is necessarily of the form (i). If it is of order $2 \cdot 3^2$, it must contain a substitution of order 2 which transforms the sub-group just given into itself. This may be taken, without loss of generality, to be (12)(45); and then H is the group

$$\text{(ii)} \quad \{(123),\ (456),\ (12)(45)\}.$$

If H is of order 3^2, it is the group

$$\text{(iii)} \quad \{(123),\ (456)\}.$$

Next, let H contain no substitutions which leave 4, 5 and 6 unchanged. Then H is simply isomorphic with a group of degree 3, and therefore it must be of the form

$$\text{(iv)} \quad \{(123)(456),\ (12)(45)\},$$

or

$$\text{(v)} \quad \{(123)(456)\}.$$

Now $\dfrac{G}{H}$ is of order 2; therefore G must have a substitution of order 2 or 4, which interchanges the systems. If the order of H is odd, this substitution must be of order 2. When the substitution is of order 2, we may, without loss of generality, take it to be (14)(25)(36) or (14)(26)(35). If H is of the form (i), (ii), or (iv), we get for G, in each case, the same group whichever of these substitutions we take. When H is of the form (iii), we get two groups which are easily seen to be conjugate in the symmetric group. These we do not regard as distinct. When H is of the form (v), we get two distinct groups, one of which is simply isomorphic with the symmetric group of three symbols, while the other is a cyclical group. In these two latter cases, the symbols can be divided into three imprimitive systems of two each.

If the substitution which interchanges the systems is of order 4, its square must occur in H; we may therefore take it to be (1425)(36). When H is of form (i), this gives the same group as before; but when H is of either of the forms (ii) or (iv), we get new forms for G. There are therefore eight distinct forms of groups of degree 6, in which the symbols form two imprimitive systems of three symbols each.

Secondly, suppose that there are three systems of symbols, containing two each, viz.

$$1,\ 2; \quad 3,\ 4; \quad \text{and} \quad 5,\ 6.$$

The self-conjugate sub-group H is of order 2^3, 2^2, 2 or 1. Corresponding to the first three cases, the forms of H are easily seen to be

$$\begin{aligned} &\text{(i)} \quad \{(12),\ (34),\ (56)\}, \\ &\text{(ii)} \quad \{(12)(34),\ (34)(56),\ (12)(56)\}, \\ \text{and } &\text{(iii)} \quad \{(12)(34)(56)\}. \end{aligned}$$

Again, since $\frac{G}{H}$ interchanges the three systems, it must be simply isomorphic with a group of degree 3, and its order is therefore either 3 or 6. First, let its order be 3. It must then contain a substitution of order 3 which, without loss of generality, may be taken to be (135)(246); this gives, with the three above forms of H, three distinct forms for G. The form of G corresponding to the form (iii) of H is, however, the same as one of those already determined.

If $\frac{G}{H}$ is of order 6, G must contain as a self-conjugate sub-group one of the three groups just obtained. Also if $\frac{G}{H}$ were cyclical, there would be a substitution in G, not belonging to H and permutable with (135)(246). This is clearly impossible, and therefore G must contain a substitution which transforms (135)(246) into its inverse. We may take this to be (13)(24) or (14)(23)(56). With the form (i) for H, these two substitutions lead to the same group. When H is of the form (ii), they give two distinct forms for G. When H is of the form (iii), G admits the imprimitive systems 1, 3, 5, and 2, 4, 6.

Lastly, if H is the identical operation, G is necessarily of order 6; no new forms can arise.

There are therefore five distinct forms of groups of degree 6, in which the symbols form three imprimitive systems of two symbols each but do not at the same time form two imprimitive systems of three symbols each.

128. An actual test to determine whether any transitive group is primitive or imprimitive may be applied as follows. Consider the effect of the substitutions of the group G on r of the symbols which are permuted transitively by it. Those substitutions, which permute the r symbols, say

$$a_1,\ a_2,\ \ldots,\ a_r$$

among themselves, form a sub-group H. Now suppose that every substitution, which changes a_1 into one of the r symbols, belongs to H. Then if S is a substitution, which does not permute the r symbols among themselves, it must change them into a new set

$$b_1, b_2, \ldots, b_r,$$

which has no symbol in common with the previous set; and every operation of the set HS changes all the a's into b's. Moreover, since G is transitive, H must permute the a's transitively; and therefore the set HS must contain substitutions changing a_1 into each one of the b's.

Suppose now, if possible, that the group contains a substitution S', which changes some of the a's into b's, and the remainder into new symbols. We may assume that S' changes a_1 into b_1, and a_2 into a new symbol c_2. Among the set HS there is at least one substitution, T, which changes a_1 into b_1. Hence TS'^{-1} changes a_1 into itself and some new symbol into a_2. This however contradicts the supposition that every substitution, which changes a_1 into one of the set of a's, belongs to H. Hence no substitution such as S' can belong to G; and every substitution, which changes one of the a's into one of the b's, must change all the a's into b's.

If the substitutions of the group are not thus exhausted, there must be another set of r symbols

$$c_1, c_2, \ldots, c_r,$$

which are all distinct from the previous sets, such that some substitution changes all the a's into c's. We may now repeat the previous reasoning to shew that every substitution, which changes an a into a c, must change all the a's into c's. By continuing this process, we finally divide the symbols into a number of distinct sets of r each, such that every substitution of the group must change the a's either into themselves or into some other set: and therefore also must change every set either into itself or into some other set. The group must therefore be imprimitive. Hence:—

THEOREM IV. *If, among the symbols permuted by a transitive group, it is possible to choose a set such that every substitution of the group, which changes a chosen symbol of the set either into itself or into another of the set, permutes all the symbols of the set among themselves; then the group is imprimitive, and the set of symbols forms an imprimitive system.*

Corollary I. If $a_1, a_2, \ldots, a_r$ are a part of the symbols permuted by a primitive group, there must be substitutions of the group, which replace some of this set of symbols by others of the set, and the remainder by symbols not belonging to the set.

Corollary II. A sub-group of a primitive group, which keeps one symbol unchanged, must contain substitutions which displace any other symbol.

If the sub-group H, that leaves a_1 unchanged, leaves every symbol of the set $a_1, a_2, \ldots, a_r$ unchanged, then H must be transformed into itself by every substitution which changes any one of these symbols into any other. Every substitution, which changes one of the set into another, must therefore permute the set among themselves; and the group, contrary to supposition, is imprimitive.

129. It has already been seen that, in particular cases, it may be possible to distribute the symbols, which are permuted by an imprimitive group, into imprimitive systems in more than one way. When this is possible, suppose that two systems which contain a_1 are

$$\begin{aligned} &a_1, a_2, \ldots, a_r, a_{r+1}, \ldots, a_m, \\ \text{and}\quad &a_1, a_2, \ldots, a_r, a'_{r+1}, \ldots\ldots\ldots, a'_n; \end{aligned}$$

and that the symbols common to the two systems are

$$a_1, a_2, \ldots, a_r.$$

A substitution of the group, which changes a_1 into a'_{r+1}, must change $a_1, a_2, \ldots, a_r$ into r symbols of that system of the first set which contains a'_{r+1}, while it changes the system of the second set that contains a_1 into itself. Hence the latter system contains at least r symbols of that system of the first set in which a'_{r+1} occurs. By considering the effect of the inverse substitution, it is clear that the system

$$a_1, a_2, \ldots, a_r, a'_{r+1}, \ldots, a'_n$$

cannot have more than r symbols in common with the system of the first set that contains a'_{r+1}. Hence the n symbols of this system can be divided into sets of r, such that each set is contained in some system of the first set. It follows that n, and therefore also m, must be divisible by r.

Suppose now that b_1 is any symbol which is not contained in either of the above systems. A substitution that changes a_1 into b_1 must change the two systems into two others, which have r symbols

$$b_1, b_2, \ldots, b_r$$

in common; and since no two systems of either set have a common symbol, these r symbols must be distinct from

$$a_1, a_2, \ldots, a_r.$$

Further, from the mode in which the set $b_1, b_2, \ldots, b_r$ has been obtained, any operation, which changes one of the symbols $a_1, a_2, \ldots, a_r$ into one of the symbols $b_1, b_2, \ldots, b_r$, must change all the symbols of the first set into those of the second. Hence the symbols operated on by the group can be divided into systems of r each, by taking together the sets of r symbols which are common to the various pairs of the two given sets of imprimitive systems; and the group is imprimitive in regard to this new set of systems of r symbols each. Hence:—

THEOREM V. *If the symbols permuted by a transitive group can be divided into imprimitive systems in two distinct ways, m being the number of symbols in each system of one set and n in each system of the other; and if some system of the first set has r symbols in common with some system of the second set; then* (i) *r is a factor of both m and n, and* (ii) *the symbols can be divided into a set of systems of r each, in respect of which the group is imprimitive.*

This result may also be regarded as an immediate consequence of Theorem III, § 124. For if H_1 and H_2 are two sub-groups of G each of which contains the sub-group R, and if K is the greatest common sub-group of H_1 and H_2, then K contains R. Now if R contains no self-conjugate sub-group of G, then G can be represented as a transitive group whose degree is the order of G divided by the order of R. If the respective orders of H_1 and H_2 are m times and n times the order of R, the symbols can be divided in two distinct ways into sets of imprimitive systems, the systems containing m and n symbols respectively. Also, if the order of K is r times the order of R, then r is a factor of m and of n; by considering the sub-group K, the symbols may be divided into a set of systems which contain r symbols each.

It might be expected that, just as we can form a new set of imprimitive systems by taking together the symbols which are common to pairs of systems of two given sets, so we might form another new set of systems by combining all the systems of one set which have any symbols in common with a single system of the other set. A very cursory consideration will shew however that this is not in general the case. In fact, it is sufficient to point out that, with the notation already used, the number of symbols in such a new system would be $\frac{mn}{r}$; and this number is not necessarily a factor of the degree of the group. Also, even if this number is a factor of the degree of the group, it will not in general be the case that the symbols so grouped together form an imprimitive system.

130. We may now discuss, more fully than was possible in § 106, the form of a self-conjugate sub-group of a given transitive group. Such a sub-group must clearly contain one or more operations displacing every symbol operated on by the group. For if every operation of the sub-group keeps the symbol a_1 unchanged, then since it is self-conjugate, every operation will keep $a_2, a_3, \ldots,$ unchanged: and the sub-group must reduce to the identical operation only.

Suppose now, if possible, that H is an intransitive self-conjugate sub-group of a transitive group G; and that H permutes the n symbols of G in the separate transitive sets $a_1, a_2, \ldots, a_{n_1}$; $b_1, b_2, \ldots, b_{n_2}$; .... If S is any operation of G which changes a_1 into b_1, then, since

$$S^{-1}HS = H,$$

it must change all the a's into b's; and since

$$SHS^{-1} = H,$$

S^{-1} must change all the b's into a's. Hence the number of symbols in the two sets, and therefore the number of symbols in each of the sets, must be the same.

Moreover every operation of G, since it transforms H into itself, must either permute the symbols of any set among themselves, or it must change them all into the symbols of some other set. Hence G must be imprimitive, and H must consist of those operations of G which permute the symbols of each imprimitive system among themselves.

Conversely, when G is imprimitive, it is immediately obvious that those operations of G, if any such exist, which permute the symbols of each of a set of imprimitive systems among themselves, form a self-conjugate sub-group. Hence:—

THEOREM VI. *A self-conjugate sub-group of a primitive group must be transitive; and if an imprimitive group has an intransitive self-conjugate sub-group, it must consist of the operations which permute among themselves the symbols of each of a set of imprimitive systems.*

If G is an imprimitive group of degree mn, and if there are n imprimitive systems of m symbols each, then we have seen in § 123 that G is isomorphic with a group G' of degree n. In particular instances, it may at once be evident, from the order of G, that this isomorphism cannot be simple. For example, if the order of G has a factor which does not divide $n!$, this is certainly the case: and more generally, if it is known independently that G cannot be expressed as a transitive group of degree n, then G must certainly be multiply isomorphic with G'. In such instances the self-conjugate sub-group of G, which corresponds to the identical operation of G', is that intransitive self-conjugate sub-group, which interchanges among themselves the symbols of each imprimitive system.

If G is soluble, a minimum self-conjugate sub-group of G must have for its order a power of a prime. Also, if G has an intransitive self-conjugate sub-group, it must have an intransitive minimum self-conjugate sub-group. Hence if G is soluble and has intransitive self-conjugate sub-groups, the symbols permuted by G must be capable of division into imprimitive systems, such that the number in each system is the power of a prime.

131. Let G be a k-ply transitive group of degree n $(k > 2)$, and let G_r be that sub-group of G which keeps r $(< k)$ given symbols unchanged, so that G_r is $(k-r)$-ply transitive in the remaining $n-r$ symbols. Also, let H be a self-conjugate sub-group of G, and let H_r be that sub-group of H

which keeps the same r symbols unchanged; so that H_r is the common sub-group of H and G_r. Since every operation of G_r transforms both H and G_r into themselves, every operation of G_r must be permutable with H_r; i.e. H_r is a self-conjugate sub-group of G_r. Now, if $r = k - 2$, G_{k-2} is doubly transitive in the $n - k + 2$ symbols on which it operates; it is therefore primitive. Hence, unless H_{k-2} consists of the identical operation only, it must be transitive in the $n - k + 2$ symbols.

If H_{k-2} is the identical operation, H contains no operation, except identity, which displaces less than $n - k + 3$ symbols. Suppose, first, that H contains operations, other than identity, which leave one or more symbols unchanged. Then, since H is a self-conjugate sub-group and G is k-ply transitive, it may be shewn, exactly as in § 110, that H must contain operations displacing not more than $2k - 2$ symbols. Hence H_{k-2} can consist of the identical operation alone, only if

$$n - k + 3 \ngtr 2k - 2,$$

or

$$k \nless \tfrac{1}{3}n + \tfrac{5}{3}.$$

When this inequality holds, we have seen (p. 171) that G contains the alternating group. Hence in this case, if G does not contain the alternating group, it follows that H_{k-2} is transitive in the $n - k + 2$ symbols on which it operates.

Since H is self-conjugate and G is k-ply transitive, H must contain a sub-group conjugate to H_{k-2} which keeps any other $k - 2$ symbols unchanged. Hence H_{k-3} must be doubly transitive in the $n - k + 3$ symbols on which it operates; and so on. Finally, if G is not the symmetric group (the alternating group, being simple, contains no self-conjugate sub-group) H must be $(k - 1)$-ply transitive.

Suppose, next, that H contains no operation, except identity, which leaves any symbol unchanged. Then if, with the notation of § 110, $j = k-1$ for every operation of H, the argument there used does not apply. For it is impossible to choose the operation T so that c_k is a symbol which does not occur in S.

The self-conjugate sub-group H contains a single operation changing a given symbol a_1 into any other symbol a_r. Also G contains operations which leave a_1 unchanged and change a_r into any other symbol a_s. Hence the operations of H, other than identity, form a single conjugate set in G; and therefore H must be an Abelian group of order p^m and type $(1, 1, \ldots, \text{to } m \text{ units})$; p being a prime. Further, since G is by supposition at least triply transitive, it must contain operations which transform any two operations of H, other than identity, into any other two. If p were an odd prime, and P_1 and P_2 were two of the generating operations of H, it follows that G would have an operation S such that

$$S^{-1}P_1S = P_1, \quad S^{-1}P_2S = P_1^{\alpha};$$

and this is impossible. Hence p must be 2. Further if G were more than triply transitive, and if A, B, C were three independent generating operations of H, then G would have an operation Σ such that

$$\Sigma^{-1}A\Sigma = A, \quad \Sigma^{-1}B\Sigma = B, \quad \Sigma^{-1}C\Sigma = AB.$$

This again is impossible, and therefore k must be 3. Hence:—

THEOREM VII. *A self-conjugate sub-group of a k-ply transitive group of degree n $(2 < k < n)$, is in general at least $(k-1)$-ply transitive*. The only exception is that a triply transitive group of degree 2^m may have a self-conjugate sub-group of order 2^m.*

132. For the further discussion of the self-conjugate sub-groups of a primitive group, it is necessary to consider in what forms the direct product of two groups can be represented as a transitive group.

Let G be the direct product of two groups H_1 and H_2, and suppose that G can be represented as a transitive group of degree n. When G is thus represented, we will suppose that H_1 is transitive in the n symbols that G permutes. We have seen in § 107 that every substitution of n symbols, which is permutable with each of the substitutions of a group transitive in the n symbols, must displace all the n symbols. It follows that every

*Jordan, *Traité des Substitutions*, p. 65; where, however, the exceptional case is overlooked.

substitution of H_2 must displace all the n symbols on which G operates; and that the order of H_2 is equal to or is a factor of n.

If the order of H_2 is equal to n, then H_2 is transitive in the n symbols, so that the order of H_1 cannot be greater than n. In this case, H_1 and H_2 must (§ 107) be two simply isomorphic groups of order n, which have no self-conjugate operations except identity. Further, if H_1 and H_2 in this case are not simple groups, let K be a self-conjugate sub-group of H_1. Since every operation of H_2 is permutable with every operation of H_1, K is a self-conjugate sub-group of G. Now the order of K is less than n, the degree of G; therefore K is intransitive and G is imprimitive. On the other hand, we have seen (*loc. cit.*) that, if H_1 and H_2 are simple, the sub-group of G that keeps one symbol fixed is a maximum sub-group: and therefore G is primitive. Hence:—

THEOREM VIII. *If the direct product Γ of H_1 and H_2 can be represented as a transitive group of degree n, in such a way that H_1 and H_2 are transitive sub-groups of Γ, then H_1 and H_2 must be simply isomorphic groups of order n, which have no self-conjugate operations except identity. When this condition is satisfied, Γ will be primitive if, and only if, H_1 and H_2 are simple.*

133. Suppose now that a primitive group G, of degree n, has two distinct minimum self-conjugate sub-groups H_1 and H_2. Then every operation of H_1 (or H_2) is permutable with H_2 (or H_1), and H_1, H_2 have no common operation except identity. Hence (§ 34) the group $\{H_1, H_2\}$, which we will call Γ, is the direct product of H_1 and H_2. Now Γ is a self-conjugate sub-group of G: it is therefore transitive in the n symbols which G permutes. Also H_1 and H_2, being self-conjugate sub-groups of G, are transitive. Hence, by Theorem VIII (§ 132), H_1 and H_2 are simply isomorphic, and n is equal to the order of H_1. Moreover, since H_1 is a minimum self-conjugate sub-group of G which contains no self-conjugate operations except identity, it must (Theorem IV, § 94) be either a simple group of composite order, or the direct product of several simply isomorphic simple groups of composite order. It follows that G cannot have two distinct minimum self-conjugate sub-groups unless the degree of G is equal to or is a power of the order of some simple group of composite order.

134. Let now Γ be a minimum self-conjugate sub-group of a doubly transitive group G, and suppose that Γ is the direct product of the α simply isomorphic simple groups

$$H_1,\ H_2,\ \ldots,\ H_\alpha.$$

Since G is primitive, Γ is transitive. If H_1 is a cyclical group of prime order p, the order of Γ is p^α; therefore the degree of Γ, or what is the same thing, the degree of G, is p^α.

If H_1 is a simple group of composite order, and if $\alpha > 2$, then (§ 132) H_1 cannot be transitive. The intransitive systems of H_1, since they form a set of imprimitive systems for Γ, must each contain the same number m of symbols. If m is less than the order of H_1, a sub-group of H_1 which leaves unchanged one symbol of one intransitive system will leave unchanged one symbol of each intransitive system. Now we shall see, in § 136, that the operations of an imprimitive self-conjugate sub-group of a doubly transitive group must displace all or all but one of the symbols. Hence m cannot be less than the order of H_1. We may similarly shew that, if m is equal to the order of H_1, some of the operations of Γ must keep more than one symbol fixed; and therefore, if $\alpha > 2$, the group assumed cannot exist. If $\alpha = 2$, H_1 may be transitive. But in this case $\{H_1, H_2\}$ certainly contains operations which leave more than one symbol unchanged; and again the group assumed cannot exist. Hence finally no doubly transitive group can contain a minimum self-conjugate sub-group of the type assumed.

No general law can be stated regarding self-conjugate sub-groups of simply transitive primitive groups; but for groups which are at least doubly transitive the preceding results may be summed up as follows:—

THEOREM IX. *A group G which is at least doubly transitive either must be simple or must contain a simple group H as a self-conjugate sub-group. In the latter case no operation of G, except identity, is permutable with every operation of H. The only exceptions to this statement are that a triply transitive group of degree 2^m may have a self-conjugate sub-group of order 2^m; and that a doubly transitive group of degree p^m, where p is a prime, may have a self-conjugate sub-group of order p^m.*

Corollary. If a primitive group is soluble, its degree must be the power of a prime*.

In fact, if a group is soluble, so also is its minimum self-conjugate sub-group. The latter must be therefore an Abelian group of order p^α: and since this group must be transitive, its order is equal to the degree of the primitive group.

135. As illustrating the occurrence of an imprimitive self-conjugate sub-group in a primitive group, we will construct a primitive group of degree 36 which has an imprimitive self-conjugate sub-group. For this purpose, let†

$$S = (1,2,3)(4,5,6)(7,8,9)(10,11,12)(13,14,15)(16,17,18)(19,20,21)(22,23,24)(25,26,27)(28,29,30)(31,32,33)(34,35,36),$$

and

$$A = (3,4)(5,6)(9,10)(11,12)(15,16)(17,18)(21,22)(23,24)(27,28)(29,30)(33,34)(35,36);$$

so that $\{S, A\}$ is an intransitive group of degree 36, the symbols being interchanged in 6 transitive systems containing 6 symbols each. This group is simply isomorphic with

$$\{(123)(456),\ (34)(56)\};$$

and it may be easily verified that this group is simply isomorphic with the alternating group of 5 symbols, the order of which is 60.

Also let

$$J = (2,7)(3,13)(4,19)(5,25)(6,31)(9,14)(10,20)(11,26)(12,32)(16,21)(17,27)(18,33)(23,28)(24,34)(30,35).$$

*This result, stated in a somewhat different form, is given, among many others, in the letter written by Galois to his friend Chevalier on the evening of May 29th, 1832, the day before the duel in which he was killed. The letter was first printed in the *Revue Encyclopédique* (1832), p. 568; it was reprinted in the collection of Galois's mathematical writings in *Liouville's Journal*, t. XI (1846), pp. 381–444

†The commas in the symbols for the substitutions are here used to prevent confusion among the one-digit and two-digit numbers.

Then

$$JSJ = (1,7,13)(19,25,31)(2,8,14)(20,26,32)$$
$$(3,9,15)(21,27,33)(4,10,16)(22,28,34)$$
$$(5,11,17)(23,29,35)(6,12,18)(24,30,36),$$

and

$$JAJ = (13,19)(25,31)(14,20)(26,32)(15,21)(27,33)$$
$$(16,22)(28,34)(17,23)(29,35)(18,24)(30,36);$$

and $\{JSJ, JAJ\}$ is similar to and simply isomorphic with $\{S, A\}$.

Now it may be directly verified that S and A are, each of them, permutable with JSJ and JAJ; and therefore every operation of the group $\{S, A\}$ is permutable with every operation of the group $\{JSJ, JAJ\}$. Also these two groups can have no common operation, since the symbols into which $\{S, A\}$ changes any given symbol are all distinct from those into which $\{JSJ, JAJ\}$ change it. Hence $\{S, A, JSJ, JAJ\}$ is the direct product of $\{S, A\}$ and $\{JSJ, JAJ\}$; it is therefore a group of order 3600. It is also, from its mode of formation, a transitive group of degree 36; and it interchanges the symbols in two and only two distinct sets of imprimitive systems, of which

$$1,\ 2,\ 3,\ 4,\ 5,\ 6 \quad \text{and} \quad 1,\ 7,\ 13,\ 19,\ 25,\ 31$$

may be taken as representatives.

Now J does not interchange the 36 symbols in either of these systems, and therefore it cannot occur in $\{S, A, JSJ, JAJ\}$. Further

$$J\{S, A, JSJ, JAJ\}J = \{S, A, JSJ, JAJ\};$$

and therefore $\{J, S, A\}$ is a transitive group of degree 36 and order 7200. Also, since J does not interchange the symbols in either of the two sets of imprimitive systems of $\{S, A, JSJ, JAJ\}$, it follows that $\{J, S, A\}$ is primitive.

136. We have seen that, with a single exception, a k-ply transitive group ($k > 2$) cannot have an imprimitive self-conjugate sub-group; while the example in the preceding section illustrates the occurrence of an imprimitive self-conjugate sub-group in a simply transitive primitive group. We will now consider, from a rather different point of view, the possibility of an imprimitive self-conjugate sub-group in a doubly transitive group. Let G be a doubly transitive group of degree mn, and let H be an imprimitive self-conjugate sub-group of G. Suppose that m is the smallest number, other than unity, of symbols which occur in an imprimitive system; and let

$$a_1,\ a_2,\ \ldots,\ a_m$$

form an imprimitive system. Since G is doubly transitive, it must contain a substitution S, which leaves a_1 unchanged and changes a_2 into a_{m+1}, a symbol not contained in the given set. If S changes the given set into

$$a_1,\ a_{m+1},\ \ldots,\ a_{2m-1},$$

then, since

$$S^{-1}HS = H,$$

this new set must form an imprimitive system for H. Also, since m is the smallest number of symbols that can occur in an imprimitive system, the two sets have no symbol in common except a_1.

Now a_{m+1} may be any symbol not contained in the original system. Hence it must be possible to distribute the mn symbols into sets of imprimitive systems of m each, such that every pair of symbols occurs in one system and no pair in more than one system. This implies that $mn - 1$ is divisible by $m - 1$, or that $n - 1$ is divisible by $m - 1$.

Consider now a substitution of H which leaves a_1 unchanged. It must permute among themselves the remaining $m - 1$ symbols of each of the $\dfrac{mn-1}{m-1}$ systems in which a_1 occurs. If a_r is any other symbol, a similar statement applies to it. Now no two systems have more than one symbol in common. Hence every substitution, which leaves both a_1 and a_r unchanged, must leave all the symbols unchanged. The sub-group H is therefore such that each of its substitutions displaces all or all but one of the symbols. Moreover the substitutions, which leave a_1 unchanged, permute among themselves the remaining symbols of each system in which a_1 occurs; therefore the order of H must be $mn\mu$, where μ is a factor of $m - 1$.

It has been seen in § 112 that, for certain values of mn, groups satisfying these conditions actually exist. The doubly transitive groups, of degree p^m and order $p^m(p^m - 1)$ there obtained, have imprimitive self-conjugate sub-groups of orders $p^m(p - 1)$ and p^m.

137. Ex. 1. Shew that a transitive group of order N and degree $2n$, which is imprimitive in respect of two systems of n symbols each, must have a self-conjugate sub-group of order $\frac{1}{2}N$.

Ex. 2. Shew that an imprimitive group of order Np^2 and degree np, where p is an odd prime, which has p imprimitive systems of n symbols each, must contain a self-conjugate sub-group whose order is a multiple of p.

Ex. 3. Shew that, if n is the smallest number of symbols in which a group G can be represented as a transitive group, then n^α is the smallest number of symbols in which the direct product of α groups, simply isomorphic with G, can be represented as a transitive group.

Ex. 4. Prove that if a group G, of order N, represented as a transitive substitution group, is imprimitive only when its degree is N; then either G is Abelian, or N is the product of two distinct primes. (Dyck.)

Ex. 5. Shew that if, in the tableau of p. 196, the sub-group

$$1,\ S_2,\ S_3,\ \ldots,\ S_\mu$$

is a self-conjugate sub-group H of the given group G, then, in each square compartment of the tableau, every horizontal line contains the same μ symbols.

Shew also that, if each square compartment of the tableau is regarded as a single symbol, the permutations of these symbols given by the tableau represents the group $\frac{G}{H}$ in regular form. (Dyck.)

CHAPTER X.

ON SUBSTITUTION GROUPS: TRANSITIVITY AND PRIMITIVITY: (CONCLUDING PROPERTIES.)

138. THE determination of all distinct transitive sub-groups of the symmetric group of n symbols has been carried out for values of n up to 12*. This investigation, if carried out for sufficiently great values of n, would involve the expression of all types of group in all possible transitive forms.

From the point of view of one of the chief problems of pure group-theory, namely, the determination of all distinct types of group of a given order, this analysis of the symmetric group of n symbols is not a succinct process, as it continually involves the redetermination of groups which have been already obtained. Thus a simple group of degree mn, in which the symbols are permuted in m imprimitive systems of n each, would in this analysis have been already obtained as a group of degree m. With reference then to the more restricted problem of determining all types of simple groups, it would certainly be sufficient to find all primitive sub-groups of the symmetric group.

139. We shall proceed to determine a superior limit to the order of a primitive group of degree n, other than the alternating or the symmetric group.

Let G be a primitive group of degree n, and suppose that G contains a sub-group H which leaves $n - m$ symbols unchanged and is transitive in the remaining m. Since G is primitive, H and the sub-groups conjugate to it must generate a transitive self-conjugate sub-group of G; and therefore

*Jordan, *Comptes Rendus*, t. LXXIII (1871), pp. 853–857; *ib.* t. LXXV (1872), pp. 1754–1757.

And see the *Bulletin of the New York Mathematical Society*, Cole, 1st series, Vol. II, pp. 184–190; 250–258: Miller, 1st series, Vol. III, pp. 168, 169; 242–245: 2nd series, Vol. I, pp. 67–72; 255–258: Vol. II, pp. 138–145. *Quarterly Journal of Mathematics*, Cole, Vol. XXVI, pp. 372–388: Vol. XXVII, pp. 35–50: Miller, Vol. XXVII, pp. 99–118; Vol. XXVIII, pp. 193–231. These memoirs were all published between 1892 and 1896; in them further references will be found.

there must be some sub-group H', conjugate to H, such that the m symbols operated on by H and the m operated on by H' are not all distinct. Suppose H' is chosen so that these two sets of m symbols have as great a number in common as possible, say s; and represent by

$$\alpha_1, \alpha_2, \ldots, \alpha_r, \gamma_1, \gamma_2, \ldots, \gamma_s,$$
$$\text{and} \quad \beta_1, \beta_2, \ldots, \beta_r, \gamma_1, \gamma_2, \ldots, \gamma_s,$$

where $r+s=m$, the symbols operated on by H and H' respectively. Then $\{H, H'\}$ is a transitive group in the $2r+s$ symbols α, β, and γ, which leaves unaltered all the remaining symbols of G.

If S is an operation of H which changes α_2 into α_1, $S^{-1}H'S$ does not affect α_1. Hence, unless S interchanges the α's among themselves, the m symbols operated on by H' and the m operated on by $S^{-1}H'S$ will have more than s in common. Every operation of H which changes one α into another must therefore interchange all the α's among themselves; hence H must be imprimitive.

If then H is primitive, s must be equal to $m-1$. In any case, if $s=m-1$, $\{H, H'\}$ is a doubly transitive and therefore primitive group of degree $m+1$, which leaves the remaining $n-m-1$ symbols of G unchanged. We may reason about this sub-group as we have done about H. Among the sub-groups conjugate to $\{H, H'\}$, there must be one at least which operates on m of the symbols displaced by $\{H, H'\}$. This, with $\{H, H'\}$, generates a triply transitive group of degree $m+2$, which leaves $n-m-2$ symbols unchanged. Proceeding thus, we find finally that G itself must be $(n-m+1)$-ply transitive.

140. If s is less than $m-1$, we may again deal with the sub-group $\{H, H'\}$, or H_1, exactly as we have dealt with H. It is a transitive group of degree m_1 ($>m$), which leaves $n-m_1$ symbols unchanged. If, among the sub-groups conjugate to H_1, none operates on more than s_1 of the symbols affected by H_1, and if H_1' is a suitably chosen conjugate sub-group, then $\{H_1, H_1'\}$ is a transitive group of degree $2m_1-s_1$, which leaves $n-2m_1+s_1$ symbols unchanged. Continuing this process, we must, before arriving at a group of degree n, reach a stage at which the number s_r is equal to m_r-1.

For suppose, if possible, that among the groups conjugate to K, of degree $\rho + \sigma$, none displaces more than σ of the symbols acted on by K, while at the same time $2\rho + \sigma = n$. If

$$\begin{aligned} &\alpha_1,\ \alpha_2,\ \ldots,\ \alpha_\rho,\ \gamma_1,\ \gamma_2,\ \ldots,\ \gamma_\sigma, \\ \text{and}\quad &\beta_1,\ \beta_2,\ \ldots,\ \beta_\rho,\ \gamma_1,\ \gamma_2,\ \ldots,\ \gamma_\sigma \end{aligned}$$

are the symbols affected by K and K' respectively, then since G is primitive, it must contain an operation S which changes α_1 into α_2 without at the same time changing all the α's into α's. If then we transform K' by S, the two groups K' and $S^{-1}K'S$ must operate on more than σ common symbols, contrary to supposition.

Hence G must in this case certainly contain a transitive sub-group of degree $n-1$, and therefore is itself at least doubly transitive*.

141. Returning to the case in which H is primitive and G therefore $(n-m+1)$-ply transitive, we at once obtain an inferior limit for m. We have seen, in fact, in Theorem IV, § 110, that a group of degree n, other than the alternating or the symmetric group, cannot be more than $(\frac{1}{3}n+1)$-ply transitive. Hence

$$n - m + 1 \ngtr \tfrac{1}{3}n + 1,$$

or

$$m \nless \tfrac{2}{3}n.$$

We may sum up these results as follows:—

THEOREM I. *A primitive group G of degree n, which has a sub-group H that keeps $n-m$ symbols unchanged and is transitive in the remaining m symbols, is at least doubly transitive. If H is primitive and G does not contain the alternating group, m cannot be less than $\frac{2}{3}n$, and G is $(n-m+1)$-ply transitive.*

*The results contained in §§ 139, 140 are due to Jordan (*Liouville's Journal*, Vol. XVI, 1871) and Netto (*Crelle's Journal*, Vol. CIII, 1889). They have been extended by Marggraff: "Ueber primitiven Gruppen mit transitiven Untergruppen geringeren Grades," (*Inaugural Dissertation*, Giessen, 1892). With the notation used in the text, Marggraff shews that, unless the symbols affected by H can be divided into imprimitive systems of r symbols each, in at least $r+1$ distinct ways, G will be $(n-m+1)$-ply transitive. In particular, if H is a cyclical group of degree m, G is $(n-m+1)$-ply transitive. He also shews that in any case $m \geqslant \frac{1}{2}n$.

Corollary. The order of a primitive group of degree n cannot exceed $\frac{n!}{2 \cdot 3 \ldots p}$, where 2, 3, ... p are the distinct primes which are less than $\frac{2}{3}n$.

If q^α is the highest power of a prime q that divides $n!$, the sub-groups of order q^α of the symmetric group form a single conjugate set, and each of them must contain circular substitutions of order q. Hence if $q < \frac{2}{3}n$, it follows by the theorem that no primitive group of degree n, other than the alternating or the symmetric group, can contain a sub-group of order q^α; and therefore $q^{\alpha-1}$ is the highest power of q that can divide the order of the group.

142. The ratio of $2 \cdot 3 \ldots p$ to n increases rapidly as n increases, and it is at once obvious that, when $n > 7$, this ratio is greater than unity; hence for values of n greater than 7, the symmetric group can have no primitive sub-group of order $(n-1)!$.

The order of the greatest imprimitive sub-group of the symmetric group is $a!\left(\frac{n}{a}!\right)^\alpha$, where a is the smallest factor of n. When $n > 4$, this is less than $(n-1)!$.

The order of the greatest intransitive sub-group of the symmetric group, other than the sub-groups that keep one symbol fixed, is $2!(n-2)!$. This is always less than $(n-1)!$.

Hence when $n > 7$, the only sub-groups of order $(n-1)!$ of the symmetric group are the sub-groups which each keep one symbol fixed; and these form a conjugate set of n sub-groups.

When $n = 7$, a sub-group of order $(n-1)!$ must be intransitive, and therefore the same result holds in this case; this also is true when n is 3, 4, or 5.

Lastly, when $n = 6$, there may, by the foregoing theorem, be primitive sub-groups of order 5!. That such sub-groups actually exist may be verified at once by considering the symmetric group of 5 symbols. This group contains 6 cyclical sub-groups of order 5, and each of them is self-conjugate in a sub-group of order 20. Hence, since the only self-conjugate sub-group contained in the symmetric group of degree 5 is the corresponding alternating group, the symmetric group of degree 5 can be expressed as a doubly transitive group of degree 6. The symmetric group of degree 6 therefore contains a set of 6 conjugate doubly transitive sub-groups of order 5!, which are simply isomorphic with the intransitive sub-groups that each keep one symbol fixed. Finally, if the 12 sub-groups of order 5!, which are thus accounted for, do not exhaust all the sub-groups of this order, any other would have in common with each of the 12 a sub-group of order 20; and therefore the operations of order 3 contained in it would be distinct from those in the previous 12. But these clearly contain all the operations of order 3 of the symmetric group, and therefore there can be no other sub-groups of order 5!. Hence:—

THEOREM II. *The symmetric group of degree* n $(n \neq 6)$ *contains* n *and only* n *sub-groups of order* $(n-1)!$, *which form a single conjugate set. The symmetric group of degree* 6 *contains* 12 *sub-groups of order* 5!, *which are simply isomorphic with one another and form two conjugate sets of* 6 *each.*

143. We shall now discuss certain further limitations on the order of a primitive group of given degree. Though it will be seen that these do not lead to general results, similar to that given by Theorem I, § 141, yet in many special cases they are of considerable assistance in determining the possible existence of groups of given orders and degrees.

We consider first a group G of order N and of prime degree p. If G is not cyclical, it must contain substitutions which keep only one symbol unchanged. For let P be a substitution of G of order p. The only substitutions permutable with P are its own powers (§ 107); and the only substitutions permutable with $\{P\}$ are substitutions which keep one symbol unchanged and are regular in the remaining $p-1$ symbols (§ 112)*. Now if the only substitutions permutable with $\{P\}$ are its own, then $\{P\}$ is one of $\frac{N}{p}$ con-

*It is shewn in § 112 that $\{P\}$ is permutable with a circular substitution of $p-1$ symbols, which leaves one symbol a_1 unchanged. If there are other substitutions which leave a_1 unchanged and are permutable with $\{P\}$, some such substitution will leave two symbols unchanged. This is clearly impossible. Hence the group of order $p(p-1)$ is the greatest group of the p symbols in which $\{P\}$ is self-conjugate.

jugate sub-groups; and these contain $N\left(1-\frac{1}{p}\right)$ substitutions of order p. In this case, G would contain exactly exactly $\frac{N}{p}$ substitutions whose orders are not divisible by p. But this is clearly impossible, since $\frac{N}{p}$ is the order of a sub-group which keeps one symbol unchanged, and there are p such sub-groups. Hence there must be substitutions in G, other than those of $\{P\}$, which are permutable with $\{P\}$; and each of these substitutions keeps one symbol unchanged.

It follows from § 134 that G, if soluble, must contain a self-conjugate sub-group of order p: therefore no group of prime degree p, which contains more than one sub-group of order p, can be soluble.

If $1 + kp$ is the number of sub-groups of order p contained in G, then

$$N = p\frac{p-1}{d}(1+kp),$$

where d is a factor of $p-1$; and a sub-group of order p is transformed into itself by every substitution of a cyclical sub-group of order $\frac{p-1}{d}$. When d is odd, a substitution which generates this cyclical sub-group is an odd substitution; and G then contains a self-conjugate sub-group of order $\frac{1}{2}N$.

If both p and $\frac{1}{2}(p-1)$ are primes, the order of a group of degree p, which contains more than one sub-group of order p, must be divisible by $\frac{1}{2}(p-1)$. For if the order is not divisible by $\frac{1}{2}(p-1)$, the order of the sub-group, within which a sub-group of order p is self-conjugate, must be $2p$. Now the substitutions of order 2 in this sub-group consist of $\frac{1}{2}(p-1)$ transpositions, so that they are odd substitutions. The group must therefore contain a self-conjugate sub-group in which these operations of order 2 do not occur. In such a sub-group, the only operations permutable with those of a sub-group of order p are its own; and we have seen that no such group can exist. The order of the group must therefore, as stated above, be divisible by $\frac{1}{2}(p-1)$.

144. Let G be a primitive group of degree n and order N; and let p be a prime, which is a factor of N but not of either n or $n-1$. Moreover, suppose that n is congruent to ν, (mod. p); ν being less than p. If $n < p^2$, and if p^α is the highest power of p which divides N, the sub-groups of order p^α must be Abelian groups of type $(1, 1, \ldots$ to α units). In fact, such a sub-group must be intransitive, and, since $n < p^2$, the number of symbols in each transitive system of the sub-group must be p. In any case the number of symbols left unchanged by a sub-group of order p^α is of the form $kp + \nu$.

Suppose now that, in a sub-group of order $\frac{N}{n}$ which leaves one symbol unchanged, a sub-group H of order p^α is one of $\frac{N}{p^\alpha mn}$ conjugate sub-groups. Then each of the n sub-groups that keep one symbol unchanged contains $\frac{N}{p^\alpha mn}$ sub-groups of order p^α; and each sub-group of order p^α belongs to $kp + \nu$ sub-groups that keep one symbol unchanged. Hence G contains $\frac{N}{p^\alpha m(kp+\nu)}$ sub-groups of order p^α, and any one of them, say H, is contained self-conjugately in a sub-group I of order $p^\alpha m(kp+\nu)$. This sub-group I must interchange transitively among themselves the $kp+\nu$ symbols left unchanged by H. For let a and b be any two of these symbols; and let S be an operation which changes a into b and transforms H into H'. There must be an operation T which keeps b unchanged and transforms H' into H, since in the sub-group that keeps b unchanged there is only one conjugate set of sub-groups of order p^α. Then ST changes a into b and transforms H into itself; and therefore I contains substitutions which change a into b. Now it may happen that the existence of a sub-group such as I requires that G is either the alternating or the symmetric group. When this is the case, we infer that the order of a primitive group of degree n, other than the alternating or the symmetric group, cannot be divisible by p.

145. As a simple example, we will shew that the order of a group of degree 19 cannot be divisible by 7, unless it contains the alternating group. It follows from Theorem I, Corollary, § 141, that the order of a group of degree 19, which does not contain the alternating group, cannot be divisible by a power of 7 higher than the first, and that if the group contains a substitution of order 7, the substitution must consist of two cycles of 7 symbols each. The sub-group of order 7 must therefore leave 5 symbols unchanged; hence, by § 144, it must be contained self-conjugately in a sub-group whose order is divisible by 5. Now (§ 83) a group of order 35 is necessarily Abelian; so that the group of degree 19 must contain a substitution of order 5 which is permutable with a substitution of order 7. Such a substitution of order 5 must clearly consist of a single cycle, and its presence in a group of degree 19 requires that the latter should contain the alternating group. It follows that, if a group of degree 19 does not contain the alternating group, its order is not divisible by 7.

As a second example, we will determine the possible forms for the order of a group of degree 13, which does not contain the alternating group. It follows, from Theorem I, Corollary, § 141, that the order of such a group must be of the form $2^{\alpha} \cdot 3^{\beta} \cdot 5^{\gamma} \cdot 11^{\delta} \cdot 13$; where α, β, γ, δ do not exceed 9, 4, 1, 1 respectively.

Suppose, first, that γ is unity, if possible. A substitution of order 5 must consist of two cycles of 5 symbols each; and a sub-group of order 5 must therefore be self-conjugate in a sub-group of order 15. There is then a substitution of order 3 which is permutable with a substitution of order 5. Such a substitution must, as in the last example, consist of a single cycle; and its existence would imply that the group contains the alternating group. It follows that, for the group as specified, γ must be zero.

Suppose, next, that δ is unity, if possible. The group is then (Theorem I, § 141) triply transitive; and the order of the sub-group, that keeps two symbols fixed and is transitive in the remaining 11, is $2^{\alpha} \cdot 3^{\beta} \cdot 11$; and this sub-group must contain more than one sub-group of order 11. We have seen in § 143 that no such group can exist. Therefore δ must be zero.

The two smallest numbers of the form $2^m 3^n$ which are congruent to unity, (mod. 13), are 3^3 and $2^4 3^2$; and every number of this form, which is congruent to unity, (mod. 13), can be written $(2^4 3^2)^x 3^{3y}$. Hence the order of every group of order 13, which contains no odd substitution, must be of the form $(2^4 3^2)^x \cdot 3^{3y} \cdot z \cdot 13$, where z is 2, 3 or 6. Since 3^4 is the highest power of 3 that can divide the order of the group, the only admissible values of x and y are (i) $x = 0$, $y = 1$: (ii) $x = 2$, $y = 0$: (iii) $x = 1$, $y = 0$.

Suppose, first, that $x = 0$, $y = 1$. The order of the group is $2 \cdot 3^3 \cdot 13$, $3^4 \cdot 13$, or $2 \cdot 3^4 \cdot 13$. There must be 13 sub-groups of order 3^4 (or 3^3), and since 13 is not congruent to unity, (mod. 9), there must be sub-groups of order 3^3 (or 3^2) common to some two sub-groups of order 3^4 (or 3^3). Such a sub-group must be self-conjugate (Theorem III, § 80). This case therefore cannot occur.

Next, suppose that $x = 2$, $y = 0$. Then z must be 2, and the order of the group is $2^9 \cdot 3^4 \cdot 13$. Now it is easy to verify that $2^9 \cdot 3^4$ is not a possible order either for an intransitive or for an imprimitive group of degree 12. The order of a sub-group of the group of degree 12 which keeps one symbol fixed is $2^7 \cdot 3^3$. This sub-group can have no substitution consisting of a single cycle of 3 symbols, since no such substitution can occur in the original group. Hence it must permute the 11 symbols in two transitive sets of 9 and 2 symbols respectively. It must therefore contain a self-conjugate sub-group of order $2^6 \cdot 3^3$ which keeps 3 of the 12 symbols unchanged; and this sub-group must occur self-conjugately in 3 of the 12 sub-groups which keep one symbol unchanged. This however makes the group of degree 12 imprimitive, contrary to supposition. Hence this case cannot occur.

Finally, then, the only possible values of x and y are $x = 1$, $y = 0$. The order of a group of degree 13, which has more than one sub-group of order 13 and no odd substitutions, is $2^5 \cdot 3^2 \cdot 13$, $2^4 \cdot 3^3 \cdot 13$, or $2^5 \cdot 3^3 \cdot 13$. The order of a group of order 13 with odd substitutions will be twice one of the preceding three numbers.

A further and much more detailed examination would be necessary to determine whether groups of degree 13 correspond to any or all of these orders. We shall see in Chapter XIV that there is a group of degree 13 and order $2^4 \cdot 3^3 \cdot 13$; and M. Jordan* states that this is the only group of degree 13 which contains more than one sub-group of order 13.

Ex. If n (> 3) and $2n+1$ are primes, shew that there is no triply transitive group of degree $2n+3$ which does not contain the alternating group.

146. As a further illustration, and for the actual value of the results themselves, we proceed to determine all types of primitive groups for degrees not exceeding 8.

(i) $n = 3$.

The symmetric group of 3 symbols has a single sub-group, viz. the alternating group. Both these groups are necessarily primitive.

(ii) $n = 4$.

Since a group whose order is the power of a prime cannot (§ 124) be represented in primitive form, a primitive group of degree 4 must contain

* *Comptes Rendus*, t. LXXV (1872), p. 1757.

3 as a factor of its order. Hence the only primitive groups of degree 4 are the symmetric and the alternating groups.

(iii) $n = 5$.

Since 5 is a prime, every group of degree 5 is a primitive group. The symmetric group of degree 5 contains 6 cyclical sub-groups of order 5; therefore, by Sylow's theorem, every group of degree 5 must contain either 1 or 6 sub-groups of order 5. Since the alternating group is simple, every sub-group that contains 6 sub-groups of order 5 must contain the alternating group. Hence, besides the alternating and the symmetric groups, we have only sub-groups which contain a sub-group of order 5 self-conjugately. In such a group, an operation of order 5 can be permutable with its own powers only. Hence (§ 112) the only sub-groups of the type in question other than cyclical sub-groups, are groups of orders 20 and 10. These are defined by

$$\{(12345),\ (2354)\},$$
$$\text{and}\quad \{(12345),\ (25)(34)\}.$$

(iv) $n = 6$.

If the order of a primitive group of degree 6 is not divisible by 5, the order must (§ 141) be equal to or be a factor of $2^3 \cdot 3$. The order of a sub-group that keeps one symbol fixed is equal to or is a factor of 2^2. Hence the sub-group must keep two symbols fixed, and therefore (§ 128) the group cannot be primitive. Hence the order of every primitive group of degree 6 is divisible by 5, and every such group is at least doubly transitive. The symmetric group contains 36 sub-groups of order 5; and hence, since no transitive group of degree 6 can contain a self-conjugate sub-group of order 5, every primitive group of degree 6, which does not contain the alternating group, must have 6 sub-groups of order 5.

If G is such a group, the sub-group of G that keeps one symbol fixed is a transitive group of degree 5 which has a self-conjugate sub-group of order 5. If this transitive group of degree 5 were cyclical, every operation of the doubly transitive group G of order 30 would displace all or all but one of the symbols. Since 6 is not the power of a prime, this is impossible

(§ 105). Hence the sub-group of G which keeps one symbol fixed must be of one of the two types given above; and the order of G must be 120 or 60. Now we have seen, in § 142, that the symmetric group of degree 6 has a single conjugate set of primitive sub-groups of order 120 and a single set of order 60. Hence there is a single type of primitive group of degree 6, corresponding to each of the orders 120 and 60. These are defined by

$$\{(126)(354),\ (12345),\ (2354)\},$$
$$\text{and}\quad \{(126)(354),\ (12345),\ (25)(34)\}:$$

where the last two substitutions in each case generate a sub-group that keeps one symbol unchanged.

(v) $n = 7$.

Every transitive group of degree 7 is primitive; and if it does not contain the alternating group, its order must (§ 141) be equal to or be a factor of $7 \cdot 6 \cdot 5 \cdot 4$. A cyclical sub-group of order 7 must (footnote, p. 223), in a group of degree 7 that contains more than one such sub-group, be self-conjugate in a group of order 21 or 42. Now neither 20 nor 40 is congruent to unity, (mod. 7); and therefore 5 cannot be a factor of the order of such a group. Hence the order of a transitive group of degree 7, that does not contain the alternating group, is equal to or is a factor of $7 \cdot 6 \cdot 4$. But 8 is the only factor of $7 \cdot 6 \cdot 4$ which is congruent to unity, (mod. 7); and therefore, if the group contains more than one sub-group of order 7, its order must be equal to $7 \cdot 6 \cdot 4$ and it must contain 8 sub-groups of order 7.

Such a group must be doubly transitive; for if a sub-group of order 24 interchanges the symbols in two intransitive systems, it is easily shewn that the group would contain substitutions displacing three symbols only, and therefore that it would contain the alternating group. A sub-group of degree 24, transitive in 6 symbols, must contain 4 sub-groups of order 3. For if it contained only one, it would necessarily have circular substitutions of 6 symbols, and the group of order $7 \cdot 6 \cdot 4$ would have a self-conjugate sub-group of order $7 \cdot 6 \cdot 2$; which is not the case. Hence the sub-groups of order 24 must be simply isomorphic with the symmetric group of 4 symbols.

The actual construction of the group is now reduced to a limited number of trials. A group of degree 6, simply isomorphic with the symmetric group of 4 symbols, and containing no odd substitutions, may always be represented in the form

$$\{(234)(567),\ (2763)(45)\};$$

and we have to find a circular substitution of the seven symbols 1, 2, 3, 4, 5, 6, 7 such that the group generated by it shall be permutable with this group. Moreover since, in the required group, every operation of order 3 transforms some operation of order 7 into its square, we may assume without loss of generality that the circular substitution of order 7 contains the sequence ...12... and is transformed into its own square by (234)(567). There are only three circular substitutions satisfying these conditions, viz.

$$\begin{aligned} &(1235476), \\ &(1236457), \\ \text{and}\quad &(1237465). \end{aligned}$$

It appears on trial that the group generated by the first of these is not permutable with the sub-group of order 24, while the groups generated by the other two are. There are therefore just two groups of order $7 \cdot 6 \cdot 4$ which contain the given group of order 24. Now in the symmetric group of 7 symbols, a sub-group of order $7 \cdot 6 \cdot 4$ must, from the foregoing discussion, be one of a set of 30 conjugate sub-groups. These all enter in the alternating group; and therefore, in that group, they must form two sets of 15 conjugate sub-groups each. Each of these contains 7 sub-groups of the type

$$\{(234)(567),\ (2763)(45)\};$$

and the alternating group contains a conjugate set of 105 such such sub-groups. Hence each sub-group of this set will enter in two, and only in two, sub-groups of the alternating group of order $7 \cdot 6 \cdot 4$; and in the symmetric group these two sub-groups are conjugate. Finally then, the sub-groups of order $7 \cdot 6 \cdot 4$ form a single conjugate set in the symmetric group. They are defined by

$$\{(1236457),\ (234)(567),\ (2763)(45)\},$$

the two latter substitutions giving a sub-group that keeps one symbol fixed.

These groups are simple; for since they are expressed as transitive groups of degree 7, there can be no self-conjugate sub-group whose order divides 24, while it is evident that a self-conjugate sub-group that contains an operation of order 7 must coincide with the group itself. Also since there are 8 sub-groups of order 7, these groups can be expressed as doubly transitive groups of eight symbols.

A group of degree 7, which has only one sub-group of order 7, must either be cyclical or be contained in the group of order $7 \cdot 6$ given by § 112. Such groups are defined by

$$\begin{aligned} &\{(1234567),\ (243756)\}, \\ \text{or}\quad &\{(1234567),\ (235)(476)\}, \\ \text{or}\quad &\{(1234567),\ (27)(45)(36)\}. \end{aligned}$$

The simple group of order 168, which here occurs as a transitive group of degree 7, is the only simple group of that order. For, if possible, let there be a simple group G of order 168 and of a distinct type from the above. It certainly cannot be expressed as a group of degree 7; and therefore it must have 21 sub-groups of order 8. If two of these sub-groups have a common sub-group of order 4, it must be contained self-conjugately (§ 80) in a sub-group of order 24 or 56; and this is inconsistent with the suppositions made. If on the other hand, 2 is the order of the greatest sub-group common to two sub-groups of order 8, such a common sub-group of order 2 must, on the suppositions made, be self-conjugate in a sub-group of order 12. But a group of order 12, which has a self-conjugate operation of order 2, must have a self-conjugate sub-group of order 3; and therefore G would only contain 7 sub-groups of order 3, and could be expressed as a group of degree 7; contrary to supposition. No other supposition is possible with regard to the sub-groups of order 8, since 21 is not congruent to unity, (mod. 8). Hence, finally, there is no simple group of order 168 distinct from the group of degree 7.

(vi) $n = 8$.

The order of a primitive group of degree 8, which does not contain the alternating group, cannot (§ 141) be divisible by 5. Suppose, if possible, that the order of such a group is $2^{\alpha+3} \cdot 3$ ($\alpha = 0, 1, 2, 3$). A substitution of order 3 must consist of two cycles; and therefore the sub-group of order $2^\alpha \cdot 3$, which keeps one symbol fixed, must interchange the others in two intransitive systems of 3 and 4 respectively. In this sub-group, a sub-group of order 3 must be one of four conjugate sub-groups, and therefore α is either 2 or 3. Now a group of order $2^5 \cdot 3$ or $2^6 \cdot 3$ is soluble, as is seen at once by considering the sub-groups of order 2^5 or 2^6. Hence a primitive group of order $2^5 \cdot 3$ or $2^6 \cdot 3$ must contain a transitive self-conjugate sub-group of order 8, whose operations are all of order 2.

If 7 is a factor of the order of the group, the group must be doubly transitive; and from the case of $n = 7$, it follows that the possible orders are $8 \cdot 7$, $8 \cdot 7 \cdot 2$, $8 \cdot 7 \cdot 3$, $8 \cdot 7 \cdot 6$, and $8 \cdot 7 \cdot 6 \cdot 4$. Moreover, for the orders $8 \cdot 7 \cdot 2$ and $8 \cdot 7 \cdot 6$, the group contains odd substitutions and therefore it contains self-conjugate sub-groups of order $8 \cdot 7$ and $8 \cdot 7 \cdot 3$ respectively.

A simple group of order $8 \cdot 7 \cdot 3$ is necessarily identical in type with the group of this order determined on p. 231; and a group of order $8 \cdot 7 \cdot 3$, which is not simple, is certainly soluble. Hence a composite group of order $8 \cdot 7 \cdot 3$, and a group of order $8 \cdot 7 \cdot 6$ which does not contain a simple sub-group of order $8 \cdot 7 \cdot 3$, must both, if expressible as primitive groups of degree 8, contain transitive self-conjugate sub-groups of order 8 whose operations are all of order 2. With the possible exception then of groups of order $8 \cdot 7 \cdot 6 \cdot 4$, the only primitive groups of degree 8, which do not contain a self-conjugate sub-group of order 8, are the simple group of order $8 \cdot 7 \cdot 3$ and any group of $8 \cdot 7 \cdot 6$ which contains this self-conjugately. We have seen that the simple group of order $8 \cdot 7 \cdot 3$ contains a single set of 8 conjugate sub-groups of order 21, and therefore it can be expressed in one form only as a group of degree 8. A group of degree 8 and order $8 \cdot 7 \cdot 6$, which contains this self-conjugately, can occur only in one form, if at all; for, if it exists, it must be triply transitive, and it must be given by combining the simple group with any operation of order 2 which transforms one of its operations of order 7 into its own inverse. That such a group does exist has been shewn in § 113. These two groups are actually given by

$$\{(15642378),\ (1234567),\ (243756)\},$$
$$\text{and}\quad \{(1627)(5438),\ (1234567),\ (235)(476)\};$$

where in each case the last two substitutions give a sub-group that keeps one symbol fixed.

To avoid unnecessary prolixity we shall, in dealing with the primitive groups of degree 8, which contain a transitive self-conjugate sub-group of order 8 whose operations are all of order 2, anticipate some of the results of the next Chapter. It will there be shewn that, if G is an Abelian group of order N, and L a group of isomorphisms (§ 156) of G, a group of degree N may be formed which has a transitive self-conjugate sub-group simply isomorphic with G, while at the same time the sub-group that keeps one symbol fixed is isomorphic with L (§ 158). The group of isomorphisms of a group of order 8, whose operations are all of order 2, will be shewn in Chapter XIV to be identical with the simple group of order 168. This group has a single set of conjugate sub-groups of each of the orders 7 and 21, but no sub-group of order 14 or 42. When expressed as a group of degree 7, it has a single set of conjugate sub-groups of order 12 (or 24) which leave no symbols unchanged. There are therefore primitive groups of degree 8 containing transitive self-conjugate sub-groups of order 8 corresponding to each of the orders $8 \cdot 7$, $8 \cdot 7 \cdot 3$, $2^5 \cdot 3$, $2^6 \cdot 3$, and $8 \cdot 7 \cdot 6 \cdot 4$; and in each case there is a single type of such group.

It remains to determine whether there can be any type of group, of degree 8 and order $8 \cdot 7 \cdot 6 \cdot 4$, other than that just obtained. Such a group must be one of 15 conjugate sub-groups in the alternating group of degree 8, and can therefore itself be expressed as a group of degree 14. Since it certainly cannot be expressed as a group of degree 7, the group of degree 14 must be transitive. The order of the sub-group, in this form, that keeps one symbol fixed is $2^5 \cdot 3$. If this keeps only one symbol unchanged, it must interchange the remaining symbols in four intransitive systems of 3, 3, 3 and 4 respectively, since a substitution of order 3 must clearly consist of 4 cycles. A group of order $2^5 \cdot 3$ cannot however be so expressed; and therefore the sub-group that keeps one symbol fixed must keep two fixed. The group of degree 14 is therefore imprimitive, and the group must contain a sub-group of order $2^6 \cdot 3$. Moreover, since the group cannot be expressed as a transitive group of degree 7, this sub-group of order $2^6 \cdot 3$ must contain (§ 123) a sub-group which is self-conjugate in the group itself. The order of this sub-group must be a power of 2; since the group is primitive, it cannot be less than 2^3. On the other hand, the order cannot be greater than 2^3

since the group contains a simple sub-group of order $7 \cdot 6 \cdot 4$. Hence finally, there is no type of primitive group of degree 8 and order $8 \cdot 7 \cdot 6 \cdot 4$ other than that already obtained.

There is no difficulty now in actually constructing the primitive groups of degree 8 which have a self-conjugate sub-group of order 8. They are all contained in the group of order $8 \cdot 7 \cdot 6 \cdot 4$; and this may be constructed from a sub-group keeping one symbol fixed, which is given on p. 231, by the method of § 109. It will thus be found that the group in question is given by

$$\{(81)(26)(37)(45),\ (1236457),\ (234)(567),\ (2763)(45)\};$$

while the groups of orders $8 \cdot 7 \cdot 3$ and $8 \cdot 7$ are given by omitting respectively the last and the two last of the four generating operations.

The construction of the two remaining groups, of order $2^5 \cdot 3$ and $2^6 \cdot 3$, is left as an exercise for the reader.

It may be noticed that it has been shewn incidentally, in discussing above the possibility of a second type of group of degree 8 and order $8 \cdot 7 \cdot 6 \cdot 4$, that the alternating group of degree 8 can be expressed as a doubly transitive group of degree 15.

It may similarly be shewn that the alternating group of degree 7 can be expressed as a doubly transitive group of degree 15, and the alternating group of degree 6 as a simply transitive and primitive group of degree 15.

147. We have seen in § 105 that a doubly transitive group, of degree n and order $n(n-1)$, can exist only when n is the power of a prime. For such a group, the identical operation is the only one which keeps more than one symbol unchanged. We shall now go on to consider the sub-groups of a doubly transitive group, of degree n and order $n(n-1)m$, which keep two symbols fixed. The order of any such sub-group is m; since the group contains operations changing any two symbols into any other two, the sub-groups which keep two symbols fixed must form a single conjugate set.

Suppose first that the sub-group, which keeps two symbols unchanged, displaces all the other symbols. The sub-group that keeps a and b unchanged cannot then be identical with that which keeps c and d unchanged, unless the symbols c and d are the same pair as a and b. Since there are

$\frac{1}{2}n(n-1)$ pairs of n symbols, the conjugate set contains $\frac{1}{2}n(n-1)$ sub-groups; and each sub-group of order m keeping two symbols fixed must be self-conjugate in a sub-group of order $2m$, which consists of the operations of the sub-group of order m and of those operations interchanging the two symbols that the sub-group of order m keeps fixed.

Suppose next that all the operations of a sub-group H, which keeps two symbols fixed, keep x symbols fixed, while none of the remaining $n-x$ symbols are unchanged by all the operations of H. From x symbols $\frac{1}{2}x(x-1)$ pairs can be formed, and therefore the sub-group that keeps one pair unchanged must keep $\frac{1}{2}x(x-1)$ pairs unchanged. In this case, the conjugate set contains $\dfrac{n(n-1)}{x(x-1)}$ distinct sub-groups of order m, and H is therefore self-conjugate in a group K of order $x(x-1)m$. The operations of this sub-group which do not belong to H interchange among themselves the x symbols that are left unchanged by H. Now since the group itself is doubly transitive, there must be operations which change any two of these x symbols into any other two; and any such operation being permutable with H must belong to K. Hence if we consider the effect of K on the x symbols only which are left unchanged by H, K reduces to a doubly transitive group of degree x and order $x(x-1)$. It follows that x must be a prime or the power of a prime.

148. The preceding paragraph suggests the combinatorial problem of forming from n distinct symbols $\dfrac{n(n-1)}{x(x-1)}$ sets of x symbols, such that every pair of symbols occurs in one set of x and no pair occurs in more than one.

There is one class of cases in which a solution of this problem is given immediately by the theory of Abelian groups. Let G be an Abelian group of order p^m, where p is a prime, and type $(1, 1, \ldots, \text{to } m \text{ units})$. We have seen, in § 49, that G has $\dfrac{p^m-1}{p-1}$ sub-groups of order p, and $\dfrac{p^m-1\cdot p^{m-1}-1}{p-1\cdot p^2-1}$ sub-groups of order p^2. Now any pair of sub-groups of order p generates a sub-group of order p^2, and therefore every pair of sub-groups of order p occurs in one and only one sub-group of order p^2. Moreover, every sub-group of order p^2 contains $p+1$ sub-groups of order p. When p is a prime and m any integer, it is therefore always possible to form from $\frac{p^m - 1}{p - 1}$ symbols $\frac{p^m - 1 \cdot p^{m-1} - 1}{p - 1 \cdot p^2 - 1}$ sets of $p + 1$ symbols each, such that every pair of the symbols occurs in one set of $p + 1$ and no pair occurs in more than one set.

Supposing that, for given values of n and x, such a distribution is possible, it is still of course an open question as to whether there is a doubly transitive substitution group of the n symbols, such that every substitution which keeps any two symbols unchanged keeps also unchanged the whole set of x in which they occur. When x is greater than 3, the question as to the existence of such groups is one which still remains to be investigated. There is however an important class of groups, to be considered later (Chapter XIV), that possess a closely analogous property. These groups are doubly transitive; and from the n symbols upon which they operate, we can form $\frac{n(n-1)}{x(x-1)}$ sets of x, that are interchanged transitively by the substitutions of the group: the sets being such that every pair occurs in one set and no pair in more than one set.

If $n(n-1)m$ is the order of such a group, and if H is a sub-group of order m which keeps a given pair fixed, then H must interchange among themselves the remaining $x-2$ symbols of that set of x which contains the pair kept unchanged by H. H contains, as a self-conjugate sub-group, the group h which leaves every symbol of the set of x unchanged; and if m'' is the order of this sub-group, while $m = m'm''$, then m' is the order of the group to which H reduces when we consider its effect only on the $x - 2$ symbols. Now h is self-conjugate in the group K that interchanges all the symbols of the set of x among themselves. But since the original group is doubly transitive, it must contain substitutions which change any two of the set of x into any other two, and every such substitution must belong to K. Hence K must be doubly transitive in the x symbols, and therefore finally the order of the group, to which K reduces when we consider its effect on the x symbols only, is $x(x-1)m'$. Since the order of h, which keeps unchanged each of the x symbols is m'', the order of K is $x(x-1)m$.

149. When $x = 3$, n must be of the form $6k + 1$ or $6k + 3$, since otherwise $\frac{n(n-1)}{x(x-1)}$ would not be an integer. The substitutions of a doubly transitive group of degree n, which possesses a complete set of $\frac{1}{6}n(n-1)$ triplets, must be such that every substitution which leaves two given symbols unchanged also leaves a third definite symbol unchanged.

The smallest possible value of n is 7; and the group of Ex. 2, § 35, which is one of the groups obtained in § 146, satisfies all the conditions. In fact, the group is clearly a doubly transitive group of degree 7; and since its order is $7 \cdot 6 \cdot 4$, the order of the sub-group which keeps two symbols fixed is 4. Now in the sub-group

$$\{(267)(345),\ (23)(47)\},$$

which keeps one symbol fixed, the only operations that keep 1 and 2 unchanged are

$$(35)(67),\ (36)(57),\ (37)(56).$$

These with identity form a sub-group of order 4, which must therefore be the sub-group that keeps three symbols fixed. The complete set of triplets in this case is

$$124,\ 137,\ 156,\ 235,\ 267,\ 346,\ 457.$$

The next smallest value of n is 9, and in this case again, a group with the required properties exists.

Ex. Shew that the group

$$\{(26973854),\ (456)(798)\}$$

is an imprimitive group of order 48, each imprimitive system containing two symbols; and that the sub-group, which keeps the symbols of one imprimitive system unchanged, is isomorphic with the symmetric group of three symbols. Prove that this group is permutable with

$$\{(123)(456)(789),\ (147)(258)(369)\},$$

and thence that

$$\{(123)(456)(789),\ (26973854),\ (456)(798)\},$$

is a doubly transitive group of degree 9, which possesses a complete set of 12 triplets.

The reader is not to infer from the examples given that, when n is of the form $6k+1$ or $6k+3$, there is always a doubly transitive group of degree n which possesses a complete set of triplets. It is a good exercise to verify that there is no such group when n is 13.

The case $n = 13$, $x = 4$ is the simplest case that can occur of the division of n symbols into sets of x in the manner of § 148 when x is greater than 3. We shall see in Chapter XIV that there is a doubly transitive group of degree 13 such that from the 13 symbols permuted by the group a complete set of 13 quartets can be formed, which are themselves permuted by the operations of the group. Of the operations forming a sub-group that keeps two given symbols fixed, half will keep fixed the two other symbols, which form a quartet with the two given symbols, and half will permute them.

On the question of the independent formation of a complete set of triplets of n symbols, and in certain cases of the group of degree n which interchanges the triplets among themselves, reference may be made to the memoirs mentioned in the subjoined footnote*.

150. We shall conclude the present Chapter with some applications of substitution groups, which enable us to complete and extend certain earlier results.

We have seen in § 107 that the substitutions of n symbols, which are permutable with each of the substitutions of a regular substitution group G of order n of the same n symbols, form another regular substitution group of order n; and that, if G is Abelian, the latter group coincides with G. Hence the only substitutions of n symbols, which are permutable with a circular substitution of the n symbols, are the powers of the circular substitution.

Let now S be a regular substitution of order m, in mn symbols. It must permute the symbols in n cycles of m symbols each; and so we may take

$$S = (a_{11}a_{12}\dots a_{1m})(a_{21}a_{22}\dots a_{2m})\dots(a_{n1}a_{n2}\dots a_{nm}).$$

If T is permutable with S, and if it changes a_{rp} into a_{rq}, it clearly must permute the m symbols

$$a_{r1},\ a_{r2},\ \dots,\ a_{rm}$$

among themselves; and therefore, so far as regards its effect on these m symbols, T must be a power of

$$(a_{r1}a_{r2}\dots a_{rm}).$$

*Netto: "Substitutionentheorie," pp. 220–235; "Zur Theorie der Tripelsysteme," *Math. Ann.* Vol. XLII, (1892), pp. 143–152. Moore: "Concerning triple systems," *Math. Ann.* Vol. XLIII, (1893), pp. 271–285. Heffter: "Ueber Tripelaysteme," *Math. Ann.*, Vol. XLIX, (1897), pp. 101–112.

Again, if T changes a_{rp} into a_{sq}, it must change the set

$$a_{r1},\ a_{r2},\ \dots,\ a_{rm},$$

into the set

$$a_{s1},\ a_{s2},\ \dots,\ a_{sm};$$

as otherwise it would not be permutable with S.

Now the totality of the substitutions of the mn symbols, which are permutable with S, form a group G_S. This group must, from the properties of T just stated, be imprimitive, interchanging the symbols in n imprimitive systems of m symbols each; and the symbols in any cycle of S will form an imprimitive system. Moreover, the self-conjugate sub-group H_S of this group, which permutes the symbols of each system among themselves, is the group of order m^n generated by

$$(a_{11}a_{12}\dots a_{1m}),\ (a_{21}a_{22}\dots a_{2m}),\ \dots,\ (a_{n1}a_{n2}\dots a_{nm}).$$

In fact, every substitution of this group is clearly permutable with S; and conversely, every substitution of the mn symbols, which does not permute the systems, must belong to this group.

Now $\dfrac{G_S}{H_S}$ is capable of representation as a group of degree n, for none of its operations changes every one of the n systems into itself. Hence $n!$ is the greatest possible order of $\dfrac{G_S}{H_S}$. On the other hand, every operation of the group, generated by

$$(a_{11}a_{21}\dots a_{n1})(a_{12}a_{22}\dots a_{n2})\dots(a_{1m}a_{2m}\dots a_{nm})$$

and

$$(a_{11}a_{21})(a_{12}a_{22})\dots(a_{1m}a_{2m}),$$

is clearly permutable with S; and this group, being simply isomorphic with the group

$$\{(a_1a_2\dots a_n),\ (a_1a_2)\},$$

i.e. with the symmetric group of n symbols, is of order $n!$.

Hence, finally, the order of G_S is $m^n \cdot n!$; and G_S is generated by

$$(a_{11}a_{12}\dots a_{n1})(a_{12}a_{22}\dots a_{n2})\dots(a_{1m}a_{2m}\dots a_{nm}),$$
$$(a_{11}a_{21})(a_{12}a_{22})\dots(a_{1m}a_{2m}),$$

and

$$(a_{11}a_{12}\dots a_{1m}).$$

151. Let h_r be a regular substitution group of order m in the m symbols

$$a_{r1},\ a_{r2},\ \dots,\ a_{rm},$$

and let S_{rt} be one of its substitutions. Then if for r we write in turn 1, 2, …, n, and if for each value of t from 1 to m we form the substitution

$$S_{1t}S_{2t}\dots S_{nt},$$

the set of m substitutions so formed constitute an intransitive group H in the mn symbols, simply isomorphic with h_r.

The method of § 150 can be applied directly to determine the group G_H of degree mn, each of whose substitutions are permutable with every substitution of H. The order of this group is $m^n \cdot n!$; and it can be generated by

$$(a_{11}a_{21}\dots a_{n1})(a_{12}a_{22}\dots a_{n2})\dots(a_{1m}a_{2m}\dots a_{nm}),$$
$$(a_{11}a_{21})(a_{12}a_{22})\dots(a_{1m}a_{2m}),$$

and

$$h_1';$$

where h_1' is the regular group in the symbols

$$a_{11},\ a_{12},\ \dots,\ a_{1m},$$

each of whose substitutions is permutable with every substitution of h_1.

This group will contain H if, and only if, H is an Abelian group. Moreover, the only self-conjugate substitutions of G_H are the substitutions of H contained in it. For if G_H contained other self-conjugate substitutions S_1,

$S_2, \ldots$, every operation of G_H would be permutable with every operation of the group $\{H, S_1, S_2, \ldots\}$. Now G_H is transitive, so that $S_1, S_2, \ldots$ must displace all the symbols; and therefore $\{H, S_1, S_2, \ldots\}$ has all its substitutions regular in the mn symbols. If its order is mn_1 where $n = n_1 n_2$, the order of the group formed of all the substitutions of mn symbols, which are permutable with each of its operations, is $(mn_1)^{n_2} \cdot n_2!$; and this number is less than $m^n \cdot n!$. Thus the supposition, that G_H has self-conjugate operations other than the operations of H which it contains, leads to an impossibility.

By means of this and the preceding section, the reader will have no difficulty in forming the group of n symbols, which is permutable with every operation of any given group in the n symbols.

152. If a group, whose order is a power of a prime p, be expressed as a transitive substitution group, its degree must also be a power of p (§ 123). Moreover such a group, since it has self-conjugate operations, must necessarily be imprimitive.

The greatest value of m, for which a group of order p^m can be expressed as a transitive group of degree p^n, where n is regarded as given, is determined at once by considering the symmetric group of degree p^n. The highest power of p that divides $p^n!$ is p^ν, where

$$\nu = p^{n-1} + p^{n-2} + \ldots + p + 1.$$

Hence the symmetric group of degree p^n contains a set of conjugate sub-groups of order p^ν and it contains no groups whose order is a higher power of p. Also, these groups are transitive in the p^n symbols; for any one of them must contain a circular substitution of order p^n. There are therefore groups of order p^ν which can be expressed as transitive groups of degree p^n; but no group of order $p^{\nu'}$ $(\nu' > \nu)$ can be so expressed. Moreover, in order that a group of order p^m may be capable of representation as a transitive group of degree p^n, it must be simply isomorphic with a transitive sub-group of the above substitution group of order p^ν.

This group may be constructed synthetically as follows. Since the sub-group of order $p^{\nu-n}$, that leaves one symbol unchanged, is contained in a sub-group of order $p^{\nu-n+1}$ (§ 55), there must be imprimitive systems containing p symbols each. If then we distribute the p^n symbols into p^{n-1} sets of p each, and with each set of p form a circular substitution, the p^{n-1} permutable and independent circular substitutions will generate an intransitive group of order $p^{p^{n-1}}$. It will be the self-conjugate sub-group of the group of order p^{ν}, which permutes the symbols of each system among themselves.

Again, the systems of p symbols each may be taken in sets of p to form systems of p^2 symbols each, since the previously considered sub-group of order $p^{n-\nu+1}$ must be contained in a sub-group of order $p^{n-\nu+2}$. Hence we may form systems of p^2 symbols by combining the previous systems in sets of p; and then with each p^2 symbols we can form a circular substitution, whose pth power is the product of the p circular substitutions of order p, which have been previously formed from the p^2 symbols. The symbols of any set of p^2 will then be interchanged by a transitive group of order p^{p+1}; and since there are p^{n-2} such sets, we obtain in this way an intransitive group of $p^{p^{n-1}+p^{n-2}}$. The group thus formed is that self-conjugate sub-group of the original group, which interchanges among themselves the symbols of each system of p^2. This process may be continued, taking greater and greater systems, till at the last step we combine the p systems of p^{n-1} symbols each into a single system by means of a circular substitution of order p^n. The order of the resulting group is clearly p^{ν}, as it should be.

The self-conjugate operations of this group form a sub-group of order p.

For suppose, if possible, they form a sub-group of order p^r. Every operation of this sub-group displaces all the symbols; and therefore, when expressed as a substitution group in the p^n symbols, it must interchange them transitively in p^{n-r} sets of p^r each.

Now (§ 151) those substitutions of the p^n symbols, which are permutable with every operation of this sub-group, form a group of order $p^{rp^{n-r}} \cdot p^{n-r}!$; this number is only divisible by p^{ν}, as it must be, when $r = 1$.

Ex. Shew that, for the group of degree p^2 and order p^{p+1}, the factor groups $\dfrac{H_{r+1}}{H_r}$ (of § 53) are all of type (1) except the last, which is of type $(1, 1)$.

The fact that ν is a function of p when n is given, explains why, in classifying all groups of order p^n, some of the lower primes may behave in an exceptional manner. Thus we saw, in § 73, that for certain groups of order p^4 it was necessary to consider separately the case $p = 3$. The present article makes it clear that, while there may be more than one type of group of order p^4 $(p > 3)$, which can be expressed as a transitive group of degree p^2, there is only a single type of group of order 3^4 which can be expressed transitively in 9 symbols.

153. In the memoirs referred to in the footnote on p. 175, M. Mathieu has demonstrated the existence of a remarkable group, of degree 12 and order $12 \cdot 11 \cdot 10 \cdot 9 \cdot 8$, which is quintuply transitive. The generating operations of this group have been given in the note to § 108, at the end of Chapter VIII. The verification of some of the more important properties of this group, as stated in the succeeding example, forms a good exercise on the results of this and the two preceding Chapters.

Ex. Shew that the substitutions

$$\begin{array}{ll} (1254)(3867), & (1758)(2643), \\ (12)(48)(57)(69), & (a2)(58)(46)(79), \\ (ab)(57)(68)(49), & (bc)(47)(58)(69), \end{array}$$

generate a quintuply transitive group of degree 12 and order

$$12 \cdot 11 \cdot 10 \cdot 9 \cdot 8.$$

Prove that this group is simple; that a sub-group of degree 11 and order $11 \cdot 10 \cdot 9 \cdot 8$, which leaves one symbol unchanged, is a simple group; and that a sub-group of degree 10 and order $10 \cdot 9 \cdot 8$, which leaves two symbols unchanged, contains a self-conjugate sub-group simply isomorphic with the alternating group of degree 6.

Shew also that the group of degree 12 contains (i) 1728 sub-groups of order 11 each of which is self-conjugate in a group of order 55: (ii) 2376 sub-groups of order 5, each of which is self-conjugate in a group of order 40: (iii) 880 sub-groups of order 27, each of which is self-conjugate in a group of order 108: (iv) 1485 sub-groups of order 64.

Prove further that the group is a maximum sub-group of the alternating group of degree 12.

CHAPTER XI.

ON THE ISOMORPHISM OF A GROUP WITH ITSELF.

154. IT is shewn in § 24 that, if all the operations of a group are transformed by one of themselves, which is not self-conjugate, a correspondence is thereby established among the operations of the group which exhibits the group as simply isomorphic with itself.

In an Abelian group every operation is self-conjugate, and the only correspondence established in the manner indicated is that in which every operation corresponds to itself. If however in an Abelian group we take, as the operation which corresponds to any given operation S, its inverse S^{-1}, then to ST will correspond $S^{-1}T^{-1}$ or $(ST)^{-1}$; and the correspondence exhibits the group as simply isomorphic with itself. For this particular correspondence, a group of order 2 is the only one in which each operation corresponds to itself.

It is therefore possible for every group, except a group of order 2, to establish a correspondence between the operations of the group, which shall exhibit the group as simply isomorphic with itself. Moreover, we shall see that in general there are such correspondences which cannot be established by either of the processes above given. We devote the present Chapter to a discussion of the isomorphism of a group with itself. It will be seen that, for many problems of group-theory, and in particular for the determination of the various types of group which are possible when the factor-groups of the composition-series are given, this discussion is most important.

155. Definition. A correspondence between the operations of a group, such that to every operation S there corresponds a single operation S', while to the product ST of two operations there corresponds the product $S'T'$ of the corresponding operations, is said to define an *isomorphism of the group with itself.*

That isomorphism in which each operation corresponds to itself is called the *identical isomorphism.* In every isomorphism of a group with itself, the identical operation corresponds to itself; and the orders of two corresponding operations are the same. For if 1 and S were corresponding operations, so also would be $1 \cdot 1$ and S^2; and therefore more than one operation would correspond to 1. Again, if S and S', of orders n and n', are corresponding operations, so also are S^n and S'^n; and therefore n must be a multiple of n'. Similarly n' must be a multiple of n; and therefore n and n' are equal.

If the operations of a group of order N are represented by

$$1,\ S_1,\ S_2,\ \ldots,\ S_{N-1},$$

and if, for a given isomorphism of the group with itself, S'_r is the operation that corresponds to S_r $(r = 1, 2, \ldots, N-1)$, the isomorphism will be completely represented by the symbol

$$\begin{bmatrix} 1, S_1, S_2, \ldots, S_{N-1} \\ 1, S'_1, S'_2, \ldots, S'_{N-1} \end{bmatrix}.$$

In this symbol, two operations in the same vertical line are corresponding operations. When no risk of confusion is thereby introduced, the simpler symbol

$$\begin{bmatrix} S \\ S' \end{bmatrix}$$

is used.

156. An isomorphism of a group with itself, thus defined, is not an operation. The symbol of an isomorphism however defines an operation. It may, in fact, be regarded as a substitution performed upon the N letters which represent the operations of the group. Corresponding to every isomorphism there is thus a definite operation; and it is obvious that the operations, which correspond to two distinct isomorphisms, are themselves distinct. The totality of these operations form a group. For let

$$\begin{bmatrix} S \\ S' \end{bmatrix} \quad \text{and} \quad \begin{bmatrix} S' \\ S'' \end{bmatrix}$$

be any two isomorphisms of the group with itself. Then if, as hitherto, we use curved brackets to denote a substitution, we have

$$\begin{pmatrix} S \\ S' \end{pmatrix}\begin{pmatrix} S' \\ S'' \end{pmatrix} = \begin{pmatrix} S \\ S'' \end{pmatrix}.$$

But since $\begin{bmatrix} S \\ S' \end{bmatrix}$ is an isomorphism, the relation

$$S_p S_q = S_r$$

requires that

$$S'_p S'_q = S'_r.$$

And since $\begin{bmatrix} S' \\ S'' \end{bmatrix}$ is an isomorphism, the relation

$$S'_p S'_q = S'_r$$

requires that

$$S''_p S''_q = S''_r.$$

Hence if

$$S_p S_q = S_r,$$

then

$$S''_p S''_q = S''_r;$$

and therefore $\begin{bmatrix} S \\ S'' \end{bmatrix}$ is an isomorphism.

The product of the substitutions which correspond to two isomorphisms is therefore the substitution which corresponds to some other isomorphism.

The set of substitutions which correspond to the isomorphisms of a given group with itself, therefore form a group.

Definition. A group, which is simply isomorphic with the group thus derived from a given group, is called *the group of isomorphisms* of the given group.

It is not, of course, necessary always to regard this group as a group of substitutions performed on the symbols of the operations of the given group. But however the group of isomorphisms may be represented, each one of its operations corresponds to a definite isomorphism of the given group. To avoid an unnecessarily cumbrous phrase, we may briefly apply the term "isomorphism" to the operations of the group of isomorphisms.

So long, at all events, as we are dealing with the properties of a group of isomorphisms, no risk of confusion is thereby introduced. Thus we shall use the phrase "the isomorphism $\begin{pmatrix} S \\ S' \end{pmatrix}$" as equivalent to "the operation of the group of isomorphisms which corresponds to the isomorphism $\begin{bmatrix} S \\ S' \end{bmatrix}$."

157. If Σ is some operation of a group G, while for S each operation of the group is put in turn, the symbol

$$\begin{pmatrix} S \\ \Sigma^{-1} S \Sigma \end{pmatrix}$$

defines an isomorphism of the group. For if

$$S_p S_q = S_r,$$

then

$$\Sigma^{-1} S_p \Sigma \cdot \Sigma^{-1} S_q \Sigma = \Sigma^{-1} S_p S_q \Sigma = \Sigma^{-1} S_r \Sigma;$$

and $\Sigma_{-1} S_r \Sigma$ is an operation of the group. An isomorphism of a group, which is thus formed on transforming the operations of the group by one of themselves, is called a *cogredient* isomorphism. All others are called *contragredient* isomorphisms*. If $\begin{pmatrix} S \\ S' \end{pmatrix}$ is any contragredient isomorphism, the isomorphisms

$$\begin{pmatrix} S \\ S' \end{pmatrix} \begin{pmatrix} S \\ \Sigma^{-1} S \Sigma \end{pmatrix},$$

when for Σ each operation of the group is taken successively, are said to form a *class* of contragredient isomorphisms.

THEOREM I. *The totality of the cogredient isomorphisms of a group G form a group isomorphic with G; this group is a self-conjugate sub-group of the group of isomorphisms of G*†.

---

*Klein, "Vorlesungen über das Ikosaeder und die Auflösung der Gleichungen vom fünften Grade" (1884), p. 232. Hölder, *Math. Ann.*, Vol. XLIII (1893), p. 314.

†Hölder, "Bildung zusammengesetzter Gruppen," *Math. Ann.*, Vol. XLVI (1895), p. 326.

The product of the isomorphisms

$$\begin{pmatrix} S \\ \Sigma^{-1}S\Sigma \end{pmatrix} \text{ and } \begin{pmatrix} S \\ \Sigma'^{-1}S\Sigma' \end{pmatrix}$$

is given by

$$\begin{aligned}\begin{pmatrix} S \\ \Sigma^{-1}S\Sigma \end{pmatrix}\begin{pmatrix} S \\ \Sigma'^{-1}S\Sigma' \end{pmatrix} &= \begin{pmatrix} S \\ \Sigma^{-1}S\Sigma \end{pmatrix}\begin{pmatrix} \Sigma^{-1}S\Sigma \\ \Sigma'^{-1}\Sigma^{-1}S\Sigma\Sigma' \end{pmatrix} \\ &= \begin{pmatrix} S \\ \Sigma'^{-1}\Sigma^{-1}S\Sigma\Sigma' \end{pmatrix} \\ &= \begin{pmatrix} S \\ \Sigma''^{-1}S\Sigma'' \end{pmatrix},\end{aligned}$$

where

$$\Sigma\Sigma' = \Sigma''.$$

The product of two cogredient isomorphisms is therefore another cogredient isomorphism; hence the cogredient isomorphisms form a group. Moreover, if we take the isomorphism

$$\begin{pmatrix} S \\ \Sigma^{-1}S\Sigma \end{pmatrix}$$

as corresponding to the operation Σ of the group G, then to every operation of G there will correspond a definite cogredient isomorphism, so that to the product of any two operations of G there corresponds the product of the two corresponding isomorphisms. The group G and its group of cogredient isomorphisms are therefore isomorphic. If G contains no self-conjugate operation, identity excepted, no two isomorphisms corresponding to different operations of G can be identical; and therefore, in this case, G is simply isomorphic with its group of cogredient isomorphisms. If however G contains self-conjugate operations, forming a self-conjugate sub-group H, then to every operation of H there corresponds the identical isomorphism; and the group of cogredient isomorphisms is simply isomorphic with $\frac{G}{H}$.

Let now

$$\begin{pmatrix} S \\ S' \end{pmatrix}$$

be any isomorphism. Then

$$\begin{aligned}\begin{pmatrix} S \\ S' \end{pmatrix}^{-1} \begin{pmatrix} S \\ \Sigma^{-1} S \Sigma \end{pmatrix} \begin{pmatrix} S \\ S' \end{pmatrix} &= \begin{pmatrix} S' \\ S \end{pmatrix} \begin{pmatrix} S \\ \Sigma^{-1} S \Sigma \end{pmatrix} \begin{pmatrix} S \\ S' \end{pmatrix} \\ &= \begin{pmatrix} S' \\ \Sigma^{-1} S \Sigma \end{pmatrix} \begin{pmatrix} \Sigma^{-1} S \Sigma \\ \Sigma'^{-1} S' \Sigma' \end{pmatrix} \\ &= \begin{pmatrix} S' \\ \Sigma'^{-1} S' \Sigma' \end{pmatrix}.\end{aligned}$$

The isomorphism $\begin{pmatrix} S \\ S' \end{pmatrix}$ therefore transforms every cogredient isomorphism into another cogredient isomorphism. It follows that the group of cogredient isomorphisms is self-conjugate within the group of isomorphisms.

158. Let G be a group of order N, whose operations are

$$1,\ S_1,\ S_2,\ \ldots,\ S_{N-1};$$

and let L be the group of isomorphisms of G. We have seen in § 20 that G may be represented as a transitive group of substitutions performed on the N symbols

$$1,\ S_1,\ S_2,\ \ldots,\ S_{N-1};$$

and that, when it is so represented, the substitution which corresponds to the operation S_x is

$$\begin{pmatrix} 1\ , & S_1\ , & S_2\ , \ldots, & S_{N-1} \\ S_x, & S_1 S_x, & S_2 S_x, \ldots, & S_{N-1} S_x \end{pmatrix},$$

or more shortly

$$\begin{pmatrix} S \\ S S_x \end{pmatrix}.$$

When G is thus represented, we will denote it by G'. We have already seen that L can be represented as an intransitive substitution group of the same N symbols; a typical substitution of L, when it is so represented, is

$$\begin{pmatrix} 1, S_1, S_2, \ldots, S_{N-1} \\ 1, S'_1, S'_2, \ldots, S'_{N-1} \end{pmatrix},$$

or more shortly

$$\begin{pmatrix} S \\ S' \end{pmatrix}.$$

When L is thus represented, we will denote it by L'. It is clear that the two substitution groups G' and L' have no substitution in common except identity. For every substitution of L' leaves the symbol 1 unchanged; and no substitution of G', except identity, leaves 1 unchanged.

Now

$$\begin{aligned}\begin{pmatrix} S \\ S' \end{pmatrix}^{-1}\begin{pmatrix} S \\ SS_x \end{pmatrix}\begin{pmatrix} S \\ S' \end{pmatrix} &= \begin{pmatrix} S' \\ SS_x \end{pmatrix}\begin{pmatrix} SS_x \\ S'S'_x \end{pmatrix}\\ &= \begin{pmatrix} S' \\ S'S'_x \end{pmatrix}\\ &= \begin{pmatrix} S \\ SS'_x \end{pmatrix}.\end{aligned}$$

Every operation of L' is therefore permutable with G'. Hence if M is the order of L, the group $\{G', L'\}$, which we will call K', is a transitive group of degree N and order NM, containing G' self-conjugately. Further $\begin{pmatrix} S \\ S' \end{pmatrix}$ transforms $\begin{pmatrix} S \\ SS_x \end{pmatrix}$ into $\begin{pmatrix} S \\ SS'_x \end{pmatrix}$; and these two substitutions of G' correspond to the operations S_x and S'_x of G. Hence the isomorphism, established on transforming the substitutions of G' by any substitution $\begin{pmatrix} S \\ S' \end{pmatrix}$ of L', is the isomorphism denoted by the symbol $\begin{pmatrix} S \\ S' \end{pmatrix}$.

Since $\begin{pmatrix} S \\ S_xSS_x^{-1} \end{pmatrix}$ is a substitution of L', the substitution $\begin{pmatrix} S \\ S_xSS_x^{-1} \end{pmatrix}\begin{pmatrix} S \\ SS_x \end{pmatrix}$, or $\begin{pmatrix} S \\ S_xS \end{pmatrix}$, belongs to K'. Hence K' contains the set of substitutions

$$\begin{pmatrix} S \\ S_xS \end{pmatrix}, \quad (x = 0, 1, 2, \ldots, N-1).$$

These form (§ 107) a transitive group G'', simply isomorphic with G' and such that every substitution of G'' is permutable with every substitution of G'. Moreover (*l.c.*), the substitutions of G'' are the only substitutions of the N symbols which are permutable with each of the substitutions of G'.

Suppose now that Σ is any substitution of the N symbols which is permutable with G'. When the substitutions of G' are transformed by Σ, the resulting isomorphism is identical with that given by some substitution, say $\begin{pmatrix} S \\ S' \end{pmatrix}$, of L'. Hence $\Sigma \begin{pmatrix} S \\ S' \end{pmatrix}^{-1}$ is a substitution of the N symbols which is permutable with every substitution of G'. It therefore belongs to G''; and hence Σ belongs to K'. It follows that K' contains every substitution of the N symbols which is permutable with G'.

The only substitutions common to G' and G'' are the self-conjugate substitutions of either. The factor group $\dfrac{K'}{\{G', G''\}}$ is simply isomorphic with $\dfrac{L}{g}$, where g is the group of cogredient isomorphisms of G contained in L. The groups G' and G'' are identical only when G' is Abelian; in this case, g consists of the identical operation alone.

Definition. A group K, simply isomorphic with the substitution group K' which has just been constructed, we shall call the *holomorph* of G.

159. An isomorphism must change any set of operations, which are conjugate to each other, into another set which are conjugate. For if

$$\begin{pmatrix} S \\ S' \end{pmatrix}$$

be the isomorphism, and if

$$S_z^{-1} S_x S_z = S_y,$$

then

$$S_z'^{-1} S_x' S_z' = S_y',$$

so that S_x' and S_y' are conjugate operations when S_x and S_y are conjugate. A cogredient isomorphism changes every set of conjugate operations into itself; and all the members of a class of contragredient isomorphisms permute the conjugate sets in the same way. If

$$\begin{pmatrix} S \\ S' \end{pmatrix}$$

is an isomorphism which changes every conjugate set of operations of G into itself, and if

$$\begin{pmatrix} S \\ S'' \end{pmatrix}$$

is any isomorphism of G, then the isomorphism

$$\begin{pmatrix} S \\ S'' \end{pmatrix}^{-1} \begin{pmatrix} S \\ S' \end{pmatrix} \begin{pmatrix} S \\ S'' \end{pmatrix}$$

changes every conjugate set into itself. It follows that those isomorphisms, which change every conjugate set of operations into itself, form a self-conjugate sub-group of the complete group of isomorphisms. This sub-group clearly contains the group of cogredient isomorphisms and may be identical with it.

If now $\begin{pmatrix} S \\ S' \end{pmatrix}$ is any isomorphism of G of order n, the substitutions

$$\begin{pmatrix} S \\ S' \end{pmatrix} \text{ and } \begin{pmatrix} S \\ SS_x \end{pmatrix}, \quad (x = 0, 1, \ldots, N-1),$$

generate a group of order N_n. When J is used to represent the isomorphism, this group may be denoted by $\{J, G\}$; as shewn above, it contains G self-conjugately. Suppose that n is prime and not a factor of N. The operation J is not permutable with every operation of G; and therefore there must be operations S of G which are permutable with no operation of the conjugate set to which J belongs. The number of operations which in $\{J, G\}$ are conjugate to such an operation S must be a multiple of n; and since n is not a factor of N, this conjugate set of operations must be made up of n distinct conjugate sets of operations in G. The isomorphism J must therefore interchange some of the conjugate sets of G.

The same result is clearly true if the order n of J has any prime factor not contained in N. Hence:—

THEOREM II. *An isomorphism of a group G, whose order contains a prime factor which does not occur in the order of G, must interchange some of the conjugate sets of G.*

160. If the isomorphism $\begin{pmatrix} S \\ S' \end{pmatrix}$ or J leaves no operation except identity unchanged, it must in $\{J, G\}$ be one of N conjugate operations. For if

$$S_x^{-1} J S_x = S_y^{-1} J S_y,$$

J would be permutable with $S_y S_x^{-1}$, which is not the case.

These N conjugate operations are

$$J, \ JS_1, \ JS_2, \ \ldots, \ JS_{N-1},$$

and since the first transforms every operation of G, except identity, into a different one, the same must be true of all the set. If now J transformed any operation S into a conjugate operation $\Sigma^{-1} S \Sigma$, $J\Sigma^{-1}$ would transform S into itself; hence J must transform every conjugate set of G into a different conjugate set.

The special case in which the order of J is two may here be considered. Representing the N operations conjugate to J by

$$J, \ J_1, \ J_2, \ \ldots, \ J_{N-1},$$

the N operations of G are

$$J^2, \ JJ_1, \ JJ_2, \ \ldots, \ JJ_{N-1}.$$

Now

$$J^{-1} \cdot JJ_x \cdot J = J_x J = (JJ_x)^{-1},$$

so that J transforms every operation of G into its own inverse. But if

$$S' = S^{-1},$$

and

$$T' = T^{-1},$$

then

$$S'T' = S^{-1}T^{-1} = (TS)^{-1}.$$

Now as $S'T'$ is the operation into which the isomorphism transforms ST, it must be $(ST)^{-1}$, and therefore

$$ST = TS.$$

The group G is therefore an Abelian group of odd order.

161. Any sub-group H of G is transformed by an isomorphism into a simply isomorphic sub-group H'; but H and H' are not necessarily conjugate within G. If however the set of conjugate sub-groups

$$H_1,\ H_2,\ \ldots,\ H_m,$$

are the only sub-groups of G of a given type, every isomorphism must interchange them among themselves; and if no isomorphism transforms each one of the set into itself, the group of isomorphisms can be represented as a transitive group of degree m.

Suppose now that no operation of G is permutable with each of the conjugate sub-groups

$$H_1,\ H_2,\ \ldots,\ H_m,$$

so that G, or what in this case is the same thing (since G can have no self-conjugate operation except identity) the group of cogredient isomorphisms of G, can be expressed as the transitive group of m symbols that arises on transforming the set of m sub-groups by each operation of G. Let J be any operation, of order μ, that transforms G and each of the set of m conjugate sub-groups, into itself. Then J^μ is the lowest power of J that can occur in G, since no operation of G transforms each of the m sub-groups into itself. Now in $\{J, G\}$, the greatest sub-group that contains H_r self-conjugately is $\{J, I_r\}$, I_r being the greatest sub-group of G that contains H_r self-conjugately. Also, in $\{J, G\}$ the set of sub-groups $\{J, I_r\}$, $(r = 1, 2, \ldots, m)$, is a complete conjugate set. Now the set of groups

$$I_1,\ I_2,\ \ldots,\ I_m$$

have by supposition no common operation except identity; and therefore the greatest common sub-group of

$$\{J, I_1\},\ \{J, I_2\},\ \ldots,\ \{J, I_m\}$$

is $\{J\}$. Hence $\{J\}$ is a self-conjugate sub-group of $\{J, G\}$; and since G is also a self-conjugate sub-group of $\{J, G\}$, while $\{J\}$ and G have no common operation except identity, J must be permutable with every operation of G. Every operation therefore which is permutable with G, and with each of the sub-groups

$$H_1,\ H_2,\ \ldots,\ H_m,$$

is permutable with every operation of G. Thus finally, no contragredient isomorphism can transform each of the sub-groups H_r $(r = 1, 2, \ldots, m)$ into itself. Hence:—

THEOREM III. *If the conjugate set of m sub-groups*

$$H_1,\ H_2,\ \ldots,\ H_m$$

contains all the sub-groups of G of a given type, and if no operation of G is permutable with each sub-group of the set, the group of isomorphisms of G can be represented as a transitive group of degree m.

Corollary I. If G contains $kp + 1$ sub-groups of order p^α, where p^α is the highest power of a prime p that divides the order of G; and if the greatest sub-group I, that contains a sub-group of order p^α self-conjugately, contains no self-conjugate sub-group of G; then the group of isomorphisms of G can be represented as a transitive group of degree $kp + 1$.

For it has been seen that the groups of order p^α contained in G form a single conjugate set.

Corollary II. If the conjugate set of m sub-groups of G

$$H_1,\ H_2,\ \ldots,\ H_m$$

is changed into itself by every isomorphism of G, and if no operation of G is permutable with every one of these sub-groups; then the group of isomorphisms of G can be expressed as a transitive group of degree m.

In fact, under the conditions stated, the reasoning applied to prove the theorem may be used to shew that no isomorphism of G can transform each of the m sub-groups into itself.

162. Definition. Any sub-group of a group G which is transformed into itself by every isomorphism of G, is called* a *characteristic sub-group* of G.

A characteristic sub-group of a group G is necessarily a self-conjugate sub-group of G; but a self-conjugate sub-group is not necessarily characteristic. A simple group, having no self-conjugate sub-groups, can have no characteristic sub-groups. Let now G be any group, and let K be the holomorph of G. A characteristic sub-group of G is then a self-conjugate sub-group of K; and conversely, every self-conjugate sub-group of K which is contained in G is a characteristic sub-group of G.

Suppose now a chief-series of K formed which contains G. If G has no characteristic sub-group, it must be the last term but one of this series, the last term being identity. It follows by § 94 that G must be the direct product of a number of simply isomorphic simple groups. Hence:—

THEOREM IV. *A group, which has no characteristic sub-group, must be either a simple group or the direct product of simply isomorphic simple groups.*

The converse of this theorem is clearly true.

163. Suppose now that G is a group which has characteristic sub-groups; and let

$$G,\ G_1,\ \dots,\ G_r,\ G_{r+1},\ \dots,\ 1$$

be a series of such sub-groups, each containing all that follow it and chosen so that, for each consecutive pair G_r and G_{r+1}, there is no characteristic sub-group of G contained in G_r and containing G_{r+1}, except G_{r+1} itself. Such a series is called† a *characteristic series* of G.

It may clearly be possible to choose such a series in more than one way. If

$$G,\ G'_1,\ \dots,\ G'_r,\ G'_{r+1},\ \dots,\ 1$$

*Frobenius, "Ueber endliche Gruppen," *Berliner Sitzungsberichte*, 1895, p. 183.

†Frobenius, "Ueber auflösbare Gruppen, II," *Berliner Sitzungsberichte*, 1895, p. 1027.

be a second characteristic series of G, then

$$\begin{array}{rl} & K,\ J,\ \ldots,\ H,\ G,\ G_1,\ \ldots,\ G_r,\ G_{r+1},\ \ldots,\ 1 \\ \text{and} & K,\ J,\ \ldots,\ H,\ G,\ G'_1,\ \ldots,\ G'_r,\ G'_{r+1},\ \ldots,\ 1 \end{array}$$

are two chief-series of K. In fact, if K had a self-conjugate sub-group contained in G_r and containing G_{r+1}, then G would have a characteristic sub-group contained in G_r and containing G_{r+1}. The two chief-series of K coincide in the terms from K to G inclusive. Hence the two sets of factor-groups

$$\frac{G}{G_1},\ \frac{G_1}{G_2},\ \ldots,\ \frac{G_r}{G_{r+1}},\ \ldots$$

and

$$\frac{G}{G'_1},\ \frac{G'_1}{G'_2},\ \ldots,\ \frac{G'_r}{G'_{r+1}},\ \ldots$$

must be equal in number and, except possibly as regards the sequence in which they occur, identical in type. Moreover, each factor-group must be either a simple group or the direct product of simply isomorphic simple groups.

164. We will now shew how to determine a characteristic series for a group whose order is the power of a prime*.

First, let the group G be Abelian; and suppose that it is generated by a set of independent operations, of which n_s are of the order p^m, $(s = 1, 2, \ldots, r)$, while

$$m_1 > m_2 > \ldots > m_r.$$

The sub-group G_μ (§ 42), formed of the operations of G which satisfy the relation

$$S^{p^\mu} = 1,$$

is clearly a characteristic sub-group. As a first step towards forming the characteristic series, we may take the set of groups

$$G_{m_1} (= G),\ G_{m_1-1},\ G_{m_1-2},\ \ldots,\ G_2,\ G_1,\ 1;$$

*Frobenius, *l.c.*, pp. 1028, 1029.

for this is a set of characteristic sub-groups such that each contains the one that follows it.

Now the sub-group H_ν (§ 45), formed of the distinct operations that remain when every operation of G is raised to the power p^ν, is also a characteristic sub-group; and since the operations common to two characteristic sub-groups also form a characteristic sub-group, the sub-group $K_{\mu,\nu}$ (common to G_μ and H_ν) is characteristic. It follows from this that G_1 will be a characteristic sub-group only when $r = 1$. If $r > 1$, G_1 is not contained in $H_{m_{r-1}-1}$, and the common sub-group $K_{1,m_{r-1}-1}$ of these two is characteristic. If $r > 2$, this sub-group again is not contained in $H_{m_{r-2}-1}$; and the common sub-group $K_{1,m_{r-2}-1}$ of G_1 and $H_{m_{r-2}-1}$ is a characteristic sub-group contained in $K_{1,m_{r-1}-1}$. Continuing thus, we form between G_1 and 1 the series

$$G_1,\ K_{1,m_{r-1}-1},\ K_{1,m_{r-2}-1},\ \ldots,\ K_{1,m_1-1},\ 1.$$

In a similar way, between G_α and $G_{\alpha-1}$ we introduce such of the series

$$\{G_{\alpha-1}, K_{\alpha,m_{r-1}-\alpha}\},\ \{G_{\alpha-1}, K_{\alpha,m_{r-2}-\alpha}\},\ \ldots,\ \{G_{\alpha-1}, K_{\alpha,m_1-\alpha}\}$$

as are distinct, the symbol $m_s - \alpha$ being replaced by zero where it is negative.

From the original series we thus form a new one, in which again each group is characteristic and contains the following. This series may be shewn to be a characteristic series.

Let

$$P_{m_s,1},\ P_{m_s,2},\ \ldots,\ P_{m_s,n_s}$$

be the n_s generating operations of G, whose orders are p^{m_s}. Then if $\{G_{\alpha-1}, K_{\alpha,m_s-\alpha}\}$ and $\{G_{\alpha-1}, K_{\alpha,m_{s-1}-\alpha}\}$ are distinct, the generating operations of the latter differ only from those of the former in containing the set

$$P_{m_s,x}^{p^{m_s-\alpha+1}},\quad (x = 1, 2, \ldots, n_s),$$

in the place of

$$P_{m_s,x}^{p^{m_s-\alpha}},\quad (x = 1, 2, \ldots, n_s).$$

Now any permutation of the n_s generating operations

$$P_{m_s,x}, \quad (x = 1, 2, \ldots, n_s),$$

among themselves, the remaining generating operations being unaltered, must clearly give an isomorphism of G with itself; and therefore no sub-group of G, contained in $\{G_{\alpha-1}, K_{\alpha,m_s-\alpha}\}$ and containing $\{G_{\alpha-1}, K_{\alpha,m_{s-1}-\alpha}\}$, can be a characteristic sub-group. This result being true for every pair of distinct groups which succeed each other in the series that has been formed, it follows that the series is a characteristic series. It may be noticed that, if Γ and Γ' are any two consecutive sub-groups in a characteristic series of G, the order of $\frac{\Gamma}{\Gamma'}$ must be p^ν, where ν is one of the r numbers n_s.

Secondly, suppose that G is not Abelian. We may first consider the series of sub-groups

$$G,\ H_n,\ H_{n-1},\ \ldots,\ H_1,\ 1$$

of § 53. Each of these is clearly a characteristic sub-group, and each contains the succeeding. Moreover, $\frac{H_{r+1}}{H_r}$ is an Abelian group; and, by the process that we have just investigated, a characteristic series may be formed for it. To each group in this series will correspond a characteristic sub-group of H_{r+1} containing H_r; and the set of groups so obtained forms part of a complete characteristic series of G. When between each consecutive pair of groups in the above series the groups thus formed are interpolated, the resulting series of groups is a characteristic series for G.

165. The isomorphisms of a given group with itself are closely connected with the composition of every composite group in which the given group enters as a self-conjugate sub-group. Let G be any composite group and H a self-conjugate sub-group of G. Then since every operation of G transforms H into itself, to every such operation will correspond an isomorphism of H with itself. If S is an operation of G not contained in H, and if the isomorphism of H arising on transforming its operations by S is contragredient, so also is the isomorphism arising from each of the set of operations SH. In this case, no one of this set of operations is permutable with every operation of H. If however the isomorphism arising from S is cogredient, there must be some operation h of H which gives the same isomorphism as S; and then Sh^{-1} is permutable with every operation of H. In this case, the set of operations SH will give all the cogredient isomorphisms of H.

Suppose now that H_1 is that sub-group of G which is formed of all the operations of G that are permutable with every operation of H. Then to every operation of G, not contained in $\{H, H_1\}$, must correspond a contragredient isomorphism of H; and to every operation of the factor group $\dfrac{G}{\{H, H_1\}}$ corresponds a class of contragredient isomorphisms. If then L is the group of isomorphisms of H, and if L_1 is that self-conjugate sub-group of L which gives the cogredient isomorphisms of H, $\dfrac{G}{\{H, H_1\}}$ must be simply isomorphic with a sub-group of $\dfrac{L}{L_1}$.

If now H contains no self-conjugate operation except identity, H and H_1 can contain no common operation except identity; and since each of them is a self-conjugate sub-group of G, every operation of H is permutable with every operation of H_1. In this case, $\{H, H_1\}$ is the direct product of H and H_1.

If, further, L coincides with H, so that H admits of no contragredient isomorphisms, $\dfrac{G}{\{H, H_1\}}$ must reduce to identity. In this case, G is the direct product of H and H_1.

Definition. A group, which contains no self-conjugate operation except identity and which admits of no contragredient isomorphism, is called* a *complete* group.

The result of the present paragraph may be expressed in the form:—

THEOREM V. *A group, which contains a complete group as a self-conjugate sub-group, must be the direct product of the complete group and*

*Hölder, "Bildung zusammengesetzter Gruppen," *Math. Ann.*, Vol. XLVI (1895), p. 325.

*some other group**.

166. THEOREM VI. *If G is a group with no self-conjugate operations except identity; and if the group of cogredient isomorphisms of G is a characteristic sub-group of L, the group of isomorphisms of G; then L is a complete group*†.

With the notation of § 158, the operations of L may be represented by the substitutions

$$\begin{pmatrix} S \\ S' \end{pmatrix}.$$

The group of cogredient isomorphisms, which we will call G', is given by the substitutions

$$\begin{pmatrix} S \\ S_x^{-1} S S_x \end{pmatrix}, \quad (x = 0, 1, \ldots, N-1);$$

it is simply isomorphic with G.

Now

$$\begin{pmatrix} S \\ S' \end{pmatrix}^{-1} \begin{pmatrix} S \\ S_x^{-1} S S_x \end{pmatrix} \begin{pmatrix} S \\ S' \end{pmatrix} = \begin{pmatrix} S' \\ S_x'^{-1} S' S_x' \end{pmatrix},$$

$$= \begin{pmatrix} S \\ S_x'^{-1} S S_x' \end{pmatrix}:$$

and therefore no operation of L is permutable with every operation of G'. Hence every isomorphism of G' is given on transforming its operations by those of L. Suppose now that J is an operation which transforms L into itself. Since G' is by supposition a characteristic sub-group of L, the operation J transforms G' into itself. If J does not belong to L, we may assume that J is permutable with every operation of G'. For if it is not, it must give the same isomorphism of G' as some operation S of L; and then JS^{-1} is permutable with every operation of G', and is not contained in L. Now J being permutable with every operation of G', we have

$$J^{-1} s^{-1} g s J = s^{-1} g s,$$

* *Ibid.* p. 325.

† Hölder (*loc. cit.* p. 331) gives a theorem which is similar but not quite equivalent to Theorem VI.

where s is any operation of L, and g any operation of G'.

Moreover

$$JgJ^{-1} = g,$$

and therefore

$$J^{-1}s^{-1}JgJ^{-1}sJ = s^{-1}gs.$$

Hence s and $J^{-1}sJ$ give the same isomorphism of G'. Now no two distinct operations of L give the same isomorphism of G', so that s and $J^{-1}sJ$ must be identical; in other words, J is permutable with every operation of L. Hence L admits of no contragredient isomorphisms. Moreover, G' has no self-conjugate operations, and no operation of L is permutable with every operation of G'; hence L has no self-conjugate operations. It is therefore a complete group.

Corollary. If G is a simple group of composite order, or if it is the direct product of a number of isomorphic simple groups of composite order, the group of isomorphisms L of G is a complete group.

For suppose, if possible, in this case that G' is not a characteristic sub-group of L; and that, by a contragredient isomorphism of L, G is transformed into G''. Then G'' is a self-conjugate sub-group of L, and each of the groups G' and G'' transforms the other into itself. Hence (§ 34) either every operation of G' is permutable with every operation of G'', or G' and G'' must have a common sub-group. The former supposition is impossible since no operation of L is permutable with every operation of G. On the other hand, if G' and G'' have a common sub-group, it is a self-conjugate sub-group of L and it therefore is a characteristic sub-group of G'. Now (§ 162) G has no characteristic sub-groups, and therefore the second supposition is also impossible. It follows that, in this case, G' is a characteristic sub-group of L, and that L is a complete group.

167. THEOREM VII. *If G is an Abelian group of odd order, and if K is the holomorph of G; then when G is a characteristic sub-group of K, the latter group is a complete group.*

If N is the order of G, then K can be expressed (§ 158) as a transitive group of degree N. When K is so expressed, those operations of K which leave one symbol unchanged form a sub-group H, which is simply isomorphic with the group of isomorphisms of G. Now (§ 160) an Abelian group of odd order admits of a single isomorphism of order two, which changes every operation into its own inverse. The corresponding substitution of H is a self-conjugate substitution in H, and is one of N conjugate substitutions in K. These are the only substitutions of K which transform every substitution of G into its inverse. If G is a characteristic sub-group of K, every isomorphism of K must transform G, and therefore also the set of N conjugate substitutions of order two, into itself. Also, no substitution of K can be permutable with each one of these N substitutions, since each of them keeps just one symbol unchanged. Hence (Theorem III, Cor. II, § 161) the group of isomorphisms of K can be expressed as a transitive group of degree N, which contains G as a transitive self-conjugate sub-group. But when the group of isomorphisms of K is so expressed, K itself consists of all the substitutions of the N symbols which are permutable with G; and at the same time, every isomorphism of K transforms G into itself. Hence the group of isomorphisms of K must coincide with K itself; i.e. K admits of no contragredient isomorphisms. Also K obviously contains no self-conjugate operation except identity; hence it is a complete group.

Corollary. The holomorph of an Abelian group of order p^m, where p is an odd prime, and type $(1, 1, \ldots, \text{to } m \text{ units})$, is a complete group.

For if, in this case, G is not a characteristic sub-group of K, let G' be a sub-group of K which, in the group of isomorphisms of K, is conjugate to G. Then G and G' being both self-conjugate in K must have a common sub-group, since K cannot contain their direct product. But the common sub-group of G and G', being self-conjugate in K, is a characteristic sub-group of G. This is impossible (§ 162); hence G is a characteristic sub-group of K.

Ex. Shew that the holomorph of an Abelian group of degree 2^m and type $(1, 1, \ldots, \text{to } m \text{ units})$ is a complete group.

168. We shall now discuss the groups of isomorphisms of certain special groups; and we shall begin with the case of a cyclical group G, of prime order p, generated by an operation P. Every isomorphism of such a group must interchange among themselves the $p-1$ operations

$$P,\ P^2,\ \ldots,\ P^{p-1};$$

and if any isomorphism replaces P by P^{α}, it must replace P^2 by $P^{2\alpha}$, and so on. Moreover, the symbol

$$\begin{pmatrix} P\ , & P^2\ , \ldots, & P^{p-1} \\ P^{\alpha}, & P^{2\alpha}, \ldots, & P^{(p-1)\alpha} \end{pmatrix}$$

does actually represent an isomorphism whatever number α may be from 1 to $p-1$; for each operation occurs once in the second line, and the change indicated leaves the multiplication-table of the group unaltered. If $\alpha = p$, the symbol does not represent an isomorphism: if $\alpha = p + \alpha'$, the symbol represents the same isomorphism as

$$\begin{pmatrix} P\ , & P^2\ , \ldots, & P^{p-1} \\ P^{\alpha'}, & P^{2\alpha'}, \ldots, & P^{(p-1)\alpha'} \end{pmatrix}.$$

The group of isomorphisms of a group of prime order p is therefore a group of order $p-1$. Now the nth power of the isomorphism

$$\begin{pmatrix} P \\ P^{\alpha} \end{pmatrix}$$

is

$$\begin{pmatrix} P \\ P^{\alpha^n} \end{pmatrix}.$$

Hence if α is a primitive root of the congruence

$$\alpha^{p-1} - 1 \equiv 0,\ (\text{mod. } p),$$

the group of isomorphisms is a cyclical group generated by the isomorphism

$$\begin{pmatrix} P \\ P^{\alpha} \end{pmatrix}.$$

Finally, if S is an operation satisfying the relations

$$S^{p-1} = 1, \quad S^{-1}PS = P^{\alpha},$$

where α is a primitive root of p, $\{S, P\}$ is the holomorph of G.

The reader will at once observe that this group of order $p(p-1)$ is identical with the doubly transitive group of § 112. It is a complete group.

169. We shall consider next the case of any cyclical group.

Suppose, first, that G is a cyclical group of order p^n, where p is an odd prime; and let it be generated by an operation S. The group contains $p^{n-1}(p-1)$ operations of order p^n; if S' is any one of these,

$$\begin{pmatrix} S \\ S' \end{pmatrix}$$

defines a distinct isomorphism. The group of isomorphisms is therefore a group of order $p^{n-1}(p-1)$. Moreover, since the congruence

$$\alpha^{p^{n-1}(p-1)} - 1 \equiv 0, \text{ (mod. } p^n),$$

has primitive roots, the group of isomorphisms is a cyclical group. The holomorph of G is defined by

$$S^{p^n} = 1, \quad J^{p^{n-1}(p-1)} = 1, \quad J^{-1}SJ = S^\alpha,$$

where α is a primitive root of the congruence

$$\alpha^{p^{n-1}(p-1)} - 1 \equiv 0, \text{ (mod. } p^n).$$

If G is a cyclical group of order 2^n, it follows, in the same way, that the group of isomorphisms is an Abelian group of order 2^{n-1}. In this case, however, the congruence

$$\alpha^{2^{n-1}} - 1 \equiv 0, \text{ (mod. } 2^n), \quad n > 2,$$

has no primitive root, and therefore the group of isomorphisms is not cyclical. The congruence

$$\alpha^{2^{n-2}} - 1 \equiv 0, \text{ (mod. } 2^n)$$

however has primitive roots, and a primitive root α of this congruence can always be found to satisfy the condition

$$\alpha^{2^{n-3}} \equiv 1 + 2^{n-1}, \text{ (mod. } 2^n).$$

The powers of the isomorphism

$$\begin{pmatrix} S \\ S^\alpha \end{pmatrix}$$

then form a cyclical group of order 2^{n-2}; and the only isomorphism of order 2 contained in it is

$$\begin{pmatrix} S \\ S^{1+2^{n-1}} \end{pmatrix}.$$

Hence

$$\begin{pmatrix} S \\ S^{\alpha} \end{pmatrix} \text{ and } \begin{pmatrix} S \\ S^{-1} \end{pmatrix},$$

the latter not being contained in the sub-group generated by the former, are two permutable and independent isomorphisms of orders 2^{n-2} and 2. They generate an Abelian group of order 2^{n-1} which is the group of isomorphisms of G. The corresponding holomorph is given by

$$S^{2^n} = 1, \quad J_1^{2^{n-2}} = 1, \quad J_2^2 = 1, \quad J_1J_2 = J_2J_1,$$
$$J_1^{-1}SJ_1 = S^{\alpha}, \quad J_2SJ_2 = S^{-1},$$

where α satisfies the conditions given above.

If G is a cyclical group of order 4, its group of isomorphisms is clearly a group of order 2.

170. It is now easy to construct the group of isomorphisms of any cyclical group G, and the corresponding holomorph. If the order of G is $2^n p_1^{m_1} p_2^{m_2} \dots$, where $p_1, p_2, \dots$ are odd primes, G is the direct product of cyclical groups of orders 2^n, $p_1^{m_1}$, $p_2^{m_2}$, ...; and every isomorphism of G transforms each of these groups into itself. Hence if the groups of isomorphisms of these cyclical groups be formed, and their direct product be then constructed, every operation of the group so formed will give a distinct isomorphism of the group G. Moreover, the order

$$2^{n-1} p_1^{m_1-1}(p_1-1)p_2^{m_2-1}(p_2-1)\dots$$

of this group is equal to the number of operations of G whose order is $2^n p_1^{m_1} p_2^{m_2} \dots$, or in other words to the number of isomorphisms of which G is capable. The group thus formed is therefore the group of isomorphisms of G. The corresponding holomorph is clearly the direct product of the holomorphs of the cyclical groups of orders 2^n, $p_1^{m_1}$, etc.

When the order of G is odd, the holomorph K is easily shewn to be a complete group. Suppose it to be expressed transitively, as in § 158, in N symbols, where N is the order of G; if G is not a characteristic sub-group of K, let G' be a group into which G is transformed by a contragredient isomorphism of K. Then G' is a self-conjugate sub-group of K; and since G' is cyclical, every sub-group of G' is a self-conjugate sub-group of K. Hence a generating operation of G' must be a circular substitution of N symbols. The N operations of order 2 which transform each operation of G' into its own inverse therefore each keep one symbol fixed; hence each of them must transform every operation of G into its inverse. But there is only one such set of operations of order 2, and therefore G' cannot differ from G. It follows by Theorem VII, § 167 that, as G is a characteristic sub-group of K, the group K itself is complete.

If the order of G is even, K must contain a self-conjugate operation other than identity, namely the operation of order 2 contained in G. Moreover, K admits of a contragredient isomorphism whose square is cogredient. From the mode of formation of K, it is clearly sufficient to verify this when the order of G is a power of 2. The holomorph K is given by

$$S^{2^n} = 1, \quad J_1^{2^{n-2}} = 1, \quad J_2^2 = 1, \quad J_1J_2 = J_2J_1,$$
$$J_1^{-1}SJ_1 = S^{\alpha}, \quad J_2SJ_2 = S^{-1};$$

where α is a primitive root of

$$\alpha^{2^{n-2}} - 1 \equiv 0 \ (\text{mod. } 2^n),$$

such that

$$\alpha^{2^{n-3}} + 1 \not\equiv 0 \ (\text{mod. } 2^n).$$

In this group, J_2 is one of 2^{n-1} conjugate operations

$$J_2,\ J_2S^2,\ J_2S^4,\ \dots.$$

Now it may be directly verified that K admits the isomorphism represented by

$$\begin{pmatrix} S, & J_1 & , J_2 \\ S, & J_1S^{\frac{1}{2}(1-\alpha)} & , J_2S \end{pmatrix},$$

Moreover, since this isomorphism changes J_2 into J_2S, it cannot be a cogredient isomorphism. Finally, the square of this isomorphism is

$$\begin{pmatrix} S, & J_1 & , J_2 \\ S, & J_1S^{1-\alpha} & , J_2S^2 \end{pmatrix},$$

or

$$\begin{pmatrix} S, & J_1 & , J_2 \\ S, & S^{-1}J_1S & , S^{-1}J_2S \end{pmatrix},$$

which is a cogredient isomorphism.

171. We shall next consider the group of isomorphisms of an Abelian group of order p^n and type $(1, 1, \ldots, \text{to } n \text{ units})$. Such a group is generated by n independent permutable operations of order p, say

$$P_1, \ P_2, \ \ldots, \ P_n.$$

Since every operation of the group is self-conjugate and of order p, while the group contains no characteristic sub-group, there must be isomorphisms transforming any one operation of the group into any other. We may therefore begin by determining under what conditions the symbol

$$\begin{pmatrix} P_r \\ P_1^{a_{1r}} P_2^{a_{2r}} \ldots P_n^{a_{nr}} \end{pmatrix} \quad (r = 1, 2, \ldots, n),$$

defines an isomorphism. This symbol replaces the operation $P_1^{x_1} P_2^{x_2} \ldots P_n^{x_n}$ by $P_1^{y_1} P_2^{y_2} \ldots P_n^{y_n}$, where

$$\left.\begin{array}{l} y_1 \equiv a_{11}x_1 + a_{12}x_2 + \ldots + a_{1n}x_n, \\ y_2 \equiv a_{21}x_1 + a_{22}x_2 + \ldots + a_{2n}x_n, \\ \ldots\ldots\ldots\ldots\ldots\ldots\ldots\ldots\ldots\ldots \\ y_n \equiv a_{n1}x_1 + a_{n2}x_2 + \ldots + a_{nn}x_n, \end{array}\right\} \text{(mod. } p).$$

Unless the p^n operations $P_1^{y_1} P_2^{y_2} \ldots P_n^{y_n}$ thus formed are all distinct, when for $P_1^{x_1} P_2^{x_2} \ldots P_n^{x_n}$ is put successively each of the p^n operations of the group, the symbol does not represent an isomorphism. On the other hand, when this condition is satisfied, the symbol represents a permutation of the operations among themselves which leaves the multiplication table of the group unchanged; it is therefore an isomorphism.

If this condition is satisfied, $x_1, \ x_2, \ \ldots, \ x_n$ must be definite numbers (mod. p), when $y_1, \ y_2, \ \ldots, \ y_n$ are given; and therefore the above set of n simultaneous congruences must be capable of definite solution with respect to the x's. The necessary and sufficient condition for this is that the determinant

$$\begin{vmatrix} a_{11}, & a_{12}, & \ldots, & a_{1n} \\ a_{21}, & a_{22}, & \ldots, & a_{2n} \\ \ldots & \ldots & \ldots & \ldots \\ a_{n1}, & a_{n2}, & \ldots, & a_{nn} \end{vmatrix}$$

should not be congruent to zero (mod. p).

Every distinct set of congruences of the above form, for which this condition is satisfied, represents a distinct isomorphism of the group, two sets being regarded as distinct if the congruence

$$a_{rs} \equiv a'_{rs} \text{ (mod. } p)$$

does not hold for each corresponding pair of coefficients. Moreover, to the product of two isomorphisms will correspond the set of congruences which results from carrying out successively the operations indicated by the two sets that correspond to the two isomorphisms.

The group of isomorphisms is therefore simply isomorphic with the group of operations defined by all sets of congruences

$$\left.\begin{array}{l} y_1 \equiv a_{11}x_1 + a_{12}x_2 + \ldots + a_{1n}x_n, \\ y_2 \equiv a_{21}x_1 + a_{22}x_2 + \ldots + a_{2n}x_n, \\ \ldots\ldots\ldots\ldots\ldots\ldots\ldots\ldots\ldots \\ y_n \equiv a_{n1}x_1 + a_{n2}x_2 + \ldots + a_{nn}x_n, \end{array}\right\} \text{(mod. } p)$$

for which the relation

$$\begin{vmatrix} a_{11}, & a_{12}, & \ldots, & a_{1n} \\ a_{21}, & a_{22}, & \ldots, & a_{2n} \\ \ldots & \ldots & \ldots & \ldots \\ a_{n1}, & a_{n2}, & \ldots, & a_{nn} \end{vmatrix} \not\equiv 0 \text{ (mod. } p)$$

is satisfied.

172. The group thus defined is of great importance in many branches of analysis. It is known as the *linear homogeneous* group. In a subsequent Chapter we shall consider some of its more important properties; we may here conveniently determine its order. Let this be represented by N_n, so that N_r represents the number of distinct solutions of the congruences

$$\begin{vmatrix} a_{11}, & a_{12}, & \ldots, & a_{1r} \\ a_{21}, & a_{22}, & \ldots, & a_{2r} \\ \ldots & \ldots & \ldots & \ldots \\ a_{r1}, & a_{r2}, & \ldots, & a_{rr} \end{vmatrix} \not\equiv 0 \text{ (mod. } p)$$

Since the group of isomorphisms of the Abelian group of order p^n transforms every one of its operations (identity excepted) into every other, it can be represented as a transitive substitution group of $p^n - 1$ symbols, and therefore, if M is the order of the sub-group that keeps P_1 unchanged,

$$N_n = (p^n - 1)M.$$

Now, in the congruences corresponding to an operation that keeps P_1 unchanged, we have

$$a_{11} \equiv 1,\ a_{12} \equiv a_{13} \equiv \ldots \equiv a_{1n} \equiv 0,\ (\text{mod. } p).$$

Hence M is equal to the number of solutions of the congruences

$$\begin{vmatrix} 1, & 0, & \ldots, & 0 \\ a_{21}, & a_{22}, & \ldots, & a_{2n} \\ \ldots & \ldots & \ldots & \ldots \\ a_{n1}, & a_{n2}, & \ldots, & a_{nn} \end{vmatrix} \not\equiv 0 \ (\text{mod. } p)$$

The value of the determinant does not depend on the values of $a_{21}, a_{31}, \ldots, a_{n1}$, and therefore

$$M = p^{n-1}N_{n-1}.$$

Hence

$$N_n = (p^n - 1)p^{n-1}N_{n-1},$$

and therefore immediately

$$N_n = (p^n - 1)(p^n - p)\ldots(p^n - p^{n-1}).$$

The reader will notice that an independent proof of this result has already been obtained in § 48. The discussion there given of the number of distinct ways, in which independent generating operations of an Abelian group of type $(1, 1, \ldots, 1)$ may be chosen, is clearly equivalent to a determination of the order of the corresponding group of isomorphisms.

The holomorph of an Abelian group, of order p^n and type $(1, 1, \ldots,$ to n units$)$, can similarly be represented as a group of linear transformations to the prime modulus p. Consider, in fact, the set of transformations

$$\left.\begin{array}{l} y_1 \equiv a_{11}x_1 + a_{12}x_2 + \ldots + a_{1n}x_n + b_1, \\ y_2 \equiv a_{21}x_1 + a_{22}x_2 + \ldots + a_{2n}x_n + b_2, \\ \ldots\ldots\ldots\ldots\ldots\ldots\ldots\ldots \\ y_n \equiv a_{n1}x_1 + a_{n2}x_2 + \ldots + a_{nn}x_n + b_n, \end{array}\right\} (\text{mod. } p);$$

where the coefficients take all integral values (mod. p) consistent with

$$\begin{vmatrix} a_{11}, & a_{12}, & \ldots, & a_{1n} \\ a_{21}, & a_{22}, & \ldots, & a_{2n} \\ \ldots & \ldots & \ldots & \ldots \\ a_{n1}, & a_{n2}, & \ldots, & a_{nn} \end{vmatrix} \not\equiv 0.$$

The set of transformations clearly forms a group whose order is $N_n p^n$. The sub-group formed by all the transformations

$$y_1 \equiv x_1 + b_1,\ y_2 \equiv x_2 + b_2,\ \ldots,\ y_n \equiv x_n + b_n,\ (\text{mod. } p),$$

is an Abelian group of order p^n and type $(1, 1, \ldots, \text{to } n \text{ units})$, and it is a self-conjugate sub-group. Moreover, the only operations of the group, which are permutable with every operation of this self-conjugate sub-group, are the operations of the sub-group itself; and, since the order of the group is equal to the order of the holomorph of the Abelian group, it follows that the group of transformations must be simply isomorphic with the holomorph of the Abelian group. In the simplest instance, where p^n is 2^2, the holomorph is simply isomorphic with the alternating group of four symbols. In any case the holomorph, when expressed as in § 158, is a doubly transitive group of degree p^n.

173. It has been seen in § 142 that, except when $n = 6$, the symmetric group of degree n has n and only n sub-groups of order $n - 1!$, which form a conjugate set. Hence by Theorem III, § 161, the group of isomorphisms of the symmetric group of degree n can be expressed, except when $n = 6$, as a transitive group of degree n. The symmetric group of n symbols however consists of all possible substitutions that can be performed on the n symbols, and therefore it must coincide with its group of isomorphisms. Hence*:—

THEOREM VIII. *The symmetric group of n symbols is a complete group, except when $n = 6$.*

Corollary. Except when $n = 6$, the alternating group of n symbols admits of one and only one class of contragredient isomorphisms.

For with this exception, the alternating group of degree n has just n sub-groups of order $\frac{1}{2}(n - 1)!$.

*Hölder, *Math. Ann.* Vol. XLVI, (1895), p. 345.

174. The alternating group of degree 6 occurs as a special case of another class of groups of which we will determine the groups of isomorphisms. These are the doubly and the triply transitive groups that have been defined in §§ 112, 113.

The doubly transitive group of degree p^m and order $p^m(p^m - 1)$ there considered has a single set of p^m conjugate cyclical sub-groups of order $p^m - 1$. Its group of isomorphisms can therefore be expressed as a transitive group of degree p^m. Let $p^m(p^m - 1)\mu$ be the order of the group of isomorphisms. The order of a sub-group that keeps two symbols fixed is μ; and every operation of this sub-group must transform a cyclical sub-group of order $p^m - 1$ into itself. With the notation of § 112, we will consider the sub-group which keeps 0 and i fixed. Every operation of this sub-group must transform the cyclical sub-group generated by the congruence*

$$x' \equiv ix$$

into itself; and no operation of the sub-group can be permutable with the given operation. If S is an operation of the sub-group which transforms

$$x' \equiv ix$$

into

$$x' \equiv i^{\alpha}x,$$

then S, when represented as a substitution, is given by

$$\begin{pmatrix} i^2 , & i^3 , & i^4 & ,\ldots \\ i^{\alpha+1}, & i^{2\alpha+1}, & i^{3\alpha+1}, & \ldots \end{pmatrix}.$$

Now S must transform the sub-group of order p^m into itself; it must therefore be permutable with that operation of this sub-group which changes 0 into i.

This operation is given by

$$x' \equiv x + i,$$

and if we denote it by T, then $S^{-1}TS$ changes i^{y+1} into $i^{\alpha z+1}$, where

$$i^{\frac{y}{\alpha}+1} + i \equiv i^{z+1}.$$

*This and all subsequent congruences in the present section are to be taken, (mod. p).

Now T changes i^{y+1} into $i^{y+1}+i$. Hence, since S and T are permutable, we must for all values of y have the simultaneous congruences

$$i^{\frac{y}{\alpha}+1}+i \equiv i^{z+1},$$

and

$$i^{y+1}+i \equiv i^{\alpha z+1}.$$

Eliminating z, the congruence

$$1+i^y \equiv (1+i^{\frac{y}{\alpha}})^\alpha$$

must hold for all values of y from 0 to p^m-1. If α is not a power of p, this congruence involves an identity of the form

$$a_1 i^{\alpha_1}+a_2 i^{\alpha_2}+\ldots \equiv 0,$$

where all the indices are less than p^m-1; and this is impossible. Hence the only possible values of α are p, p^2, p^3, ...; and the greatest possible value of μ is m.

Now the congruence

$$x' \equiv x^p$$

defines a substitution performed on the p^m symbols permuted by the group, and this substitution is permutable with the group. For if we denote this operation by J, and any operation

$$x' \equiv \alpha x+\beta$$

of the group by Σ, then $J^{-1}\Sigma J$ is

$$x' \equiv \alpha^p x+\beta^p,$$

another operation of the group. Moreover, J clearly transforms

$$x' \equiv ix$$

into its pth power.

The group of isomorphisms of the doubly transitive group of order p^m is therefore the group defined by

$$x' \equiv ix, \quad x' \equiv x+i, \quad x' \equiv x^p;$$

where i is a primitive root of the congruence

$$i^{p^m-1} \equiv 1.$$

From this it immediately follows that the group of isomorphisms of the triply transitive group, of degree p^m+1 and order

$$(p^m+1)p^m(p^m-1),$$

defined by the congruences

$$x' \equiv \frac{\alpha x+\beta}{\gamma x+\delta},$$

where α, β, γ, δ satisfy the conditions of § 113, is the group of order $(p^m+1)p^m(p^m-1)m$ obtained by combining the previous congruences with

$$x' \equiv x^p.$$

In fact, it may be immediately verified that the operation given by this congruence is permutable with the group, and does not give a cogredient isomorphism of the group. Moreover, by Theorem III, § 161, the group of isomorphisms of the given group can be expressed as a transitive group of degree p^m+1; and therefore, among a class of contragredient isomorphisms, there must be some transforming into itself a sub-group which keeps one symbol fixed. Hence the order of the group of isomorphisms cannot exceed

$$(p^m+1)p^m(p^m-1)m.$$

When p is an odd prime, the triply transitive group of degree p^m+1, which may be defined by

$$x' \equiv -\frac{1}{x}, \quad x' \equiv ix, \quad x' \equiv x+i,$$

contains as a self-conjugate sub-group a doubly transitive group defined by

$$x' \equiv -\frac{1}{x}, \quad x' \equiv i^2x, \quad x' \equiv x+i.$$

It will be shewn in Chapter XIV that this is a simple group. When p^m is equal to 3^2, it is easy to verify that this group is simply isomorphic with the alternating group of degree 6.

The group of isomorphisms of this simple group can be expressed as a transitive group of degree p^m+1, and must clearly contain the triply transitive group of order $(p^m+1)p^m(p^m-1)$ self-conjugately. It therefore coincides with the group of order $(p^m+1)p^m(p^m-1)m$, which has just been determined. The latter group is therefore (Theorem VI, Cor. § 166) a complete group. Further, if the simple group be denoted by G and the group of isomorphisms by L, the factor group $\frac{L}{G}$ will be determined by

$$x' \equiv ix, \quad x' \equiv x^p,$$

when all operations of G are treated as the identical operation. Denoting these operations by I and J, then $J^{-1}IJ$ is I^p, which is the same as I multiplied by an operation of G. Hence $\frac{L}{G}$ is an Abelian group generated by two independent operations of orders 2 and m.

The group of isomorphisms of the alternating group of degree 6 is therefore a group of order 1440; and the symmetric group of degree 6 admits a single class of contragredient isomorphisms*.

175. Let P be any group whose order is the power of a prime, and let

$$P,\ P_1,\ P_2,\ \ldots,\ P_n,\ 1$$

of orders

$$p^\alpha,\ p^{\alpha_1},\ p^{\alpha_2},\ \ldots,\ P^{\alpha_n},\ 1$$

be a characteristic series (§ 163) of P. Every isomorphism of P must transform P_r and P_{r+1} into themselves, and therefore also $\frac{P_r}{P_{r+1}}$ into itself. Suppose now that an isomorphism I of P transforms every operation of P_{r+1} into itself and every operation of $\frac{P_r}{P_{r+1}}$ into itself. If S is any operation of P_r, not contained in P_{r+1}, I must transform S into SA, where A is some operation of P_{r+1}; so that, if p^μ is the order of A, I transforms the operations of the set SP_{r+1} cyclically among themselves in sets of p^μ. Similarly, if S' is an operation of P_r not contained in the set SP_{r+1}, I will transform the operations of the set $S'P_{r+1}$ cyclically among themselves in sets of $p^{\mu'}$. Hence the order of the isomorphism I is a multiple of p; and

*Hölder, *loc. cit.* p. 343.

any isomorphism, that transforms every operation of P_{r+1} into itself and every operation of $\frac{P_r}{P_{r+1}}$ into itself and is of order prime to p, will transform every operation of P_r into itself. Therefore, the only isomorphism of P, that transforms every operation of each of the groups

$$\frac{P}{P_1}, \frac{P_1}{P_2}, \ldots, \frac{P_{n-1}}{P_n}, P_n$$

into itself and is of order prime to p, is the identical isomorphism. Now each of these groups is an Abelian group, whose operations are all of order p; and it has been shewn that, if p^m is the order of such a group, the order of its group of isomorphisms is

$$(p^m - 1)(p^m - p) \ldots (p^m - p^{m-1}).$$

Every isomorphism of P, whose order is relatively prime to p, must therefore be such that its order is a factor of one of the expressions of the above form, obtained by writing

$$\alpha - \alpha_1, \alpha_1 - \alpha_2, \ldots, \alpha_n,$$

in succession for m. If then k is the greatest of these numbers, the order of any isomorphism of P, whose order is relatively prime to p, is a factor of

$$(p-1)(p^2-1) \ldots (p^k - 1).$$

Herr Frobenius* has introduced the symbol $\vartheta(P)$ to denote this product. If G is a group of order $p_1^{\alpha_1} p_2^{\alpha_2} \ldots p_n^{\alpha_n}$, where $p_1, p_2, \ldots, p_n$ are distinct primes, and if $P_1, P_2, \ldots, P_n$ are groups of orders $p_1^{\alpha_1}, p_2^{\alpha_2}, \ldots, p_n^{\alpha_n}$ contained in G, we shall use the symbol $\vartheta(G)$ to denote the least common multiple of

$$\vartheta(P_1), \vartheta(P_2), \ldots, \vartheta(P_n).$$

176. If P' is a sub-group of a group P of order p^α, $\vartheta(P')$ is not necessarily a factor of $\vartheta(P)$. For instance, the group of order p^4, generated

*"Ueber auflösbare Gruppen, II," *Berliner Sitzungsberichte*, 1895, p. 1028.

by the four operations A, B, C, D of order p, of which A and B are self-conjugate while

$$D^{-1}CD = CA,$$

has a characteristic sub-group $\{A, B\}$ of order p^2, and it has no characteristic sub-group of order p^3. Hence

$$\vartheta(P) = (p-1)(p^2-1).$$

The sub-group P' however, which is generated by A, B and C, is an Abelian group whose operations are all of order p; and therefore

$$\vartheta(P') = (p-1)(p^2-1)(p^3-1).$$

So again, for the Abelian group P of order p^3, generated by A and B, where

$$A^{p^2} = 1, \quad B^p = 1, \quad AB = BA,$$

we have

$$\vartheta(P) = p - 1;$$

while its sub-group P', generated by A^p and B, is an Abelian group whose operations are all of order p, so that

$$\vartheta(P') = (p-1)(p^2-1).$$

Suppose now that P is an Abelian group of order p^α, generated by ρ independent and permutable operations. For such a group, we define* a new symbol $\theta(P)$ by the equation

$$\theta(P) = (p-1)(p^2-1)\ldots(p^\rho-1).$$

In forming the characteristic series of P in § 164, we commenced with the series of groups G_r $(r = 1, 2, \ldots)$, such that G_r consists of all the operations of P satisfying the relation

$$S^{p^r} = 1.$$

*Frobenius, *loc. cit.* p. 1030.

Since P is generated by ρ independent operations, the order of G_1 is p^ρ, and the order of $\frac{G_{r+1}}{G_r}$ cannot be greater than p^ρ. If the generating operations of P are all of the same order β, the series

$$P,\ G_{\beta-1},\ G_{\beta-2},\ \ldots,\ G_1,\ 1$$

is a complete characteristic series of P, and each factor-group $\frac{G_{r+1}}{G_r}$ is of order p^ρ. In this case, therefore,

$$\vartheta(P) = \theta(P).$$

If however the generating operations are not all of the same order, G_1 will not be the last term of the complete characteristic series; nor will G_{r+1} and G_r be consecutive terms in the series, if the order of $\frac{G_{r+1}}{G_r}$ is p^ρ. Hence, in this case, $\vartheta(P)$ will be a factor of $\theta(P)$.

If now P' is any sub-group of P, then it has been seen (§ 46) that the number of independent generating operations of P' is equal to or less than ρ. Hence $\theta(P')$ is equal to or is a factor of $\theta(P)$.

177. THEOREM IX. *If a group G, of order N, is transformed into itself by an operation S, whose order is relatively prime to $N\vartheta(G)$, every operation of G is permutable with S**.

Let $p_1^{\alpha_1}$ be the highest power of a prime p_1 which divides N. The number of sub-groups of G, and therefore also of $\{S, G\}$, whose order is $p_1^{\alpha_1}$ is a factor of N; and since the order of S is relatively prime to N, one at least of them, say P_1, must be transformed into itself by S. Now the order of S is relatively prime both to p_1 and to $\vartheta(P_1)$; and therefore the isomorphism of P given on transforming its operations by S must be the identical isomorphism. In other words, S is permutable with every operation of P_1. In the same way it follows that, if $p_2^{\alpha_2}$, $p_3^{\alpha_3}$, ... are the highest powers of p_2, p_3, ... that divide N, there must be sub-groups of orders $p_2^{\alpha_2}$, $p_3^{\alpha_3}$, ... with every operation of each of which S is permutable.

*Frobenius, *loc. cit.* p. 1030.

But these groups of orders $p_1^{\alpha_1}$, $p_2^{\alpha_2}$, $p_3^{\alpha_3}$, ... generate G; therefore S is permutable with every operation of G.

Corollary. If the order of an isomorphism of G is relatively prime to N, it must be a factor of $\vartheta(G)$.

It follows immediately, from the theorem, that there is no isomorphism of G whose order contains a prime factor not occurring in $N\vartheta(G)$. Suppose, if possible, that the group of isomorphisms of G contains an operation S of order q^m, where q is a prime which does not divide N, and that the highest power of q that divides $\vartheta(G)$ is $q^{m'}$, where $m' < m$. If $p_r^{\alpha_r}$ is the highest power of any prime p_r which divides N, S must transform some sub-group P_r of order $p_r^{\alpha_r}$ into itself. Since q^m is not a factor of $\vartheta(P_r)$, some power of S must be permutable with every operation of P_r. Hence, as in the theorem, it follows that some power of S, certainly $S^{q^{m-1}}$, must be permutable with every operation of G. But no operation of the group of isomorphisms of G, except identity, is permutable with every operation of G. Hence the group of isomorphisms cannot contain an operation of order q^m, if $m > m'$; and therefore there is no isomorphism of G whose order contains a higher power of q than $q^{m'}$. If then $q^r q'^{r'} \ldots$, a number relatively prime to N, is the order of an isomorphism of G, all the numbers q^r, $q'^{r'}$, ... are factors of $\vartheta(G)$, and so also is their product.

178. The method, by which it has been shewn in § 175 that the order of any isomorphism of P_r which transforms every operation of each of the groups $\dfrac{P_r}{P_{r+1}}$ and P_{r+1} into itself is a power of p, may be used to obtain the following more general result.

THEOREM X. *If H is a self-conjugate sub-group of G, the order of an isomorphism of G, which transforms every operation of each of the groups $\dfrac{G}{H}$ and H into itself, is a factor of the order of H.*

If S is any operation of G not contained in H, the isomorphism will change S into Sh, where h is some operation of H. If then m is the order of h, the isomorphism transforms

$$S,\ Sh,\ Sh^2,\ \ldots,\ Sh^{m-1}$$

cyclically; and therefore it transforms all the operations of the set SH in cycles of m each. If S' is any operation of G not contained in SH, the isomorphism will interchange the operations of the set $S'H$ among themselves in cycles of m' each, where m' again is the order of some operation of H. The isomorphism, when expressed as a substitution performed on the operations of G, will consist of a number of cycles of m, m', ... symbols; and its order is therefore the least common multiple of m, m', ... Now if q is any prime that divides the order of H, and q^n the highest power of q that occurs as the order of a cyclical operation of H, no power of q higher than q^n can occur in any of the numbers m, m', ...; and q^n is therefore the highest power of q that can occur in their least common multiple. This least common multiple, which is the order of the isomorphism, must therefore divide the order of H.

179. Ex. 1. Shew that, for the group of order p^3 defined by

$$P^p = 1, \quad Q^p = 1, \quad R^p = 1, \quad Q^{-1}PQ = PR,$$
$$RP = PR, \quad RQ = QR,$$

the symbol

$$\begin{pmatrix} P & , & Q & , R \\ P^x Q^y R^z, & P^\alpha Q^\beta R^\gamma, & R^n \end{pmatrix}$$

gives an isomorphism if

$$\beta x - \alpha y \equiv n, \ (\text{mod. } p).$$

Hence determine the order of the group of isomorphisms.

Ex. 2. Shew that, for the group of order p^3 defined by

$$P^{p^2} = 1, \quad Q^p = 1, \quad Q^{-1}PQ = P^{1+p},$$

the symbol

$$\begin{pmatrix} P & , & Q \\ P^x Q^y, & P^{\alpha p} Q \end{pmatrix}$$

gives an isomorphism if x is not a multiple of p; and determine the order of the group of isomorphisms.

Ex. 3. Shew that the group of isomorphisms of the group of order 2^n, defined by (§ 63)

$$P^{2^{n-1}} = 1, \quad Q^2 = P^{2^{n-2}}, \quad Q^{-1}PQ = P^{-1},$$

is of order 2^{2n-3}, when $n > 3$. If $n = 3$, its order is 24 and it is simply isomorphic with the last type but one of § 84.

Ex. 4. If G is a complete group of order N, shew that the order of K, the holomorph of G, is N_2; and that the order of the holomorph of K is $2N^4$.

CHAPTER XII.

ON THE GRAPHICAL REPRESENTATION OF GROUP*.

180. OUR discussions hitherto have been confined exclusively to groups of finite order. When however, as we now propose to do, we consider a group in relation to the operations that generate it, it becomes almost necessary to deal, incidentally at least, with groups whose order is not finite; for it is not possible to say a priori what must be the number and the nature of the relations between the given generating operations, which will ensure that the order of the resulting group is finite.

Many of the definitions given in respect of finite groups may obviously be extended at once to groups containing an infinite number of operations. Among these may be specially mentioned the definitions of a sub-group, of conjugate operations and sub-groups, of self-conjugate sub-groups, of the relation of isomorphism between two groups and of the factor-group given by this relation. In regard to the last of them, the isomorphism between two groups, one at least of which is not of finite order, may be such that to one operation of the one group there correspond an infinitely great number of operations of the other. On the other hand, all the results obtained for finite groups, which depend directly or indirectly on the order of the group, necessarily become meaningless when the group is not a group of finite order.

181. Suppose that

$$S_1,\ S_2,\ \ldots,\ S_n$$

represent any n distinct operations which can be performed, directly or inversely, on a common object, and that between these operations no relations exist. Then the totality of the operations represented by

$$\ldots S_p^{\alpha} S_q^{\beta} S_r^{\gamma} \ldots,$$

*The investigations of this Chapter are due to Dyck, "Gruppentheoretische Studien," *Math. Ann.*, Vol. XX, (1882), pp. 1–44. We have followed Dyck's memoir closely except in two respects. Firstly, we have used a rather more definite geometrical operation than that of the memoir; and secondly, we have not specially considered a regular and symmetric division of a closed surface, apart from a merely regular division.

where the number of factors is any whatever and the indices are any positive or negative integers, form a group G of infinite order, which is generated by the n operations. If, moreover, whenever such a succession of factors as $S_p^\alpha S_p^\beta$ occurs in the above expression, it is replaced by $S_p^{\alpha+\beta}$, each operation of the group can be expressed in one way and in one way only by an expression of the above form, which is then called *reduced.*

It will sometimes be convenient to avoid the use of negative indices in the expression of any operation of the group. To this end we may write

$$S_1 S_2 \dots S_n S_{n+1} = 1,$$

so that S_{n+1} is a definite operation of the group; then

$$S_r^{-1} = S_{r+1} S_{r+2} \dots S_n S_{n+1} S_1 \dots S_{r-1}, \quad (r = 1, 2, \dots, n).$$

By using these relations to replace all negative powers of operations wherever they occur, we may represent every operation of the group in a single definite way by means of the $n+1$ operations

$$S_1,\ S_2,\ \dots,\ S_n,\ S_{n+1},$$

with positive indices only.

The group, thus defined and represented, is the most general group conceivable that is generated by n distinct operations. Any two such groups, for which n is the same, are simply isomorphic with each other.

Suppose now that

$$\overline{S}_1,\ \overline{S}_2,\ \dots,\ \overline{S}_n$$

represent n distinct operations, but that, instead of being entirely independent, they are connected by a relation of the form

$$\overline{S}_p^a \overline{S}_q^b \dots \overline{S}_r^c = 1,$$

which will be represented by

$$f(\overline{S}_i) = 1.$$

If $\overline{G}$ is the group generated by these operations, an isomorphism may be established between G and $\overline{G}$ by taking $\overline{S}_i$ $(i = 1, 2, \ldots, n)$ as the operation of $\overline{G}$ that corresponds to the operation S_i of G.

Then to every operation of G

$$\ldots S_p^{\alpha} S_q^{\beta} S_r^{\gamma} \ldots$$

will correspond a single definite operation

$$\ldots \overline{S}_p^{\alpha} \overline{S}_q^{\beta} \overline{S}_r^{\gamma} \ldots$$

of $\overline{G}$; for the supposition that two distinct operations of $\overline{G}$ correspond to the same operation of G leads to the result that between the generating operations of G there is a relation, which is not the case. On the other hand, to the identical operation of $\overline{G}$ there will correspond an infinite number of distinct operations of G, namely those which are formed by combining together in every possible way all operations of G of the form

$$R^{-1} f(S_i) R,$$

where R is any operation of G. These operations of G form a self-conjugate sub-group H; the corresponding factor-group $\frac{G}{H}$ is simply isomorphic with $\overline{G}$.

If between the generating operations of $\overline{G}$ there are several independent relations

$$f_1(\overline{S}_i) = 1, \quad f_2(\overline{S}_i) = 1, \quad \ldots, \quad f_m(\overline{S}_i) = 1,$$

it may be shewn exactly as before that the groups G and $\overline{G}$ are isomorphic in such a way that to the identical operation of $\overline{G}$ there corresponds that self-conjugate sub-group of G, which is formed by combining in every possible way all the operations of G of the form

$$R^{-1} f_j(S_i) R, \quad (j = 1, 2, \ldots, m).$$

182. We may at once extend the result of the preceding paragraph in the following way:—

Theorem I. *If G is the group generated by the n operations*

$$S_1,\ S_2,\ \dots,\ S_n,$$

between which the m relations

$$f_1(S_i) = 1, \quad f_2(S_i) = 1, \quad \dots, \quad f_m(S_i) = 1,$$

exist; and if $\overline{G}$ is the group generated by the n operations

$$\overline{S}_1,\ \overline{S}_2,\ \dots,\ \overline{S}_n,$$

which are connected by the same m relations

$$f_1(\overline{S}_i) = 1, \quad f_2(\overline{S}_i) = 1, \quad \dots, \quad f_m(\overline{S}_i) = 1,$$

as hold between the generating operations of G, and by the further m' relations

$$g_1(\overline{S}_i) = 1, \quad g_2(\overline{S}_i) = 1, \quad \dots, \quad g_{m'}(\overline{S}_i) = 1;$$

then $\overline{G}$ is simply isomorphic with the factor-group $\frac{G}{H}$; where H is that self-conjugate sub-group of G, which results from combining in every possible way all operations of the form

$$R^{-1} g_j(S_i) R, \quad (j = 1, 2, \dots, m'),$$

R being any operation of G.

In proving this theorem, it is sufficient to notice that, if we take $\overline{S}_i$ $(i = 1, 2, \dots, n)$ as the operation of $\overline{G}$ which corresponds to the operation S_i of G, then to each operation of G a single definite operation of $\overline{G}$ will correspond, while to the identical operation of $\overline{G}$ there corresponds the self-conjugate sub-group H of G.

The theorem just stated is of such a general nature that it is perhaps desirable to illustrate it by considering shortly some simple examples.

Let us take first the case of a group G, generated by two independent operations S_1 and S_2, subject to no relations; and let us suppose that the single relation

$$\overline{S}_1 \overline{S}_2 = \overline{S}_2 \overline{S}_1$$

holds between the generating operations of $\overline{G}$. The self-conjugate sub-group H of G then consists of all the operations

$$\dots S_1^{\alpha_n} S_2^{\beta_n} \dots$$

of G which reduce to identity if we regard S_1 and S_2 as permutable; or, in other words, of those operations of G for which the relations

$$\sum \alpha_n = 0, \quad \sum \beta_n = 0$$

simultaneously hold.

In respect of this sub-group, the operations of G can be divided into an infinite number of classes of the form

$$S_1^p S_2^q H.$$

For the operations of the class $S_1^p S_2^q H$, multiplied by those of the class $S_1^{p'} S_2^{q'} H$, give always operations of the class

$$S_1^p S_2^q S_1^{p'} S_2^{q'} H,$$

since H is a self-conjugate sub-group; and, because

$$S_1^p S_2^q S_1^{p'} S_2^{q'} = S_1^{p+p'} \cdot S_1^{-p'} S_2^q S_1^{p'} S_2^{-q} \cdot S_2^{q+q'},$$

while $S_1^{-p'} S_2^q S_1^{p'} S_2^{-q}$ belongs to H, the class $S_1^p S_2^q S_1^{p'} S_2^{q'} H$ is the same as $S_1^{p+p'} S_2^{q+q'} H$. Hence the operations of any two given classes, multiplied in either order, give the same third class; and therefore the group $\frac{G}{H}$ is an Abelian group generated by two permutable, but otherwise unrestricted, operations.

As a second illustration, we will choose a case in which $\overline{G}$ is of finite order. Let G be generated by the operations S and T, which satisfy the relations

$$S^3 = 1, \quad T^3 = 1, \quad (ST)^3 = 1;$$

and for $\overline{G}$, suppose that the generating operations satisfy the additional relation

$$(ST^2)^3 = 1.$$

Then H is formed by combining in all possible ways the operations

$$R^{-1}(ST^2)^3 R.$$

Now it may be easily verified that in G, the operation ST^2 belongs to a set of three conjugate operations

$$ST^2, \quad TST, \quad T^2S;$$

and that these three operations are permutable among themselves, while their product is identity. Hence H consists of the Abelian group

$$(ST^2)^{3\alpha}(TST)^{3\beta};$$

and in respect of H, G may be divided into 27 classes of the form

$$S^x(ST^2)^y(TST)^zH, \quad (x, y, z = 0, 1, 2).$$

The group $\overline{G}$ will be defined by the laws according to which these 27 classes combine among themselves; and the reader will have no difficulty in verifying that it is isomorphic with the non-Abelian group of order 27, whose operations are all of order 3 (§ 73).

Ex. If $S_1(=1)$, S_2, ..., S_N are the operations of a group G of finite order, prove that the totality of the operations of the form $S_p^{-1}S_q^{-1}S_pS_q$ generate a self-conjugate sub-group H; and that the group $\frac{G}{H}$ is an Abelian group. Shew also that, if H' is a self-conjugate sub-group of G and if $\frac{G}{H'}$ is Abelian, then H' contains H.

(Miller, *Quarterly Journ. of Math.*, Vol. XXVIII, (1896), p. 266.)

183. For the further discussion of a group, as defined by its generating operations and the relations between them, a suitable graphical mode of representation becomes of the greatest assistance. To this we shall now proceed.

In the simple case in which the group is generated by a single unrestricted operation, such a representation may be constructed as follows. Let C_1 and C_{-1} be two circles which touch each other; C_2 and C_{-2} the inverses of C_{-1} in C_1 and C_1 in C_{-1}; C_3 and C_{-3} the inverses of C_{-2} in C_1 and C_2 in C_{-1}, and so on. These circles (fig. 1) divide the plane in which they are drawn into an infinite number of crescent-shaped spaces. Suppose now that the space between C_1 and C_{-1} is left white, and the spaces between C_1 and C_2 and between C_{-1} and C_{-2} (on either side of this white

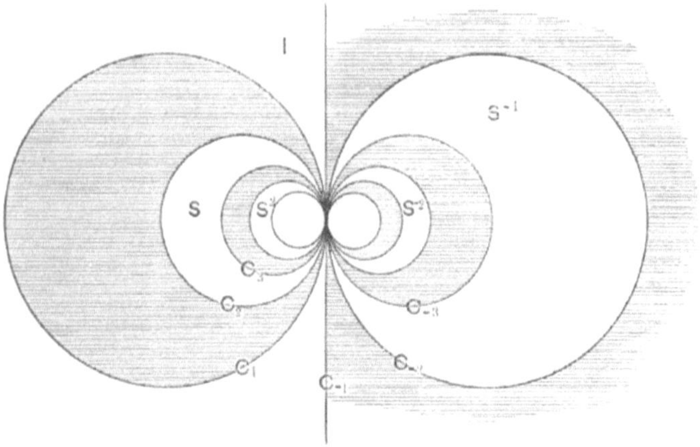

Fig. 1.

space) are coloured black; the next pair on either side left white, the next coloured black, and so on. Then any white space may be transformed into any other (and any black into any other) by an even number of inversions at the circles C_{-1} and C_1; and if S denote the operation consisting of an inversion at C_{-1} followed by an inversion at C_1, the space between C_{-1} and C_1 will be transformed into another perfectly definite white space by the operation S^n, while conversely the operation necessary to transform the space between C_{-1} and C_1 into any other given white space will be a definite power of S. Hence if one of the white spaces, say that between C_{-1} and C_1, is taken to correspond to the identical operation, there is then a unique correspondence between the white spaces and the operations of the group generated by the unrestricted operation S; and the figure that has been constructed gives a graphical representation of the group. It should be noticed that the actual geometrical process of inversion, which has been here used to construct the spaces corresponding to the operations of the group, is in no way essential to the graphical representation. It is however convenient as giving definiteness to the construction; and later, when we deal with the case of a general group, such definiteness becomes almost a necessity.

In a precisely similar manner, the group generated by a single relation S, satisfying the relation

$$S^n = 1,$$

may be treated. In this case, we take two circles C_{-1} and C_1 intersecting at an angle $\frac{\pi}{n}$, and from these form, as before, the circles obtained by successive inversions. This gives a finite series of n circles, each of which

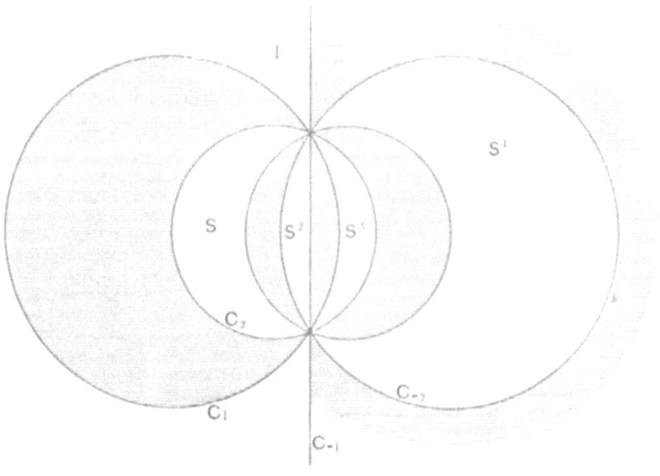

Fig. 2.

intersects the two next to it on either side at angles $\frac{\pi}{n}$, while the n circles divide the plane into $2n$ spaces. If these are left white and coloured black in alternate succession, and if one of the white is taken to correspond to the identical operation, there is a unique correspondence between the white spaces and the operations of the group generated by S, where S represents the result of successive inversions first at C_{-1} and then at C_1.

This operation obviously satisfies the relation

$$S^n = 1$$

and no simpler relation; so that the figure gives a graphical representation of a cyclical group of order n.

The systems of circles in figures 1 and 2 have a common geometrical property which may be noticed here as it will be of use in the sequel. Successive inversions at any one of the pairs C_{-1} and C_1, C_1 and C_2, C_2 and C_3 are equivalent to the operation S; and therefore successive inversions at C_{-1} and C_r are equivalent to the operation S^r. Hence the result of an even number of inversions at any of the circles in either figure is equivalent to some operation of the group that the figure represents.

184. We may now proceed to construct a graphical representation of the group which is generated by n operations subject to no relations. To this end, suppose $n+1$ circles drawn, each of which is external to all the others while each touches two and only two of the rest. Such a system can be drawn in an infinite variety of ways: we will suppose, to give definiteness and simplicity to the resulting figure, that the $n+1$ points of contact lie on a circle, which cuts the $n+1$ circles orthogonally. If these $n+1$ points taken in order are $A_1, A_2, \ldots, A_{n+1}$, the successive circles are $A_{n+1}A_1$, $A_1A_2, \ldots, A_nA_{n+1}$. We will suppose that only so much of these circles is drawn as lies within the common orthogonal circle $A_1A_2\ldots A_{n+1}$. The $n+1$ circular arcs $A_{n+1}A_1$, $A_1A_2, \ldots$ then bound a finite simply connected plane figure which we will denote provisionally by P. Suppose now that P is inverted in each of its sides, that the resulting figures are inverted in each of their new sides, and so on continually. Then from their mode of formation no two of the figures thus arising can overlap either wholly or in part; and when the process is continued without limit, every point in the interior of the orthogonal circle $A_1A_2\ldots A_{n+1}$ will lie in one and only one of the figures thus formed from P by successive inversions.

If P' is any one of the figures or polygons so formed, the set of inversions by which it is derived from P is perfectly definite. For suppose, if possible, that P' is derived from P by two distinct sets of inversions represented by Σ and Σ'. Then $\Sigma\Sigma'^{-1}$ is a set of inversions in the sides of P which transforms P into itself. But every set of inversions necessarily transforms P into some polygon lying outside it, and therefore

$$\Sigma\Sigma'^{-1} = 1;$$

or the set of inversions composing Σ is identical term for term with the

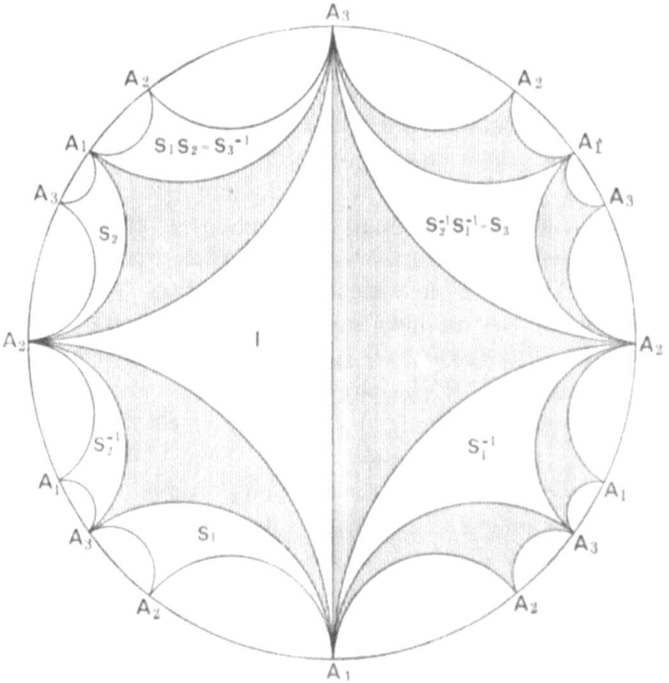

Fig. 8.

set composing Σ'. It immediately follows that the polygons can be divided into two sets, according as they are derived from P by an even or an odd number of inversions. The latter we shall suppose coloured black, and the former (including P) left white. Every white polygon will be surrounded by black polygons and vice versâ. Since there is only one definite set of inversions that will transform P into any other white polygon P', the $n+1$ corners of P' will correspond one by one to the $n+1$ corners of P; and when the perimeters of the two polygons are described in the same direction of rotation with regard to their interiors, the angular points that correspond will occur in the same cyclical order. On the other hand, in order that the corresponding angular points of a white and a black polygon may occur in the same cyclical order, their perimeters must be described in opposite directions. In consequence of these results, we may complete our figure (fig. 3) by lettering every angular point of every polygon with the same letter that occurs at the corresponding angular point of the polygon P.

185. If now $T_1, T_2, \ldots, T_{n+1}$ represent inversions at A_1A_2, $A_2A_3, \ldots,$ $A_{n+1}A_1$, the operation $T_{r-1}T_r$ leaves the corner A_r of P unchanged and it transforms P into the next white polygon which has the corner A_r in common with P, the direction of turning round A_r coinciding with the direction $A_{n+1}A_n \ldots A_1$ of describing the perimeter of P. For brevity, we shall describe this transformation of P as a positive rotation round A_r. If then we denote the operation $T_{r-1}T_r$ by the single symbol S_r, we may say that S_r produces a positive rotation of P round A_r. Let P_1 be the new polygon so obtained; and let P_1' be the polygon into which any other white polygon P' is changed by a positive rotation round the corner of P' that corresponds to A_r. Then if Σ is the set of inversions that changes P into P', it also changes P_1 into P_1': so that $\Sigma^{-1}S_r\Sigma$ changes P' into P_1', i.e. produces a positive rotation round the corner A_r of P'; and $S_r\Sigma$ changes P into P_1'.

Let us now represent the operations

$$T_{n+1}T_1,\ T_1T_2,\ \ldots,\ T_nT_{n+1},$$

by

$$S_1,\ S_2,\ \ldots,\ S_{n+1},$$

so that

$$S_1S_2 \ldots S_nS_{n+1} = 1.$$

Then every operation, consisting of a pair of inversions in the sides of P, can be represented in terms of

$$S_1,\ S_2,\ S_3,\ \ldots,\ S_n.$$

For an inversion at $A_r A_{r+1}$, followed by an inversion at $A_s A_{s+1}$, is given by $T_r T_s$; and

$$\begin{aligned} T_r T_s &= T_r T_{r+1} \cdot T_{r+1} T_{r+2} \cdot \ldots \cdot T_{s-1} T_s \\ &= S_{r+1} S_{r+2} \ldots S_s. \end{aligned}$$

If $s > r$, this is of the form required. If $s < r$, the term S_{n+1} that then occurs may be replaced by

$$S_n^{-1} S_{n-1}^{-1} \ldots S_2^{-1} S_1^{-1}.$$

Hence finally, every operation consisting of an even number of inversions in the sides of P can be expressed in terms of

$$S_1,\ S_2,\ S_3,\ \ldots,\ S_n;$$

and with a restriction to positive indices, every such operation can be expressed in terms of

$$S_1,\ S_2,\ \ldots,\ S_n,\ S_{n+1}.$$

Now it has been seen that no two operations, each consisting of a set of inversions in the sides of P, can be identical unless the component inversions are identical term for term. Hence no two reduced operations of the form

$$f(S_i)$$

are identical; in other words, the n generating operations

$$S_1,\ S_2,\ \ldots,\ S_n$$

are subject to no relations.

If then we take the polygon P to correspond to the identical operation of the group G generated by

$$S_1,\ S_2,\ \ldots,\ S_n,$$

each white polygon may be taken as associated with the operation which will transform P into it. The foregoing discussion makes it clear that in this way a unique correspondence is established between the operations of G and the system of white polygons; or in other words, that the geometrical figure gives a complete graphical representation of the group.

Moreover, since the operation $\Sigma^{-1}S_r\Sigma$ is a positive rotation round the corner A_r of the polygon Σ (calling now P the polygon 1), a simple rule may be formulated for determining by a mere inspection of the figure what operation of the group any given white polygon corresponds to.

This rule may be stated as follows. Let a continuous line be drawn inside the orthogonal circle from a point in the white polygon 1 to a point in any other white polygon, so that every consecutive pair of white polygons through which the line passes have a common corner, a positive rotation round which leads from the first to the second of the pair. This is always possible. Then if the common corners of each consecutive pair of white polygons through which the line passes, starting from 1, are A_p, A_q, ..., A_r, A_s, the final white polygon corresponds to the operation

$$S_sS_r \dots S_qS_p{}^*.$$

186. The graphical representation of a general group we have thus arrived at is only one of an infinite number that could be constructed; and we choose this in preference to others mainly because the form of the figure and the relative positions of the successive polygons are readily apprehended by the eye. As regards the mere establishment of such a representation we might, still using the process of inversion for the purpose of forming a definite figure, have started with $n+1$ circles each exterior to and having no point in common with any of the others. Taking as the figure P the space external to all the circles and inverting it continually in the circles, we should form a series of black and white spaces of which the latter would again give a complete picture of the group. It is however only necessary to begin the construction of such a figure in order to convince ourselves that it would not appeal to the eye in the same way as the figure actually chosen.

*The reader who refers to Prof. Dyck's memoir should notice that the definition of the operation S_r above given is not exactly equivalent to that used by Prof. Dyck. With his notation, the white polygon here considered would correspond to the operation $S_pS_q \dots S_rS_s$.

Moreover, as in the representation of a cyclical group, the process of inversion is in no way essential to the representation at which we arrive. Any arbitrary construction, which would give us the series of white and black polygons having, in the sense of the geometry of position, the same relative configuration as our actual figure, would serve our purpose equally well.

187. If Σ is the operation which transforms P into P', and if Q is the black polygon which has the side A_rA_{r+1} in common with P, then Σ transforms Q into the black polygon which has in common with P' the side corresponding to A_rA_{r+1}. If then we take Q to correspond to the identical operation, any black polygon will correspond to the same operation as that white polygon with which it has the side A_rA_{r+1} in common. In this way we may regard our figure as divided in a definite way into double-polygons, each of which represents a single operation of the group.

188. We have next to consider how, from the representation of a general group whose n generating operations are subject to no relations, we may obtain the representation of a special group generated by n operations connected by a series of relations

$$F_j(S_i) = 1, \quad (j = 1, 2, \dots, m).$$

It has been seen (§ 181) that to the identical operation of the special group there corresponds a self-conjugate sub-group H of the general group; or in other words, that the set of operations ΣH of the general group give one and only one operation in the special group.

Hence, to obtain from our figure for the general group one that will apply to the special group, we must regard all the double-polygons of the set ΣH as equivalent to each other; and if from each such set of double-polygons we choose one as a representative of the set, the totality of these representative polygons will have a unique correspondence with the operations of the special group.

We shall first shew that a set of representative double-polygons can always be chosen so as to form a single simply connected figure. Starting with the double-polygon, P_1, that corresponds to the identical operation of the general group as the one which shall correspond to the identical operation of the special group, we take as a representative of some other operation of the special group a double-polygon, P_2, which has a side in common with P_1. Next we take as a representative of some third operation of the special group a double-polygon which has a side in common with either P_1 or P_2; and we continue the choice of double-polygons in this way until it can be carried no further. The set of double-polygons thus arrived at of necessity forms a single simply connected figure C, bounded by circular arcs; and no two of the double-polygons belonging to it correspond to the same operation of the special group. Moreover, in C there is one double-polygon corresponding to each operation of the special group. To shew this, let C' be the figure formed by combining with C every double-polygon which has a side in common with C; and form C'' from C', C''' from C'', and so on, as C' has been formed from C. From the construction of C it follows that every polygon in C' is equivalent, in respect of the special group, to some polygon in C. Similarly, every polygon in C'' is equivalent to some polygon in C' and therefore to some polygon in C; and so on. Hence finally, every polygon in the complete figure of the general group is equivalent to some polygon in C, in respect of the special group; and therefore, since no two polygons of C are equivalent in respect of the special group, the figure C is formed of a complete set of representative double-polygons for the special group.

Suppose now that S is a double-polygon outside C, with a side $A'_r A'_{r+1}$ belonging to the boundary of C. Within C there must be just one polygon, say ST, of the set SH. If this polygon lay entirely inside C, so as to have no side on the boundary of C, every polygon having a side in common with it would belong to C. Now since S and ST are equivalent, every polygon having a side in common with S is equivalent to some polygon having a side in common with ST. Hence since C contains no two equivalent polygons, ST must have a side on the boundary of C; and if this side is $A''_r A''_{r+1}$, the operation T of H transforms $A'_r A'_{r+1}$ into $A''_r A''_{r+1}$. Moreover, no operation of H can transform $A'_r A'_{r+1}$ into another side of C; for if this were possible, C would contain two polygons equivalent to S. It is also clear that, regarded as sides of polygons within C, $A'_r A'_{r+1}$ and $A''_r A''_{r+1}$ belong to polygons of different colours. Hence a correspondence in pairs of the sides of C is established: to each portion $A'_r A'_{r+1}$ of the boundary of C, which forms a side of a white (or black) polygon of C, there corresponds another definite portion $A''_r A''_{r+1}$, forming a side of a black (or white) polygon of C, such that a certain operation of H and its inverse will change one into the other, while no other operation of H will change either into any other portion of the boundary of C.

The system of double-polygons forming the figure C, and the correspondence of the sides of C in pairs, will now give a complete graphical representation of the group. For the figure has been formed so that there is a unique correspondence between the white polygons of C and the operations of the group, such that until we arrive at the boundary the previously obtained rule will apply; and when we arrive at a polygon on the boundary, the correspondence of the sides in pairs enables the process to be continued.

189. From the mode in which the figure C has been formed, no two of the figures CH can have a polygon in common, when for H is taken in turn each operation of the self-conjugate sub-group H of the general group G; also the complete set contains every double-polygon of our original figure. This set of figures, or rather the division of the original figure into this set, will then represent in a graphical form the self-conjugate sub-group H of G. Moreover, the operations which transform corresponding pairs of sides of C into each other will, when combined and repeated, clearly suffice to transform C into any one of the figures CH and will therefore form a set of generating operations of H.

190. A simple example, in which the process described in the preceding paragraphs is actually carried out, will help to familiarize the reader with the nature of the process and will also serve to introduce a further modification of our figure. The example we propose to consider is the special group with two generating operations which are connected by the relations

$$S_1^3 = 1, \quad S_2^3 = 1, \quad S_1S_2 = S_2S_1.$$

As a first step, we will take account only of the relation

$$S_1S_2 = S_2S_1,$$

and form for this special group the figure C. All operations

$$\dots S_1^{\alpha_n} S_2^{\beta_n} \dots,$$

for which $\Sigma\alpha_n$ and $\Sigma\beta_n$ have given values, are in the special group identical. We may thus select from the figure for the general group the set of polygons

$$S_2^\alpha S_1^\beta \quad (\alpha, \beta = -\infty \text{ to } +\infty)$$

as a set of representative polygons; and a reference to the diagram* (fig. 4) makes it clear that this set of polygons forms a figure with a single bounding curve. The black polygon which corresponds to the operation Σ has here been chosen as that which has the side A_1A_2 in common with the white polygon Σ.

Each double-polygon, except those of the set S_1^m, contributes two sides to the boundary of C, one belonging to a white polygon and one to a black. The

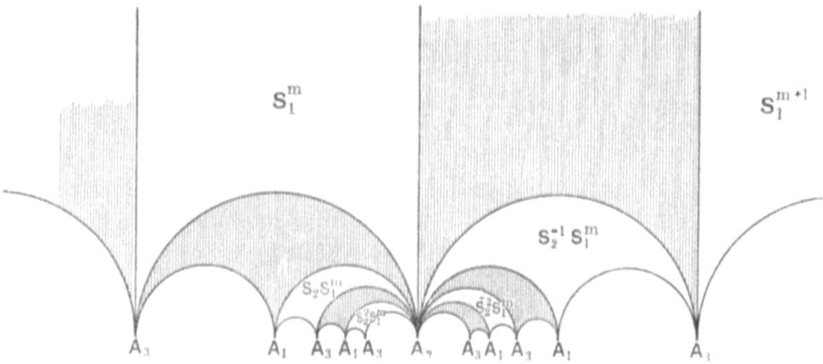

Fig. 4.

polygons, which border C and have sides in common with $S_2^\alpha S_1^\beta$, are $S_1 S_2^\alpha S_1^\beta$ and $S_1^{-1} S_2^\alpha S_1^\beta$; and these, regarded as operations of the special group, are equivalent to $S_2^\alpha S_1^{\beta+1}$ and $S_2^\alpha S_1^{\beta-1}$. Hence the correspondence between the sides of C is such that

(i) to the side A_1A_3 of the white polygon $S_2^\alpha S_1^\beta$ corresponds the side A_1A_3 of the black polygon $S_2^\alpha S_1^{\beta-1}$;

(ii) to the side A_1A_3 of the black polygon $S_2^\alpha S_1^\beta$ corresponds the side A_1A_3 of the white polygon $S_2^\alpha S_1^{\beta+1}$.

*In fig. 4 the orthogonal circle, which is not shewn, is taken to be a straight line.

When we now take account of the additional relations

$$S_1^3 = 1, \quad S_2^3 = 1,$$

the figure C is found to reduce to a set of nine double-polygons, which is completely represented by fig. 5.

In addition to the correspondences between the sides of C to which those just written simplify when the indices of S_1 and S_2 are reduced (mod. 3), we have now also the correspondences, indicated in the figure by curved lines with arrow-heads, which result from the new relations. Our figure may be further modified in such a way that its form takes direct account of these four new correspondences. Thus without in any way altering the configuration of the double-polygons, from the point of view of geometry of position, we may continuously deform the figure so that the pairs of corresponding sides indicated by the curved arrow-heads are brought to actual coincidence. When this is done, the resulting figure will have the form shewn in fig. 6. The correspondence in pairs of the sides of the boundary is indicated in the figure by full and dotted lines. The two unmarked portions $A_1A_3A_1$ correspond, as also do the two similar portions marked with a full line, and the two marked with a dotted line.

It will be noticed that, in this final form of the figure for the special group, direct account is taken of the finite order of the generating operations S_1 and S_2 and also of the operation S_1S_2. The simplification of the figure that results by thus taking account directly of the finite order of the generating operations, and the greater ease with which the eye follows this simplified representation, are immediately obvious on a comparison of figs. 5 and 6.

191. In the applications of this graphical representation of a group that we have specially in view, namely to groups of finite order, the generating operations themselves are necessarily of finite order. The generating operations

$$S_1, \ S_2, \ \ldots, \ S_n,$$

of such a group may be taken as of orders

$$m_1, \ m_2, \ \ldots, \ m_n;$$

and if

$$S_1S_2 \ldots S_nS_{n+1} = 1,$$

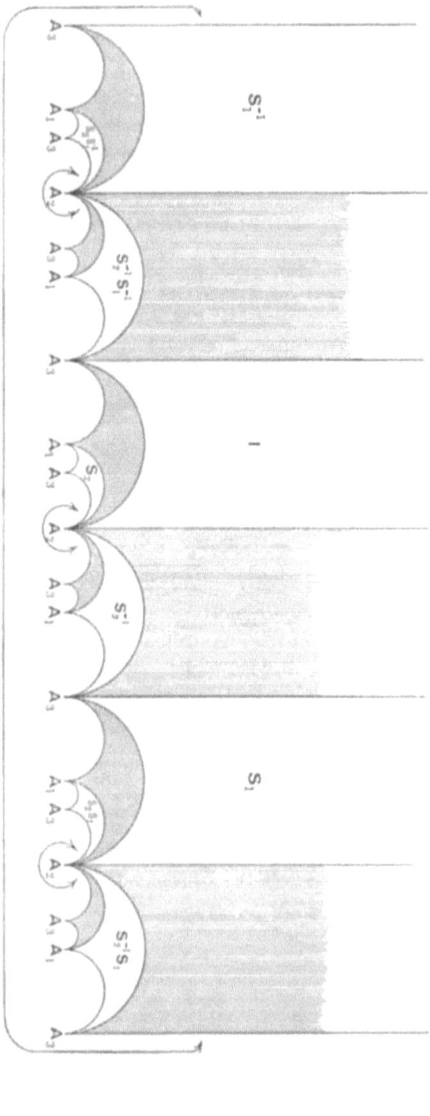

Fig. 5.

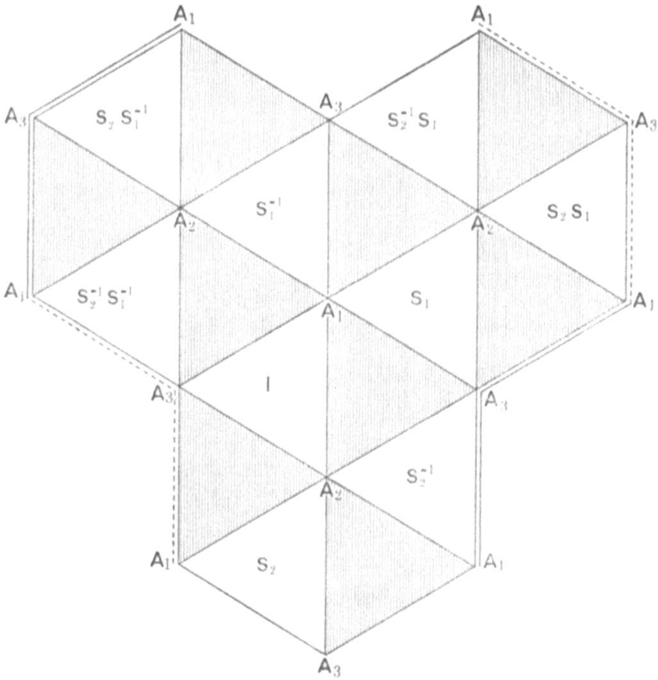

Fig. 6.

then S_{n+1} will be of finite order m_{n+1}. We shall therefore next consider a group generated by n operations which satisfy the relations

$$S_1^{m_1} = 1, \quad S_2^{m_2} = 1, \quad \ldots, \quad S_n^{m_n} = 1, \quad (S_1 S_2 \ldots S_n)^{m_{n+1}} = 1.$$

The simple example we have given makes it clear that, at least in some particular cases, relations of this form may be directly taken into account in constructing our figure; in such a way that in the complete figure, consisting of a finite or an infinite number of double-polygons, the correspondence in pairs of the sides of the boundary, if any, will depend upon further relations between the generating operations.

We may, in fact, always take account of relations of the form in question in the construction of our figure as follows.

Let us take as before $n+1$ arcs of circles

$$A_{n+1}A_1,\ A_1A_2,\ \ldots,\ A_{n-1}A_n,\ A_nA_{n+1},$$

bounding a polygonal figure P of $n+1$ corners; but now, instead of supposing the circles $A_{r-1}A_r$ and A_rA_{r+1} to touch at A_r, let them cut at an angle (measured inside P) of

$$\frac{\pi}{m_r}, \quad (r = 1, 2, \ldots, n),$$

while A_nA_{n+1} and $A_{n+1}A_1$ cut at an angle $\dfrac{\pi}{m_{n+1}}$. Such a figure can again be chosen in an infinite variety of ways: we will suppose that it is drawn so that the $n+1$ circles have a common orthogonal circle. This clearly is always possible; but it is not now necessarily the case that this orthogonal circle is real. Let the figure P be now inverted in each of its sides; let the new figures so formed be inverted in each of their new sides; and so on continually. Then since the angles of P are sub-multiples of two right angles, no two of the figures thus formed can overlap in part without coinciding entirely. Moreover, when the process is completely carried out, every point within the orthogonal circle when it is real, and every point in the plane of the figure when the orthogonal circle is evanescent or imaginary, will lie in one and in only one of the polygons thus formed from P by successive inversions.

192. Exactly as with the general group, these polygons are coloured white or black according as they are derivable from P by an even or an odd number of inversions. The corners of any white polygon correspond one by one to the corners of P; so that, when the perimeters of the polygons are described in the same direction, corresponding corners occur in the same cyclical order.

If now the operation of successive inversions at $A_{n+1}A_1$ and A_1A_2 is represented by S_1, and that of successive inversions at $A_{r-1}A_r$ and A_rA_{r+1} by S_r, $(r = 2, 3, \ldots, n)$; all operations, consisting of an even number of inversions in the sides of P, can be represented in terms of

$$S_1,\ S_2,\ \ldots,\ S_n.$$

Moreover, from the construction of the polygon P, these operations satisfy the relations

$$S_1^{m_1} = 1,\quad S_2^{m_2} = 1,\quad \ldots,\quad S_n^{m_n} = 1,\quad S_{n+1}^{m_{n+1}} = 1,$$

where

$$S_1 S_2 \ldots S_n S_{n+1} = 1.$$

Again, if P' is any white polygon of the figure, which can be derived from P by the operation Σ, a positive rotation (§ 185) of P' round its corner A'_r is effected by the operation $\Sigma^{-1} S_r \Sigma$; and, if P'' is the polygon so obtained, P'' is derived from P by the operation $S_r \Sigma$. It is to be observed that a positive rotation of a polygon round its A_r corner is now an operation of finite order m_r.

Suppose now that two operations Σ and Σ' transform P into the same polygon P', so that $\Sigma\Sigma'^{-1}$ leaves P unchanged. If this operation, written at length, is

$$S_p^{\alpha} \ldots S_q^{\beta} S_r^{\gamma} S_s^{\delta},$$

and if P is transformed into P_1 by a positive rotation round A_s repeated δ times, P_1 into P_2 by a rotation round its corner A_r repeated γ times, and so on; then the operation may be indicated by a broken line drawn from P to P_1, from P_1 to P_2, and so on, the line returning at last to P. But the operation indicated by such a line is clearly equivalent to complete rotations, (i.e. rotations each of which lead to identity), round each of the corners which the broken line includes. In other words, $\Sigma\Sigma'^{-1}$ reduces to identity when account is taken of the relations which the generating operations satisfy. Hence finally, to every white polygon P' will correspond one and only one of the operations of the group, namely that operation which transforms P into P'. The same is clearly true of the black polygons; and by taking P and a chosen black polygon which has a side in common with P as corresponding to the identical operation, the required unique correspondence is established between the complete set of double-polygons in the figure and the operations of the group, the relations which the generating operations satisfy being directly indicated by the configuration of the figure. Moreover, as with the general group (§ 185), a simple rule may be stated for determining, from an inspection of the figure, the polygon that corresponds to any given operation of the group.

193. The number of polygons in the figure and therefore the order of the group will still, in general, be infinite. We may now proceed, just as in the previous case of a quite general group, to derive from the figure representing the group G, generated by n operations satisfying the relations

$$S_1^{m_1} = 1, \quad S_2^{m_2} = 1, \quad \ldots, \quad S_n^{m_n} = 1, \quad (S_1 S_2 \ldots S_n)^{m_{n+1}} = 1,$$

a suitable representation of the more special group $\overline{G}$, generated by n operations which satisfy the above relations and in addition the further m relations

$$f_j(S_i) = 1, \quad (j = 1, 2, \ldots, m).$$

As has been seen in § 182, if H is the self-conjugate sub-group of G which is formed by combining all possible operations of the form

$$R^{-1} f_j(S_i) R,$$

and if Σ is any operation of G, then the set of operations ΣH, regarded as operations of $\overline{G}$, are all equivalent to each other. From each set of polygons ΣH in the figure of G, we may therefore choose one to represent the corresponding operation of $\overline{G}$; and, as was shewn with the general group, a complete set of such representative polygons may be selected to form a connected figure, i.e. a figure which does not consist of two or more portions which are either isolated or connected only by corners. Moreover, as in the former case, the sides of this figure C will be connected in pairs $A'_r A'_{r+1}$ and $A''_r A''_{r+1}$, which are transformed into each other by some operation T of H and its inverse, while no other operation of H will transform either $A'_r A'_{r+1}$ or $A''_r A''_{r+1}$ into any other side of C.

It is not now however necessarily the case that the figure C, as thus constructed, is simply connected. Let us suppose then that C has one or more inner boundaries as well as an outer boundary, and denote one of these inner boundaries by L. If the sides of L do not all correspond in pairs, and if $A'_r A'_{r+1}$ is a side of L such that the other side $A''_r A''_{r+1}$ corresponding to it does not belong to L, we may replace the double-polygon P'' in C of which $A''_r A''_{r+1}$ is a side by the double-polygon, not previously belonging to C, of which $A'_r A'_{r+1}$ is a side. If P'' has a side on the boundary L, the new figure C' thus obtained will have one inner boundary less than C; and if P'' has no side on the boundary L, the new inner boundary L' that is thus formed from L will contain one double-polygon less than L, while the number of inner boundaries is not increased. This process may be continued till the new inner boundary L_1 which replaces L is such that all of its sides correspond in pairs.

Let now $A_s A_{s+1}$ and $A'_s A'_{s+1}$ be a pair of corresponding sides of L_1, such that $A_s A_{s+1}$ is transformed into $A'_s A'_{s+1}$ by an operation h of the self-conjugate sub-group H. A side $A_t A_{t+1}$ of another boundary of C may be chosen such that $A_s A_{s+1}$ and $A_t A_{t+1}$ are sides of a simply connected portion, say B, of C; while no side of L_1 except $A_s A_{s+1}$ forms part of the boundary of B. The polygons of B are equivalent, in respect of the special group, to those of Bh. Moreover, since the sides of L_1 correspond in pairs, no side of Bh, except $A'_s A'_{s+1}$ can coincide with a side of L_1. Hence when B is replaced by Bh, the inner boundary L_1 will be got rid of and no new inner boundary will be formed. Finally then, C may always be chosen so as to form a single simply connected figure.

The simply connected plane figure C, which has thus been constructed, with the correspondence of the sides of its boundary in pairs, will now give a complete graphical representation of the special group. The rule already formulated will determine the operation of the group to which each white polygon corresponds; and when, in carrying out this rule, we come to a polygon on the boundary, the correspondence of the sides of the boundary in pairs will enable the process to be continued.

The correspondence of the sides of C in pairs involves a correspondence of the corners in sets of two or more. Thus if A_r is a corner of C and if, of the m_r white polygons which in the complete figure have a corner at A_r, n_1 lie within C, there must within C be $m_r - n_1$ white polygons equivalent to the remainder, and each of these must have an A_r corner on the boundary. If A'_r is a corner of C such that there are n_2 white polygons, lying within C and having a corner at A'_r, and if one of the sides of the boundary with a corner at A'_r corresponds to one of the sides of the boundary with a corner at A_r, these n_2 white polygons must be equivalent to n_2 of the white polygons, lying outside C and having a corner at A_r. If

$$n_1 + n_2 < m_r,$$

there must be a third corner A''_r, contributing n_3 more white polygons towards the set. With this we proceed as before; and the process may be continued till the whole of the m_r white polygons surrounding A_r are accounted for. The set of corners A_r, A'_r, A''_r, ... will then form a set of corresponding corners, which are equivalent to each other in respect of the special group; and the whole of the corners of C may be divided into such sets. At each set of corresponding corners A_r of C there must clearly be also m_r black polygons belonging to C; and the sum of the angles of C at a set of corresponding corners must be equal to four right angles.

194. When the order of the group is finite, we may still further so modify our figure as to take account of the correspondence of the sides of the boundary in pairs. We may, in fact, by a suitable bending and stretching of the figure, bring corresponding sides of the boundary to actual coincidence. When this is done, the figure will no longer be a piece of a plane with a single boundary, but will form a continuous surface, which is unbounded and in general is multiply connected. Every point A_r on the surface, which in the plane figure did not lie on the boundary, will be a corner common to $2m_r$ polygons alternately black and white; and, in consequence of what has just been seen in regard to the correspondence of corners of the boundary, the same is true for every point A_r on the surface which in the plane figure consisted of a set of corresponding corners of the boundary. If N is the order of the group, the continuous unbounded surface will be divided into $2N$ polygons, black and white. The configuration of the set of white polygons with respect to any one of them will, from the point of view of geometry of position, be the same as that with respect to any other; and the like is true for the black polygons. Such a division of a continuous unbounded surface is described as a *regular* division; and we have finally, as a graphical representation of any group of finite order N, a division of a continuous surface into $2N$ polygons, half black and half white, which is regular with respect to each set. The correspondence between the

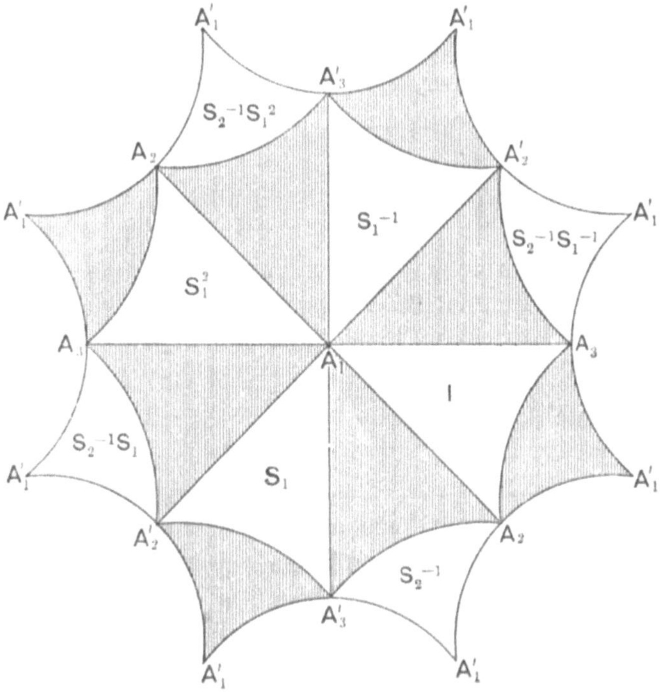

Fig. 7.

operations of the group and the white polygons on the surface is given by the rule that a single positive rotation of the white polygon Σ round its corner A_r leads to the white polygon $S_r\Sigma$.

195. We may again here illustrate this final modification of the graphical representation of a finite group by a simple example. For this purpose, we choose the group defined by

$$S_1^4 = 1, \quad S_2^4 = 1, \quad S_3^4 = 1,$$
$$S_1 S_2 S_3 = 1, \quad S_2^{-1} S_1 S_2 S_1 = 1.$$

This group (§ 74) is a non-Abelian group of order 8, containing a single operation of order 2. The reader will have no difficulty in verifying that the plane figure for this group is given by fig. 7; and that opposite sides of the octagonal boundary correspond. The single operation of order 2 is

$$S_1^2 (= S_2^2 = S_3^2);$$

this corresponds to a displacement of the triangles among themselves in which all the six corners remain fixed. If now corresponding sides of the boundary are brought to coincidence, the continuous surface formed will be a double-holed anchor-ring, or sphere with two holes through it. A view of one half of the surface divided into black and white triangles, is given in fig. 8. The half of the surface, not shewn, is divided up in a similar manner; and the operation of order 2 replaces each triangle of one half by the corresponding triangle of the other, an operation which clearly leaves the six corners of the polygons undisplaced.

196. The form of the plane figure C, which with the correspondence of its bounding sides in pairs represents the group, is capable of indefinite modification by replacing individual polygons on the boundary by equivalent polygons. If however we reckon a pair of corresponding sides of the boundary as a single side and a set of corresponding corners of the boundary as a single corner, it is clear that, however the figure may be modified, the numbers of its corners, sides and polygons remain each constant. This may be immediately verified on replacing any single boundary polygon by its equivalent.

If now A be the number of corners, and E the number of sides in the figure C when reckoned as above, $2N$ being the number of polygons, then the connectivity* $2p + 1$ of the closed surface is given by the equation

$$2p = 2 + E - 2N - A.$$

*Forsyth, *Theory of Functions*, p. 325.

Fig. 8.

When the group and its generating operations are given, the integer p is independent of the form of the plane figure C, which as has been seen is capable of considerable modification. The plane figure C however depends directly on the set of generating operations that is chosen for the group. For a given group of finite order, such a set is not in general unique; and the number of generating operations as well as their order will in general vary from one set to another. It does not necessarily follow, and in fact it is not generally the case, that the connectivity of the surface by whose regular division the group is represented, is independent of the choice of generating operations. There must however obviously be a lower limit to the number p for any given group of finite order, whatever generating operations are chosen; this we shall call the *genus* of the group*.

197. We shall now shew that there is a limit to the order of a group which can be represented by the regular division of a surface of given connectivity $2p+1$. If N is the order of such a group, generated by the n operations

$$S_1,\ S_2,\ \ldots,\ S_n,$$

which satisfy the relations

$$S_1^{m_1}=1,\quad S_2^{m_2}=1,\quad \ldots,\quad S_n^{m_n}=1,$$
$$S_1S_2\ldots S_n=1;$$

the surface will be divided in $2N$ polygons of n sides each. Let A_1, A_2, ..., A_n be the angular points of one of these polygons; and suppose that on the surface there are C_1 corners in the set to which A_1 belongs, C_2 in the set to which A_2 belongs, and so on. Round each corner A_r there are $2m_r$ polygons; and each polygon has one and only one corner of the set to which A_r belongs. Hence

$$C_r m_r = N,$$

and so

$$C_1+C_2+\ldots+C_n = N\sum_1^n \frac{1}{m_r}.$$

Again, each side belongs to two and only to two polygons, so that the number of sides is

$$Nn.$$

Using these values for A and E in the formula of § 196, we obtain the equation

$$2(p-1) = N\left(n-2-\sum_1^n \frac{1}{m_r}\right).$$

A complete discussion of this equation for the cases $p=0$ and $p=1$ will be given in the next chapter.

*Hurwitz, "Algebraische Gebilde mit eindeutigen Transformationen in sich," *Math. Ann.* XLI, (1893), p. 426.

When p is a given integer greater than unity, we can determine the greatest value that is possible for N by finding the least possible positive value of the expression

$$n - 2 - \sum_1^n \frac{1}{m_r}.$$

If $n > 4$, this quantity is not less than $\frac{1}{2}$, since m_r cannot be less than 2.

If $n = 4$, the simultaneous values

$$m_1 = m_2 = m_3 = m_4 = 2$$

are not admissible, since they make the expression zero. Its least value in this case will therefore be given by

$$m_1 = m_2 = m_3 = 2, \quad m_4 = 3;$$

and the expression is then equal to $\frac{1}{6}$.

If $n = 3$, we require the least positive value of

$$K = 1 - \frac{1}{m_1} - \frac{1}{m_2} - \frac{1}{m_3}.$$

Now the three sets of values

$$\begin{aligned} m_1 &= 3, \quad m_2 = 3, \quad m_3 = 3, \\ m_1 &= 2, \quad m_2 = 4, \quad m_3 = 4, \\ \text{and} \quad m_1 &= 2, \quad m_2 = 3, \quad m_3 = 6, \end{aligned}$$

each make K zero; and therefore no positive value of K can be less than the least of those given by

$$\begin{aligned} m_1 &= 3, \quad m_2 = 3, \quad m_3 = 4, \\ m_1 &= 2, \quad m_2 = 4, \quad m_3 = 5, \\ \text{and} \quad m_1 &= 2, \quad m_2 = 3, \quad m_3 = 7. \end{aligned}$$

These sets of values give for K the values $\frac{1}{12}$, $\frac{1}{20}$ and $\frac{1}{42}$. Hence finally, the absolutely least positive value of the expression is $\frac{1}{42}$, and therefore the greatest admissible value of N is

$$84(p-1).$$

Hence*:—

THEOREM II. *The order of a group, that can be represented by the regular division of a surface of connectivity* $2p+1$, *cannot exceed* $84(p-1)$, *p being greater than unity.*

198. If, when a group is represented by the regular division of an unbounded surface, we draw a line from any point inside the white polygon 1 (or any other polygon) returning after any path on the surface to the point from which it started, it will represent a relation between the generating operations of the group. For in following out along the line so drawn the rule that determines the operation of the group corresponding to each white polygon, some operation

$$F(S_i)$$

will be found to correspond to the final polygon; and this being the white polygon 1, it follows that

$$F(S_i) = 1.$$

If the surface is simply connected, any such line can be continuously altered till it shrinks to a point; and therefore the $n+1$ relations between the n generating operations completely define the group, since all other relations can be deduced from them.

If however the surface is of connectivity $2p+1$, there are $2p$ independent closed paths that can be drawn on the surface, no one of which can by continuous displacement either be shrunk up to a point or brought to coincidence with another; and every closed path on the surface can by continuous displacement either be brought to a point or to coincidence with a path constructed by combination and repetition of the $2p$ independent paths†. Any one of these $2p$ independent paths will give a relation between the n generating operations of the group, which cannot be deduced from the $n+1$ relations on which the angles of the polygons depend. Moreover, every relation between the generating operations can be represented by a closed path on the surface; and therefore there can be no further relation

*Hurwitz, *loc. cit.* p. 424.

†Forsyth, *Theory of Functions*, p. 330.

independent of the original relations and those obtained from the $2p$ independent paths. There cannot therefore be more than $2p$ independent relations between the n generating operations of a group, in addition to the $n+1$ relations that give the order of the generating operations and of their product; $2p+1$ being the connectivity of the surface by whose regular division into n-sided polygons the group is represented.

The $2p$ relations given by $2p$ independent paths on the surface are not, however, necessarily independent. In fact we have already had an example to the contrary in § 195. On the closed surface, by the regular division of which the group there considered is represented, four independent closed paths can be drawn. Any three of the corresponding relations can be derived from the fourth by transformation.

The only known cases in which the $2p$ relations are independent are those of a class of groups of genus one (§ 205).

Ex. Draw the figure of the group generated by S_1, S_2, S_3, where

$$S_1^2 = 1, \quad S_2^3 = 1, \quad S_3^8 = 1, \quad S_1 S_2 S_3 = 1.$$

Shew from the figure that the special group, given by the additional relation

$$(S_1 S_3^4)^2 = 1,$$

is a finite group of order 48; and that it can be represented by the regular division of a surface of connectivity 5.

Note to § 194.

If in the process of bending and stretching, described in § 194, by means of which the plane figure C is changed into an unbounded surface, the angles of the polygons all remain unaltered, the circles of the plane figure will become continuous curves on the surface. These curves on the surface, which we will still call circles, are necessarily re-entrant. It is not however necessarily the case that, on the surface, a circle will not cut itself.

In the plane figure for the general group, an inversion at any circle of the figure leaves the figure unchanged geometrically but interchanges the black and white polygons. Each circle is, in fact, a line of symmetry for the figure such that, in respect of it, there is corresponding to every white polygon a symmetric black polygon and vice versâ.

Similarly on the surface a circle which does not cut itself may be a line of symmetry, such that a reflection at it is an operation of order two which leaves the surface and its division into polygons unchanged, but interchanges black and white polygons. When this is the case, every circle on the surface will be a line of symmetry and no circle will cut itself. On the other hand no such operation can ever be connected with a circle which cuts itself.

When such lines of symmetry exist, Prof. Dyck speaks of the division of the surface as regular and symmetric.

CHAPTER XIII.

ON THE GRAPHICAL REPRESENTATION OF GROUPS: GROUPS OF GENUS ZERO AND UNITY: CAYLEY'S COLOUR GROUPS.

199. WE shall now proceed to a discussion in the cases $p = 0$ and $p = 1$ of the relation

$$2(p-1) = N\left(n - 2 - \sum_{1}^{n} \frac{1}{m_r}\right),$$

which connects the number and the orders of the generating operations of a group with the order of the group itself; and to the consideration of the corresponding groups.

For any given value of p, other than $p = 1$, we may regard this relation as an equation connecting the positive integers N, n, m_1, m_2, ..., m_n. It does not however follow from the investigations of the last Chapter that there is always a group or a set of groups corresponding to a given solution of the equation. In fact, for values of p greater than 1, this is not necessarily the case. We shall however find that, when $p = 0$, there is a single type of group corresponding to each solution of the equation; and that, when $p = 1$, there is an infinite number of types of group, all characterized by a common property, corresponding to each solution of the equation. When $p = 0$, the groups are (§ 196) of genus zero; and all possible groups of genus zero are found by putting $p = 0$ in the equation. The groups thus obtained are of special importance in many applications of group-theory; for this reason, they will be dealt with in considerable detail.

200. When $p = 0$, the equation may be written in the form

$$2\left(1 - \frac{1}{N}\right) = \sum_{1}^{n}\left(1 - \frac{1}{m_r}\right);$$

in this form, it is clear that the only admissible values of n are 2 and 3.

First, let $n = 2$. The only possible solution then is

$$N = m_1 = m_2 = n,$$

n being any integer. The corresponding group is a cyclical group of order n.

Secondly, let $n = 3$. In this case, one at least of the three integers m_1, m_2, m_3 must be equal to 2, as otherwise the right-hand side of the equation would be not less than 2. We may therefore without loss of generality put $m_1 = 2$. If now both m_2 and m_3 were greater than 3, the right-hand side would still be not less than 2; and therefore we may take m_2 to be either 2 or 3. When m_1 and m_2 are both 2, the equation becomes

$$\frac{2}{N} = \frac{1}{m_3};$$

giving

$$m_3 = n, \quad N = 2n,$$

where n is any integer.

When m_1 is 2 and m_2 is 3, the equation is

$$\frac{2}{N} + \frac{1}{6} = \frac{1}{m_3}.$$

This has three solutions in positive integers; namely,

$$\begin{aligned} m_3 &= 3, \quad N = 12; \\ m_3 &= 4, \quad N = 24; \\ \text{and} \quad m_3 &= 5, \quad N = 60. \end{aligned}$$

The solutions of the equation for the case $p = 0$ may therefore be tabulated in the form:—

	m_1	m_2	m_3	N
I	n	n		n
II	2	2	n	$2n$
III	2	3	3	12
IV	2	3	4	24
V	2	3	5	60

201. That a single type of group actually exists, corresponding to each of these solutions, may be seen at once by returning to our plane figure. The sum of the internal angles of the triangle $A_1A_2A_3$ formed by circular arcs is, in each of these cases, greater than two right angles; and the common orthogonal circle is therefore imaginary. The complete figure will therefore divide the whole plane into black and white triangles, so that there are no boundaries to consider. Moreover, the number of white triangles in each case must be equal to the corresponding value of N; for the preceding investigation shews that this is a possible value, and on the other hand the process, by which the figure is completed from a given original triangle, is a unique one. There is therefore a group corresponding to each solution; and the correspondence which has been established in any case between the operations of a group and the polygons of a figure, proves that there cannot be two distinct types of group corresponding to the same solution.

202. The plane figure for $p = 0$ does not, in fact, differ essentially from the figure drawn on a continuous simply connected surface in space. The former may be regarded as the stereographic projection of the latter. The five distinct types are represented graphically by the following figures.

The first is a cyclical group, and the figure (fig. 9) does not differ essentially from fig. 2 in § 183.

The group given by the second solution of the equation is called the dihedral group. It is represented by fig. 10.

The group given by the third solution of the equation is represented in fig. 11. It is known as the tetrahedral group.

To the fourth solution of the equation corresponds the group represented by fig. 12. It is known as the octohedral group.

To the fifth solution of the equation corresponds the group represented in fig. 13. It is known as the icosahedral group.

The four last groups are identical with the groups of rotations which will bring respectively a double pyramid on an n-sided base, a tetrahedron, an octohedron, and an icosahedron to coincidence with itself in each case*.

---

*Klein, "Vorlesungen über das Ikosaeder," Chap. I.

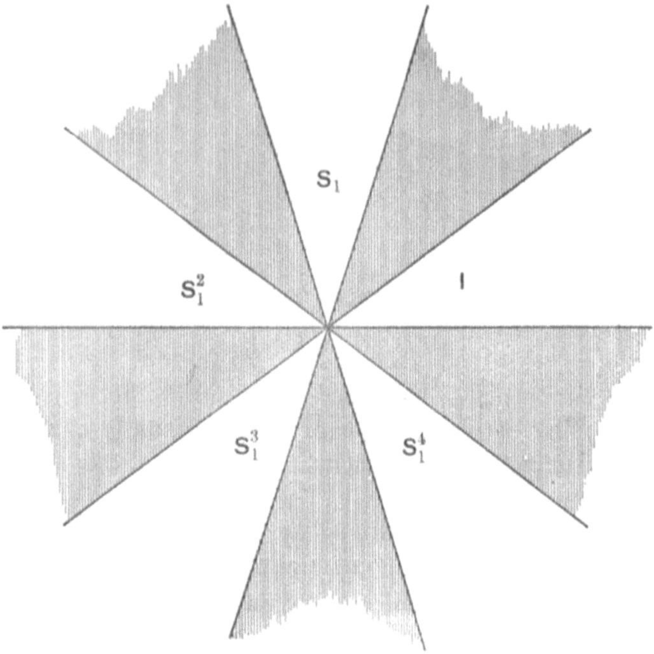

Fig. 9.

When the figures are drawn on a sphere, and the three circles of the original triangle and therefore also all the circles of the figure are taken to be great-circles of the sphere, the actual displacements of the triangles among themselves which correspond to the operations of the group can be effected by real rotations about diameters of the sphere; thus the statement of the preceding sentence may be directly verified.

203. In terms of their generating operations, the five types of group

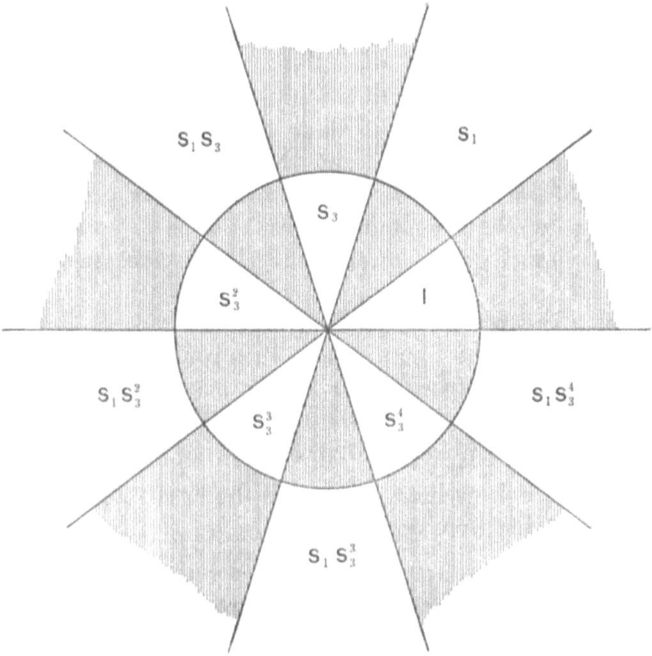

Fig. 10.

of genus zero are given by the relations:—

$$\begin{array}{lllll} \text{I.} & S_1^n = 1, & S_2^n = 1, & & S_1 S_2 = 1; \\ \text{II.} & S_1^2 = 1, & S_2^2 = 1, & S_3^n = 1, & S_1 S_2 S_3 = 1; \\ \text{III.} & S_1^2 = 1, & S_2^3 = 1, & S_3^3 = 1, & S_1 S_2 S_3 = 1; \\ \text{IV.} & S_1^2 = 1, & S_2^3 = 1, & S_3^4 = 1, & S_1 S_2 S_3 = 1; \\ \text{V.} & S_1^2 = 1, & S_2^3 = 1, & S_3^5 = 1, & S_1 S_2 S_3 = 1. \end{array}$$

The first of these does not require special discussion.

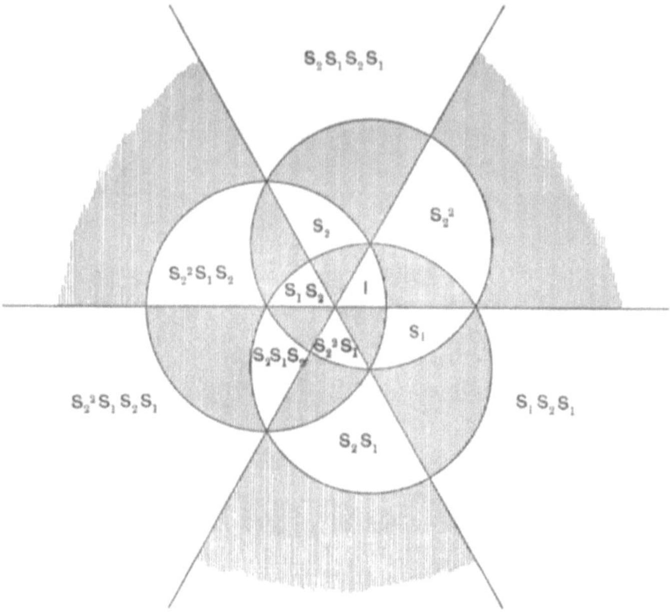

Fig. 11.

In the dihedral group, we have

$$S_1S_3S_1 = S_1S_2 = S_3^{-1}.$$

The dihedral group of order $2n$ therefore contains a cyclical sub-group of order n self-conjugately; and every operation of the group which does not belong to this self-conjugate sub-group is of order 2. The operations of the group are given, each once and once only, by the form

$$S_1^{\alpha}S_3^{\beta}, \quad (\alpha = 0, 1;\ \beta = 0, 1, 2, \ldots, n-1).$$

When $n = 3$, this group is simply isomorphic with the symmetric group of three symbols.

Fig. 12.

In the tetrahedral group, since

$$(S_1S_2)^3 = 1,$$
$$S_2^{-1}S_1S_2S_1 = S_2S_1S_2^{-1},$$

and therefore S_1, $S_2^{-1}S_1S_2$, $S_2S_1S_2^{-1}$ are permutable with each other. These operations of order 2 (with identity) form a self-conjugate sub-group of order 4; and the 12 operations of the group are therefore given by the form

$$S_2^{\alpha}(S_2^{-1}S_1S_2)^{\beta}S_1^{\gamma},$$

Fig. 13.

or

$$S_2^{\alpha} S_1^{\beta} S_2 S_1^{\gamma}, \quad (\alpha = 0, 1, 2;\ \beta, \gamma = 0, 1).$$

If

$$S_1 = (12)(34), \quad S_2 = (123),$$

then

$$S_1 S_2 = (134);$$

and therefore the tetrahedral group is simply isomorphic with the alternating group of four symbols.

If, in the octohedral group, we write

$$S_3^2 = S',$$

then

$$S_2 S' = S_1 S_3 = S_1 S_2^2 S_1;$$

and therefore

$$(S_2 S')^3 = 1.$$

Hence S_2 and S' generate a tetrahedral group.

Again

$$S_1 S_2 S_1 = (S_2 S')^2,$$

and

$$S_1 S' S_1 = S_2 S' S_2^{-1},$$

so that this is a self-conjugate sub-group. The operations of the group are given, each once and once only, in the form

$$S_1^{\alpha} S_2^{\beta} S_3^{2\gamma} S_2 S_3^{2\delta}, \quad (\alpha, \gamma, \delta = 0, 1;\ \beta = 0, 1, 2).$$

If

$$S_1 = (12), \quad S_2 = (234),$$

then

$$S_1 S_2 = (1342);$$

and therefore the octohedral group is simply isomorphic with the symmetric group of four symbols.

The icosahedral group is simple. It is, in fact, simply isomorphic with the alternating group of five symbols which has been shewn (§ 111) to be a simple group. Thus if

$$S_1 = (12)(34), \quad S_2 = (135),$$

then

$$S_1 S_2 = (12345);$$

so that the substitutions S_1 and S_2 satisfy the relations

$$S_1^2 = 1, \quad S_2^3 = 1, \quad (S_1S_2)^5 = 1.$$

They must therefore generate an icosahedral group or one of its sub-groups. On the other hand, from the substitutions S_1 and S_2 all the even substitutions of five symbols may be formed, and these are 60 in number. The group therefore cannot be a sub-group of the icosahedral group; the only alternative is that the two are identical.

As the icosahedral group has no self-conjugate group, we cannot in this case so easily construct a form which will represent each operation of the group just once in terms of the generating operations. It is however not difficult to verify that this is true of the set of forms*

$$\begin{array}{l} S_3^\alpha, \quad S_3^\alpha S_1 S_3^\beta, \quad S_3^\alpha S_1 S_3^2 S_1 S_3^\beta, \\ \qquad S_3^\alpha S_1 S_3^2 S_1 S_3^3 S_1, \end{array} \quad (\alpha, \beta = 0, 1, 2, 3, 4).$$

204. We shall next deal with the equation in the case $p = 1$. In this case alone, the order of the group disappears from the equation, which merely gives a relation between the number and order of the generating operations. This may be written in the form

$$2 = \sum_1^n \left(1 - \frac{1}{m_r}\right);$$

and n must therefore be either 4 or 3.

When n is 4, the equation becomes

$$2 = \frac{1}{m_1} + \frac{1}{m_2} + \frac{1}{m_3} + \frac{1}{m_4},$$

and the only solution is clearly

$$m_1 = m_2 = m_3 = m_4 = 2.$$

*Dyck, "Gruppentheoretische Studien," *Math. Ann.* XX, (1882), p. 35, and Klein, *loc. cit.* p. 26.

When n is 3, the equation takes the form

$$1 = \frac{1}{m_1} + \frac{1}{m_2} + \frac{1}{m_3},$$

and is easily seen to have three solutions, viz.

$$\begin{aligned} m_1 = 3, \quad m_2 = 3, \quad m_3 = 3; \\ m_1 = 2, \quad m_2 = 4, \quad m_3 = 4; \\ \text{and} \quad m_1 = 2, \quad m_2 = 3, \quad m_3 = 6. \end{aligned}$$

205. Take first the solution

$$n = 4, \quad m_1 = m_2 = m_3 = m_4 = 1.$$

The corresponding general group is defined by the relations

$$S_1^2 = 1, \quad S_2^2 = 1, \quad S_3^2 = 1, \quad S_4^2 = 1,$$
$$S_1 S_2 S_3 S_4 = 1.$$

If we proceed to form the plane figure representing this group, the sum of the internal angles of the quadrilateral $A_1A_2A_3A_4$ is equal to four right angles, and the four circles that form it therefore pass through a point. If this point be taken at infinity, the four circles (and therefore all the circles of the figure) become straight lines. The plane figure will now take the form given in fig. 14, and the four generating operations are actual rotations through two right angles about lines through A_1, A_2, A_3 and A_4, perpendicular to the plane of the figure.

Every operation of the group is therefore, in this form of representation, either a rotation through two right angles about a corner of the figure or a translation; and it will clearly be the former or the latter according as it consists of an odd or an even number of factors, when expressed in terms of the generating operations. The operations which correspond to translations form a sub-group; for if two operations each consist of an even number of factors, so also does their product. Moreover, this sub-group is self-conjugate, since the number of factors in $\Sigma^{-1}S\Sigma$ is even if the

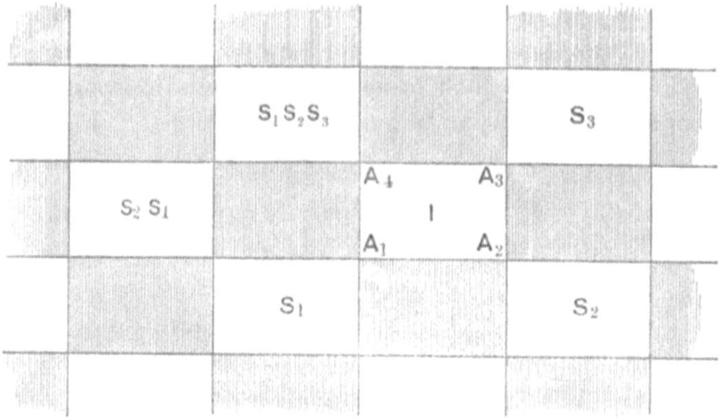

Fig. 14.

number in S is even. This self-conjugate sub-group is generated by the two operations

$$S_1S_2 \quad \text{and} \quad S_2S_3;$$

for

$$S_1S_3 = S_1S_2 \cdot S_2S_3,$$
$$S_2S_1 = (S_1S_2)^{-1}, \quad S_3S_2 = (S_2S_3)^{-1}, \quad S_3S_1 = (S_2S_3)^{-1}(S_1S_2)^{-1};$$

and therefore every operation containing an even number of factors can be represented in terms of S_1S_2 and S_2S_3. Lastly, these two operations are permutable with each other; for

$$S_2S_3 \cdot S_1S_2 = S_2S_3 \cdot S_4S_3 = S_1S_3 = S_1S_2 \cdot S_2S_3;$$

and therefore every operation of the group is contained, once and only once, in the form

$$S_1^{\alpha}(S_1S_2)^{\beta}(S_2S_3)^{\gamma}, \quad (\alpha = 0, 1: \ \beta, \gamma = -\infty, \ldots, 0, \ldots, \infty).$$

The results thus arrived at may also be verified very simply by purely kinematical considerations. If a group generated by S_1, S_2, S_3 and S_4 is of finite order, there must, since it is of genus 1, be either one or two additional relations between the generating operations; and any such relation is expressible by equating the symbol of some operation of the general group to unity. Such a relation is therefore either of the form

$$S_1(S_1S_2)^b(S_2S_3)^c = 1,$$

or

$$(S_1S_2)^b(S_2S_3)^c = 1.$$

The operation $S_1(S_1S_2)^b(S_2S_3)^c$ of the general group, consisting of an odd number of factors, must be a rotation round some corner of the figure, say a rotation round the corner A_r of the white quadrilateral Σ; it is therefore identical with $\Sigma^{-1}S_r\Sigma$.

Now the relation

$$\Sigma^{-1}S_r\Sigma = 1,$$

gives

$$S_r = 1.$$

A relation of the first of the two forms is therefore inconsistent with the supposition that the group is actually generated by S_1, S_2 and S_3. It, in fact, reduces the generating operations and the relations among them to

$$S_1^2 = 1, \quad S_2^2 = 1, \quad S_3^2 = 1, \quad S_1S_2S_3 = 1,$$

which define a group of genus zero.

The only admissible relations for a group of genus 1 are therefore those of the form

$$(S_1S_2)^b(S_2S_3)^c = 1.$$

A single relation of this form reduces the operations of the general group to those contained in

$$S_1^\alpha(S_1S_2)^\beta(S_2S_3)^\gamma,$$
$$(\alpha = 0, 1;\ \beta = 0, 1, \ldots, b-1;\ \gamma = -\infty, \ldots, 0, \ldots, \infty);$$

and the group so defined is still of infinite order.

Finally, two independent relations

$$(S_1S_2)^b(S_2S_3)^c = 1,$$
$$(S_1S_2)^{b'}(S_2S_3)^{c'} = 1,$$

where

$$\frac{b}{b'} \neq \frac{c}{c'},$$

must necessarily lead to a group of finite order. If m is the greatest common factor of b and b', so that

$$b = b_1m, \quad b' = b_1'm,$$

where b_1 and b_1' are relatively prime; and if

$$b_1x - b_1'y = 1;$$

the two relations give

$$(S_2S_3)^{cb_1'-c'b_1} = 1,$$

and

$$(S_1S_2)^m = (S_2S_3)^{c'y-cx}.$$

Every operation of the group is now contained, once and only once, in the form

$$S_1^\alpha(S_1S_2)^\beta(S_2S_3)^\gamma,$$
$$(\alpha = 0, 1;\ \beta = 0, 1, \ldots, m-1;\ \gamma = 0, 1, \ldots, cb_1' - b_1c' - 1);$$

and the order of the group is $2(bc' - b'c)$.

206. Corresponding to the solution

$$n = 3, \quad m_1 = m_2 = m_3 = 3,$$

we have the general group generated by S_1, S_2, S_3, where

$$S_1^3 = 1, \quad S_2^3 = 1, \quad S_3^3 = 1,$$
$$S_1S_2S_3 = 1.$$

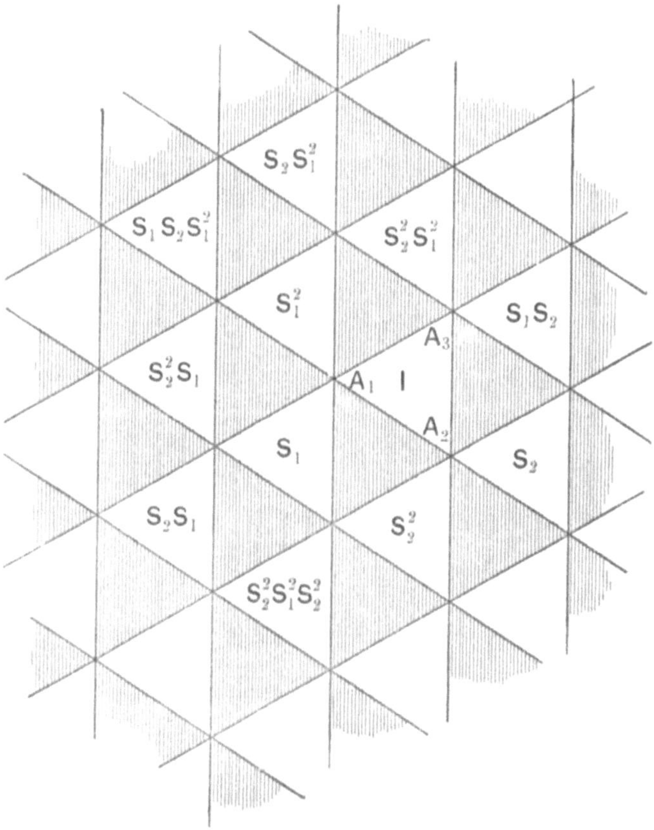

Fig. 15.

The sum of the three angles of the triangle $A_1 A_2 A_3$ is two right angles, and therefore again the circles in the plane figure (fig. 15) may be taken as straight lines. When the figure is thus chosen, the generating operations are rotations through $\frac{2}{3}\pi$ about the angles of an equilateral triangle; and every operation of the group is either a translation or a rotation.

The three operations

$$S_1S_2^2, \quad S_2S_1S_2, \quad S_2^2S_1,$$

when transformed by S_2, are interchanged among themselves. When transformed by S_1 they become

$$S_2^2S_1, \quad S_1^2S_2S_1S_2S_1, \quad S_1^2S_2^2S_1^2,$$

and since

$$(S_1S_2)^3 = 1,$$

the two latter are $S_1S_2^2$ and $S_2S_1S_2$ respectively.

Hence the three operations generate a self-conjugate sub-group; and since

$$S_1S_2^2 \cdot S_2S_1S_2 \cdot S_2^2S_1 = 1,$$

this sub-group is generated by $S_1S_2^2$ and $S_2S_1S_2$.

These two operations are permutable; for

$$S_1S_2^2 \cdot S_2S_1S_2 = S_1^2S_2 = S_2S_1S_2S_1S_2 \cdot S_2 = S_2S_1S_2 \cdot S_1S_2^2.$$

Hence finally, every operation of the group is represented, once and only once, by the form

$$S_1^\alpha(S_1S_2^2)^\beta(S_2S_1S_2)^\gamma, \quad (\alpha = 0, 1, 2;\ \beta, \gamma = -\infty, \ldots, 0, \ldots, \infty).$$

This result might also be arrived at by purely kinematical considerations; for an inspection of the figure shews that the two simplest translations are

$$S_1S_2^2 \quad \text{and} \quad S_2S_1S_2,$$

and that every translation in the group can be obtained by the combination and repetition of these. Every operation in which the index α is not zero must be a rotation through $\frac{2}{3}\pi$ or $\frac{4}{3}\pi$ about one of the angles of the figure; it is therefore necessarily identical with an operation of the form

$$\Sigma^{-1}S_r^\alpha\Sigma.$$

If now the group generated by S_1, S_2, S_3 is of finite order, there must be either one or two additional relations between them. A relation of the form

$$S_1^a(S_1S_2^2)^b(S_2S_1S_2)^c = 1,$$

where a is either 1 or 2, is equivalent to

$$\Sigma^{-1}S_r^a\Sigma = 1,$$

so that

$$S_r = 1.$$

Such a relation would reduce the group to a cyclical group of order 3. This is not admissible, if the group is actually to be generated by two distinct operations S_1 and S_2.

A relation

$$(S_1S_2^2)^b(S_2S_1S_2)^c = 1,$$

gives, on transformation by S_2^{-1},

$$(S_2S_1S_2)^b(S_2^2S_1)^c = 1.$$

Now

$$S_2^2S_1 = (S_1S_2^2)^{-1}(S_2S_1S_2)^{-1},$$

so that

$$(S_1S_2^2)^{-c}(S_2S_1S_2)^{b-c} = 1.$$

If m is the greatest common factor of b and c, so that

$$b = b'm, \quad c = c'm,$$

where b' and c' are relatively prime; and if

$$b'x - c'y = 1,$$

the two relations

$$(S_1S_2^2)^b(S_2S_1S_2)^c = 1,$$

and

$$(S_1S_2^2)^{-c}(S_2S_1S_2)^{b-c} = 1,$$

lead to

$$(S_2S_1S_2)^{m(b'^2-b'c'+c'^2)}=1,$$

and

$$(S_1S_2)^m=(S_2S_1S_2)^{m\{(c'-b')y-c'x\}};$$

and every operation of the group is contained, once and only once, in the form

$$S_1^\alpha(S_1S_2^2)^\beta(S_2S_1S_2)^\gamma,$$

where

$$\alpha=0,1,2;\ \beta=0,1,\ldots,m-1;\ \gamma=0,1,\ldots,m(b'^2-b'c'+c'^2)-1.$$

Thus the group is of finite order $3(b^2-bc+c^2)$. In this case then, unlike the previous one, a single additional relation is sufficient to ensure that the group is of finite order. Any further relation, which is independent, must of necessity reduce the group to a cyclical group of order 3 or to the identical operation.

207. The two remaining solutions may now be treated in less detail. The general group corresponding to the solution

$$n=3,\quad m_1=2,\quad m_2=m_3=4,$$

is given by

$$S_1^2=1,\quad S_2^4=1,\quad S_3^4=1,$$
$$S_1S_2S_3=1,$$

and is represented graphically by fig. 16.

All the translations of the group can be generated from the two operations

$$S_1S_2^2,\quad S_2S_1S_2;$$

and every operation of the group is given, once and only once, by the form

$$S_2^\alpha(S_1S_2^2)^\beta(S_2S_1S_2)^\gamma,\quad(\alpha=0,1,2,3;\ \beta,\gamma=-\infty,\ldots,0,\ldots,\infty).$$

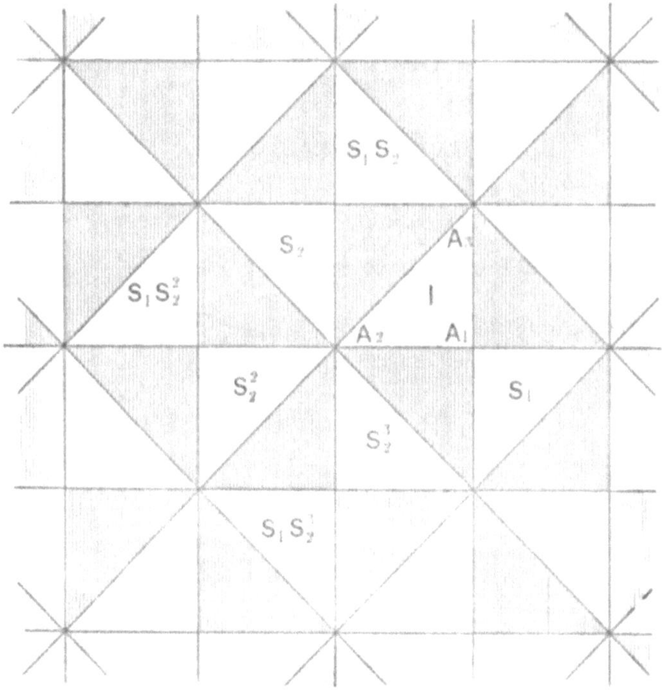

Fig. 16.

An additional relation of the form

$$S_2^a(S_1S_2^2)^b(S_2S_1S_2)^c - 1,$$

where a is 1, 2 or 3, leads either to

$$S_1 = 1, \quad S_2 = 1 \quad \text{or} \quad S_3 = 1,$$

and is therefore inconsistent with the supposition that the group is generated by two distinct operations.

An additional relation

$$(S_1S_2^2)^b(S_2S_1S_2)^c = 1$$

gives

$$(S_1S_2^2)^{-c}(S_2S_1S_2)^b = 1:$$

and if

$$b = b_1m, \quad c = c_1m,$$
$$b_1x + c_1y = 1,$$

where b_1 and c_1 are relatively prime, these relations are equivalent to

$$(S_2S_1S_2)^{m(b_1^2+c_1^2)} = 1,$$
$$(S_1S_2^2)^m = (S_2S_1S_2)^{m(b_1y-c_1x)}.$$

Every operation of the group is then contained, once and only once, in the form

$$S_2^\alpha(S_1S_2^2)^\beta(S_2S_1S_2)^\gamma,$$
$$(\alpha = 0, 1, 2, 3;\ \beta = 0, 1, \ldots, m-1;\ \gamma = 0, 1, \ldots, m(b_1^2+c_1^2)-1);$$

and the order of the group is $4(b^2 + c^2)$.

208. Lastly, the general group corresponding to the solution

$$n = 3, \quad m_1 = 2, \quad m_2 = 3, \quad m_3 = 6,$$

is given by

$$S_1^2 = 1, \quad S_2^3 = 1, \quad S_3^6 = 1,$$
$$S_1S_2S_3 = 1;$$

and it is represented graphically by fig. 17.

Now it may again be verified, either from the generating relations or from the figure, that the two operations

$$S_2^2S_3^2 \quad \text{and} \quad S_3S_2^2S_3,$$

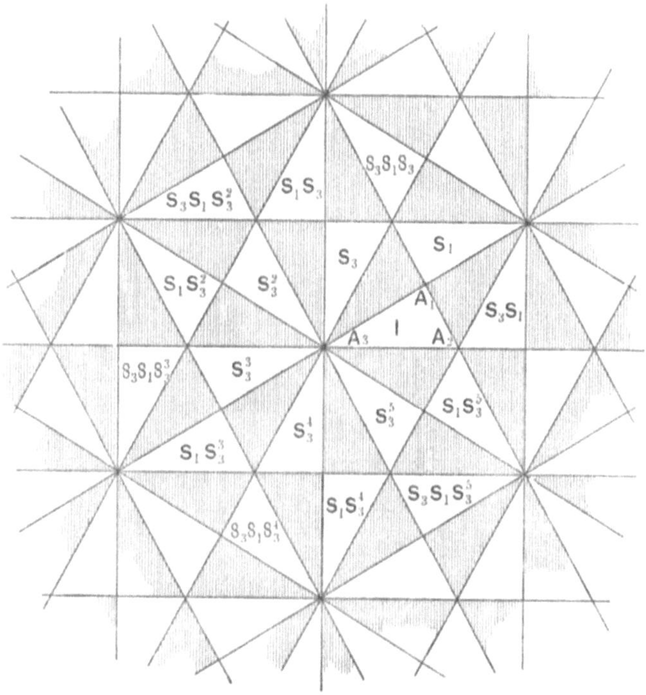

Fig. 17.

which are permutable with each other, generate all the operations which in the kinematical form of the group are translations; and that every operation of the group is represented, once and only once, by the form

$$S_3^\alpha (S_2^2 S_3^2)^\beta (S_3 S_2^2 S_3)^\gamma,$$
$$(\alpha = 0, 1, \ldots, 5;\ \beta, \gamma = -\infty, \ldots, 0, \ldots, \infty).$$

Also as before, any further relation, which does not reduce the group to a cyclical group, is necessarily of the form

$$(S_2^2 S_3^2)^b (S_3 S_2^2 S_3)^c = 1.$$

On transforming this relation by S_3^{-1} we obtain

$$(S_3 S_2^2 S_3)^b (S_3^2 S_2^2)^c = 1.$$

Now

$$S_3^2 S_2^2 = (S_2^2 S_3^2)^{-1} (S_3 S_2^2 S_3);$$

so that

$$(S_2^2 S_3^2)^{-c} (S_3 S_2 S_3)^{b+c} = 1.$$

If then

$$b = b_1 m, \quad c = c_1 m,$$
$$b_1 x + c_1 y = 1,$$

where b_1 and c_1 are relatively prime, it follows that

$$(S_3 S_2^2 S_3)^{m(b_1^2 + b_1 c_1 + c_1^2)} = 1,$$

and

$$(S_2^2 S_3^2)^m = (S_3 S_2 S_3)^{m\{b_1 y + c_1 (y-x)\}}.$$

Every operation of the group is then contained, once and only once, in the form

$$S_3^\alpha (S_2^2 S_3^2)^\beta (S_3 S_2^2 S_3)^\gamma,$$
$$(\alpha = 0, 1, \ldots, 5;\ \beta = 0, 1, \ldots, m-1;\ \gamma = 0, 1, \ldots, m(b_1^2 + b_1 c_1 + c_1^2) - 1);$$

and the order of the group is $6(b^2 + bc + c^2)$.

209. There are thus four distinct classes of groups* of genus 1, which are defined in terms of their generating operations by the following sets of relations:—

I. $$S_1^2 = 1, \quad S_2^2 = 1, \quad S_3^2 = 1, \quad (S_1 S_2 S_3)^2 = 1,$$
$$(S_1 S_2)^a (S_2 S_3)^b = 1, \quad (S_1 S_2)^{a'} (S_2 S_3)^{b'} = 1, \quad (ab' - a'b > 0);$$
$$N = 2(ab' - a'b).$$

*Dyck, "Ueber Aufstellung und Untersuchung von Gruppe und Irrationalität regulärer Riemann'scher Flächen," *Math. Ann.* XVII, (1880), pp. 501–509.

II. $S_1^3 = 1, \quad S_2^3 = 1, \quad (S_1S_2)^3 = 1,$

$$(S_1^2S_2^2)^a(S_2S_1S_2)^b = 1;$$

$$N = 3(a^2 - ab + b^2).$$

III. $S_1^2 = 1, \quad S_2^4 = 1, \quad (S_1S_2)^4 = 1,$

$$(S_1S_2^2)^a(S_2S_1S_2)^b = 1;$$

$$N = 4(a^2 + b^2).$$

IV. $S_1^6 = 1, \quad S_2^3 = 1, \quad (S_1S_2)^2 = 1,$

$$(S_1^2S_2^2)^a(S_1S_2^2S_1)^b = 1;$$

$$N = 6(a^2 + ab + b^2).$$

For special values of a and b, some of these groups may be groups of genus zero; for instance, in Class I, if $ab' - a'b$ is a prime, the group is a dihedral group. It is left as an exercise to the reader to determine all such exceptional cases.

Ex. Prove that the number of distinct types of group, of genus two, is three; viz. the groups defined by

(i) $A^4 = 1, \quad B^2 = A^2, \quad B^{-1}AB = A^{-1};$

(ii) $A^8 = 1, \quad B^2 = 1, \quad B^{-1}AB = A^3;$

(iii) $A^8 = 1, \quad B^2 = 1, \quad (AB)^3 = 1, \quad (A^4B)^2 = 1.$

210. As a final illustration of the present method of graphical representation, we will consider the simple group of order 168 (§ 146), given by

$$\{(1236457), (234)(567), (2763)(45)\}.$$

The operations of this group are of orders 7, 4, 3 and 2; and it is easy to verify that three operations of orders 2, 3, and 7 can be chosen such that their product is identity.

In fact, if

$$S_2 = (16)(34), \quad S_3 = (253)(476), \quad S_7 = (1673524);$$

then

$$S_2S_3S_7 = 1.$$

Moreover, these three operations generate the group. The connectivity of the corresponding surface, by the regular division of which the group can be represented, is $2p+1$; where

$$2p-2 = 168(3-2-\tfrac{1}{2}-\tfrac{1}{3}-\tfrac{1}{7}).$$

This gives

$$p = 3;$$

it follows from Theorem II, § 197, that the genus of the group is 3.

The figure for the general group, generated by S_2, S_3, and S_7, where

$$S_2^2 = 1, \quad S_3^3 = 1, \quad S_7^7 = 1,$$
$$S_2 S_3 S_7 = 1,$$

acquires as symmetrical a form as possible, by taking the centre of the orthogonal circle for that angular point of the triangle 1 at which the angle is $\frac{1}{7}\pi$. In fig. 18 a portion of the general figure, which is contained between two radii of the orthogonal circle inclined at an angle $\frac{2}{7}\pi$, is shewn. The remainder may be filled in by inversions at the different portions of the boundary.

The operations, which correspond to the white triangles of the figure, are given by the following table:—

1	1	9	$S_2S_7^5S_2$	17	$S_2S_7^5S_2S_7^3S_2$
2	S_2	10	$S_7S_2S_7^5S_2$	18	$S_7^6S_2S_7^3S_2$
3	S_7S_2	11	$S_7^2S_2S_7^5S_2$	19	$S_2S_7^3S_2$
4	$S_7^2S_2$	12	$S_7^3S_2S_7^5S_2$	20	$S_7S_2S_7^3S_2$
5	$S_7^3S_2$	13	$S_7^6S_2S_7^4S_2$	21	$S_2S_7^2S_2$
6	$S_7^4S_2$	14	$S_2S_7^4S_2$	22	$S_7^6S_2S_7^2S_2$
7	$S_7^5S_2$	15	$S_7S_2S_7^4S_2$	23	$S_7^5S_2S_7^2S_2$
8	$S_7^6S_2$	16	$S_7^2S_2S_7^4S_2$	24	$S_2S_7^4S_2S_7^2S_2$.

The representation of the special group is derived from this general figure by retaining only a set of 168 white (and corresponding black) triangles, which are distinct when S_2, S_3 and S_7 are replaced by the corresponding substitutions given on p. 337. When each white triangle is thus marked

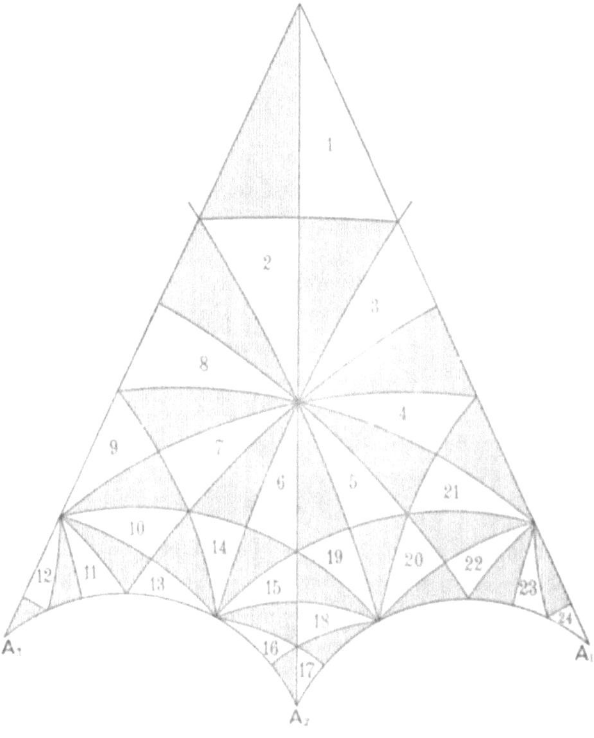

Fig. 18.

with the corresponding substitution, it is found that a complete set of 168 distinct white (and black) triangles is given by the portion of the figure actually drawn and the six other distinct portions obtained by rotating it round the centre of the orthogonal circle through multiples of $\frac{2}{7}\pi$.

To complete the graphical representation of the group of order 168, it is necessary to determine the correspondence in pairs of the sides of the boundary. This is facilitated by noticing that the angular points A_1, A_2, A_3, ... of the boundary must correspond in sets. Now the white triangle, which has an angle at A_1 and lies inside the polygon, is given by

$$S_2S_7^4S_2S_7^2S_2.$$

This must be equivalent to a white triangle, which lies outside the polygon, and has a side on the boundary and an angle at one of the points A_2, A_3, ...

The triangle, which satisfies these conditions and has an angle at A_{2n+1}, is given by

$$S_7S_2S_7^4S_2S_7^2S_2S_7^n;$$

while the triangle, which satisfies the conditions and has an angle at A_{2n+2}, is given by

$$S_7^6S_2S_7^5S_2S_7^3S_2S_7^n.$$

When (16)(34) and (1673524) are written for S_2 and S_7, we find that

$$S_2S_7^4S_2S_7^2S_2 = S_7S_2S_7^4S_2S_7^2S_2S_7^5.$$

The white triangle with an angle at A_1 inside the polygon is therefore equivalent to the white triangle with an angle at A_{11}, which lies outside the polygon and has a side on the boundary. It follows, from the continuity of the figure, that the arcs A_1A_2 and $A_{11}A_{10}$ of the boundary correspond. Since the operation S_7 changes the figure and the boundary into themselves, it follows that A_3A_4, $A_{13}A_{12}$; A_5A_6, A_1A_{14}; A_7A_8, A_3A_2; A_9A_{10}, A_5A_4; $A_{11}A_{12}$, A_7A_6; and $A_{13}A_{14}$, A_9A_8; are pairs of corresponding sides. Hence the above single condition is sufficient to ensure that the general group shall reduce to the special group of order 168.

By taking account of the relation

$$(S_7S_2)^3 = 1, \quad \text{or} \quad (S_2S_7)^3 = 1,$$

the form of the condition may be simplified. Thus it may be written

$$\begin{aligned} S_7^4S_2S_7^2S_2 &= S_2S_7S_2S_7 \cdot S_7^3S_2S_7^2S_2S_7^5 \\ &= S_7^6S_2S_7^3S_2S_7^2S_2S_7^5, \end{aligned}$$

or

$$S_2 S_7^2 S_2 = S_7^2 S_2 S_7^3 S_2 S_7^2 S_2 S_7^5.$$

Now

$$\begin{aligned} S_2 S_7 \cdot S_7 S_2 &= S_7^6 S_2 S_7^6 S_2 \cdot S_2 S_7^6 S_2 S_7^6 \\ &= S_7^6 S_2 S_7^5 S_2 S_7^6. \end{aligned}$$

Hence

$$S_7^4 S_2 S_7^5 S_2 S_7 = S_2 S_7^3 S_2 S_7^2 S_2,$$

or

$$S_7^4 S_2 S_7^4 \cdot S_2 S_7^6 S_2 = S_2 S_7^3 S_2 S_7^2 S_2,$$

or

$$S_7^4 S_2 S_7^4 S_2 S_7^4 = S_2 S_7^3 S_2,$$

or finally

$$(S_7^4 S_2)^4 = 1.$$

The simple group of order 168 is therefore defined abstractly by the relations*

$$S_2^2 = 1, \quad S_7^7 = 1, \quad (S_7 S_2)^3 = 1, \quad (S_7^4 S_2)^4 = 1.$$

Ex. Shew that the symmetric group of degree five is a group of genus four; and that it is completely defined by the relations

$$S_2^2 = 1, \quad S_5^5 = 1, \quad (S_2 S_5)^4 = 1, \quad (S_5^{-1} S_2 S_5 S_2)^3 = 1.$$

211. The regular division of a continuous surface into $2N$ black and white polygons is only one of many methods that may be conceived for representing a group graphically.

We shall now describe shortly another such mode of representation, due to Cayley†, who has called it the method of *colour-groups*. As given

*This agrees with the result as stated by Dyck, "Gruppentheoretische Studien," *Math. Ann.* Vol. XX, (1882), p. 41.

†*American Journal of Mathematics*, Vol. I, (1878), pp. 174–176, Vol. XI, (1889), pp. 139–157; *Proceedings of the London Mathematical Society*, Vol. IX, (1878), pp. 126–133.

by Cayley, this method is entirely independent of the one we have been hitherto dealing with; but there is an intimate relation between them, and the new method can be most readily presented to the reader by deriving it from the old one.

Let

$$1,\ S_1,\ S_2,\ \ldots,\ S_{N-1}$$

be the operations of a group G of order N. We may take the $N-1$ operations other than identity as a set of generating operations. Their continued product

$$S_1 S_2 \ldots S_{N-1}$$

is some definite operation of the group. If it is the identical operation, the only modification in the figure, which represents the group by the regular division of a continuous surface, will be that the Nth corner of the polygon has an angle of two right-angles.

With this set of generating operations, the representation of the group is given by a regular division of a continuous surface into N white and N black polygons $A_1A_2\ldots A_N$, the angle at A_r being $\dfrac{2\pi}{m_r}$, m_r being the order of S_r. Suppose now that in each white polygon we mark a definite point. From the marked point in the polygon Σ, draw a line to the marked point in the polygon derived from it by a positive rotation round its angle A_r. Call this line an S_r-line, and denote the direction in which it is drawn by means of an arrow. Carry out this construction for each polygon Σ, and for each of its angles except A_N. We thus form a figure which, disregarding the original surface, consists of N points connected by $N(N-1)$ directed lines, two distinct lines joining each pair of points. Now if the line drawn from a to b, where a and b are two of the points, is an S-line, then the line drawn from b to a is from the construction an S^{-1}-line. We may then at once modify our diagram, in the direction of simplification, by dropping out one of the two lines between a and b, say the S^{-1}-line, on the understanding that the remaining line, with the arrow-head reversed, will give the line omitted. If S is an operation of order 2, S and S^{-1} are identical, and two arrow-heads may be drawn on such a line in opposite directions. The modified figure will now consist of N points connected by $\frac{1}{2}N(N-1)$ lines. From the construction it follows at once that, for every value of r, a single S_r-line ends at each point of the figure and a single S_r-line begins at each point of the figure; these two lines being identical when the order of S_r is 2.

We may pass from one point of the figure to another along the lines in various ways; but any path between two points of the figure will be specified completely by such directions as: follow first an S_r-line, then an S_s-line, then an S_t^{-1}-line, and so on. Such a set of directions is said to define a route. It is an immediate consequence of the construction that, if starting from some one particular point a given route leads back to the starting point, then it will lead back to the starting point from whatever point we begin. In fact, a route will be specified symbolically by a symbol

$$S_u \dots S_t^{-1} S_s S_r,$$

and if

$$S_u \dots S_t^{-1} S_s S_r \Sigma = \Sigma,$$

then

$$S_u \dots S_t^{-1} S_s S_r = 1,$$

and therefore

$$S_u \dots S_t^{-1} S_s S_r \Sigma' = \Sigma',$$

whatever operation Σ' may be.

212. If the diagram of N points connected by $\frac{1}{2}N(N-1)$ directed lines is to appeal readily to the eye, some method must be adopted of easily distinguishing an S_r-line from an S_s-line. To effect this purpose, Cayley suggested that all the S_r-lines should be of one colour, all the S_s-lines of another, and so on. Suppose now that, independently of any previous consideration, we have a diagram of N points connected by $\frac{1}{2}N(N-1)$ coloured directed lines satisfying the following conditions:—

(i) all the lines of any one colour have either (a) a single arrow-head denoting their directions: or (b) two opposed arrow-heads, in which case each may be regarded as equivalent to two coincident lines in opposite directions;

(ii) there is a single line of any given colour leading to every point in the diagram, and a single line of the colour leading from every point: if the colour is one with double arrow-heads, the two lines are a pair of coincident lines;

(iii) every route which, starting from some one given point in the diagram, is closed, i.e. leads back again to the given point, is closed whatever the starting point.

Then, under these conditions, the diagram represents in graphical form a definite group of order N.

It is to be noticed that the first two conditions are necessary in order that the phrase "a route" used in the third shall have a definite meaning. Suppose that R and R' are two routes leading from a to b. Then RR'^{-1} is a closed route and will lead back to the initial point whatever it may be. Hence if R leads from c to d, so also must R'; and therefore R and R' are equivalent routes in the sense that from any given starting point they lead to the same final point. There are then, with identity which displaces no point, just N non-equivalent routes on the diagram, and the product of any two of these is a definite route of the set. The N routes may be regarded as operations performed on the N points; on account of the last property which has been pointed out, they form a group. Moreover, the diagram gives in explicit form the complete multiplication table of the group, for a mere inspection will immediately determine the one-line route which is equivalent to any given route; i.e. the operation of the group which is the same as the product of any set of operations in any given order.

From a slightly different point of view, every route will give a permutation of the N points, regarded as a set of symbols, among themselves; no symbol remaining unchanged unless they all do. To the set of N independent routes, there will correspond a set of N substitutions performed on N symbols; and we can therefore immediately from the diagram represent the group as a transitive substitution group of degree N.

213. It cannot be denied that, even for groups of small order, the diagram we have been describing would not be easily grasped by the eye. It may however still be considerably simplified since, so far as a graphical definition of the group is concerned, a large number of the lines are always redundant.

If in the diagram consisting of N points and $\frac{1}{2}N(N-1)$ coloured lines, which satisfies the conditions of § 212, all the lines of one or more colours are omitted, two cases may occur. We may still have a figure in which it is possible to pass along lines from one point to any other; or the points may break up into sets such that those of any one set are connected by lines, while there are no lines which enable us to pass from one set to another.

Suppose, to begin with, that the first is the case. There will then, as before, be N non-equivalent routes in the figure, which form a group when they are regarded as operations; it is obviously the same group as is given by the general figure. The sole difference is that there will not now be a one-line route leading from every point to every other point, and therefore the diagram will no longer give directly the result of the product of any number of operations of the group.

If on the other hand the points break up into sets, the new diagram will no longer represent the same group as the original diagram. Some of the routes of the original diagram will not be possible on the new one, but every route on the new one will be a route on the original diagram. Hence the new diagram will give a sub-group; and since it is still the case that no route, except identity, can leave any point unchanged, the number of points in each of the sets must be the same. The reader may verify that the sub-group thus obtained will be self-conjugate, only if the omitted colours interchange these sets bodily among themselves.

214. The simplest diagram that will represent the group will be that which contains the smallest number of colours and at the same time connects all the points. To each colour corresponds a definite operation of the group (and its inverse). Hence the smallest number of colours is the smallest number of operations that will generate the group. It may be noticed that this simplified diagram can be actually constructed from the previously obtained representation of the group by the regular division of a surface, the process being exactly the same as that by which the general diagram was obtained. For if

$$S_1,\ S_2,\ \ldots,\ S_n$$

are a set of independent generating operations, and if

$$S_1 S_2 \ldots S_n S_{n+1} = 1,$$

we may represent the group by the regular division of a surface into $2N$ black and white $(n+1)$-sided polygons. When we draw on this surface the S_1-, S_2-, ..., S_n-lines, the N points will be connected by lines in a single set, since from

$$S_1,\ S_2,\ \ldots,\ S_n$$

every operation of the group can be constructed; and the set of points and directed coloured lines so obtained is clearly the diagram required.

As an illustration of this form of graphical representation, we may consider the octohedral group (§ 201), defined by

$$S_1^2 = 1, \quad S_2^3 = 1, \quad S_3^4 = 1,$$
$$S_1 S_2 S_3 = 1.$$

On the diagram already given (p. 318), we may at once draw S_1-, S_2- and S_3-lines. These in the present figure are coloured respectively red, yellow, and green. On the lines which correspond to operations of order two (in this case the red lines), the double arrow-heads may be dispensed with. By omitting successively the red, the yellow, and the green lines we form from this the three simplest colour diagrams which will represent the group*.

*For further illustrations, the reader may refer to Young, *Amer. Journal*, Vol. XV, (1893), pp. 164–167; Maschke, *Amer. Journal*, Vol. XVIII, (1896), pp. 156–188.

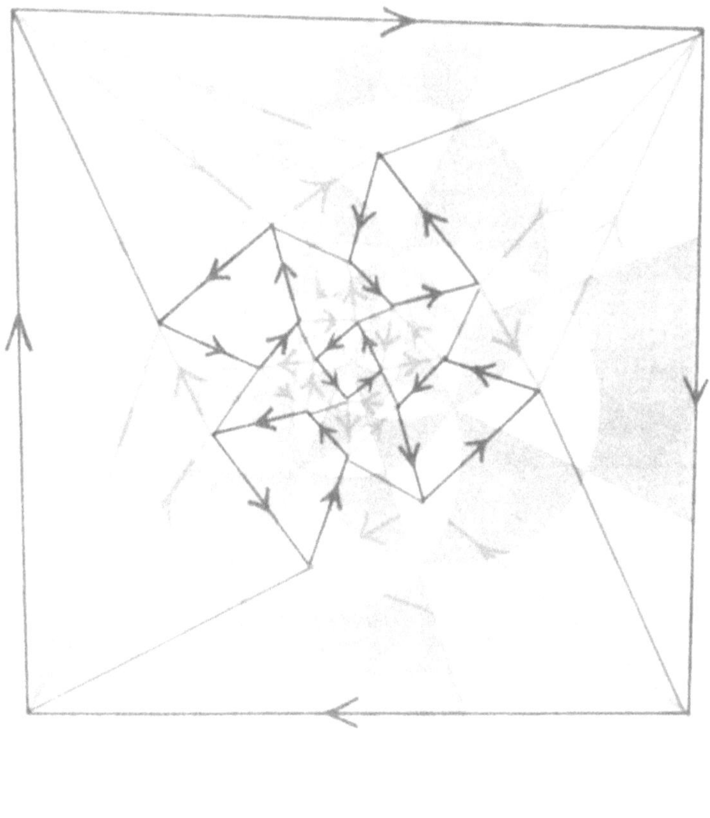

CHAPTER XIV.

ON THE LINEAR GROUP*.

215. WE shall now, in illustration of the general principles that have been developed in the preceding chapters, proceed to discuss and give an analysis of certain special groups. The first that we choose for this purpose is the group of isomorphisms of an Abelian group of order p^n and type $(1, 1, \ldots, \text{to } n \text{ units})$. This group has been defined and its order determined in §§ 171, 172. It is there shewn that the group is simply isomorphic with the homogeneous linear group defined by all sets of congruences

$$\left.\begin{array}{l} y_1 \equiv a_{11}x_1 + a_{12}x_2 + \ldots + a_{1n}x_n, \\ y_2 \equiv a_{21}x_1 + a_{22}x_2 + \ldots + a_{2n}x_n, \\ \ldots\ldots\ldots\ldots\ldots\ldots\ldots\ldots\ldots\ldots \\ y_n \equiv a_{n1}x_1 + a_{n2}x_2 + \ldots + a_{nn}x_n, \end{array}\right\} \pmod{p},$$

whose determinants are not congruent to zero; and that its order is

$$N = (p^n - 1)(p^n - p) \ldots (p^n - p^{n-1}).$$

*The homogeneous linear group and its sub-groups forms the subject of the greater part of Jordan's *Traité des Substitutions.* The investigation of its composition-series, given in the text, is due to Jordan.

The complete analysis of the fractional linear group, defined by

$$y \equiv \frac{\alpha x + \beta}{\gamma x + \delta}, \pmod{p},$$

where

$$\alpha\delta - \beta\gamma \equiv 1,$$

is due originally to Gierster, "Die Untergruppen der Galois'schen Gruppe der Modulargleichungen für den Fall eines primzahligen Transformationsgrades," *Math. Ann.* Vol. XVIII, (1881), pp. 319–365. With a few unimportant modifications, the investigation in the text follows the lines of Gierster's memoir.

A similar analysis of the simple groups of order $2^n(2^{2n} - 1)$, which can be expressed as triply transitive groups of degree $2^n + 1$, has been given by the author, "On a class of groups defined by congruences," *Proc. L. M. S.* Vol. XXV, (1894), pp. 132–136.

The operation given by the above set of congruences will be denoted in future by the symbol

$$(a_{11}x_1 + \ldots + a_{1n}x_n,\ a_{21}x_1 + \ldots + a_{2n}x_n,\ \ldots,\ a_{n1}x_1 + \ldots + a_{nn}x_n).$$

216. It may be easily verified by direct calculation that, if D and D' are the determinants of two operations S and S' of the group, then DD' is the determinant of the operation SS', all the numbers involved being reduced, (mod. p). Hence it immediately follows that those operations of the group, whose determinant is unity, form a sub-group. If this sub-group is denoted by Γ, the group itself being G, then Γ is a self-conjugate sub-group of G. For if Σ is any operation of Γ and S any operation of G whose determinant is D, the determinant of $S^{-1}\Sigma S$ is $D^{-1}D$ or unity; and therefore $S^{-1}\Sigma S$ belongs to Γ. Suppose now that S is an operation* of G whose determinant is z, a primitive root of the congruence

$$z^{p-1} \equiv 1 \text{ (mod. } p).$$

Then the determinant of every operation of the set

$$S^r\Gamma$$

is z^r; and therefore, if r and s are not congruent (mod. p), the two sets

$$S^r\Gamma \quad \text{and} \quad S^s\Gamma$$

can have no operation in common. Moreover, if S' is any operation of G whose determinant is z^r, then $S^{-r}S'$ belongs to Γ, and therefore S' belongs to the set $S^r\Gamma$. Hence finally, the sets

$$\Gamma,\ S\Gamma,\ S^2\Gamma,\ \ldots,\ S^{p-2}\Gamma$$

are all distinct, and they include every operation of G; so that

$$G = \{S, \Gamma\}.$$

*The operation $(zx_1, x_2, \ldots, x_n)$ has z for its determinant.

The factor-group $\frac{G}{\Gamma}$ is therefore cyclical and of order $p-1$.

217. It may be very readily verified that the operations of the cyclical sub-group generated by

$$(zx_1, zx_2, \ldots, zx_n)$$

are self-conjugate operations of G. To prove that these are the only self-conjugate operations of G, we will deal with the case $n = 3$: it will be seen that the method is perfectly general. Suppose then that

$$T = (\alpha x_1 + \beta x_2 + \gamma x_3,\ \alpha' x_1 + \beta' x_2 + \gamma' x_3,\ \alpha'' x_1 + \beta'' x_2 + \gamma'' x_3)$$

is a self-conjugate operation of G, while

$$S = (ax_1 + bx_2 + cx_3,\ a' x_1 + b' x_2 + c' x_3,\ a'' x_1 + b'' x_2 + c'' x_3)$$

is any operation. The relation

$$ST = TS$$

involves the nine simultaneous congruences*

$$\begin{aligned} a\alpha + b\alpha' + c\alpha'' &\equiv \alpha a + \beta a' + \gamma a'', \\ a\beta + b\beta' + c\beta'' &\equiv \alpha b + \beta b' + \gamma b'', \\ a\gamma + b\gamma' + c\gamma'' &\equiv \alpha c + \beta c' + \gamma c'', \\ \text{etc.,} \qquad & \qquad \text{etc.;} \end{aligned}$$

and these must be satisfied for all possible values of the coefficients of S. Now

$$b \equiv c \equiv a' \equiv 0$$

is a possible relation between the coefficients of S, whether regarded as an operation of G or Γ; and therefore

$$\gamma \equiv 0.$$

---

*These and all succeeding congruences are to be taken (mod. p), unless the contrary is stated.

In the same way, it may be shewn that

$$\beta \equiv \alpha' \equiv \gamma' \equiv \alpha'' \equiv \beta'' \equiv 0,$$

and that

$$\alpha \equiv \beta' \equiv \gamma'';$$

so that T is a power of the operation

$$(zx_1, zx_2, zx_3).$$

The only self-conjugate operations of G are therefore the powers of A, where A denotes

$$(zx_1, zx_2, \ldots, zx_n);$$

and the only self-conjugate operations of Γ are those operations of this cyclical sub-group which are contained in Γ. Now the order of A is $p-1$ and its determinant is z^n. Hence the self-conjugate operations of Γ form a cyclical sub-group D of order d, where d is the greatest common factor of $p-1$ and n; and this sub-group is generated by $A^{\frac{p-1}{d}}$.

218. To determine completely the composition-series of G, it is necessary to find whether Γ has a self-conjugate sub-group greater than and containing D. A simple calculation will shew that, from

$$(x_1 + x_s, x_2, x_3, \ldots, x_n)$$

and its conjugate operations, all the operations of Γ may be generated; and hence no self-conjugate sub-group of Γ which is different from Γ itself can contain an operation of this form. If then it is shewn that any self-conjugate sub-group of Γ, distinct from D, necessarily contains operations of this form, it follows that D is a maximum self-conjugate sub-group of Γ.

We shall first deal with the case $n = 2$.

If $p = 2$, the orders of G, Γ and D are 6, 6 and 1. In this case, Γ is simply isomorphic with the symmetric group of three symbols, which has a self-conjugate sub-group of order 3. The successive factor-groups of the composition-series of G are therefore cyclical groups of orders 2 and 3.

If $p = 3$, the orders of G, Γ and D are 48, 24 and 2. The factor-group $\frac{\Gamma}{D}$ has 12 for its order, and cannot therefore be a simple group. The reader will have no difficulty in verifying that, in this case, the successive factor-groups of G have orders 2, 3, 2, 2 and 2. We may therefore, in dealing with the case $n = 2$, assume that p is not less than 5.

Let us suppose now that Γ has a self-conjugate sub-group I that contains D; and let S or

$$(ax_1 + bx_2,\ a'x_1 + b'x_2)$$

be one of its operations, not contained in D.

If b is different from zero, Γ contains Σ, where Σ denotes

$$\left(\alpha a x_1 + \alpha b x_2, -\frac{1+\alpha^2 a^2}{\alpha b} x_1 - \alpha a x_2\right),$$

and therefore I contains $\Sigma^{-1} S \Sigma S$, which is

$$\left(-\alpha^{-2} x_1, -\frac{(1+\alpha^2)(b' + a\alpha^2)}{b\alpha^2} x_1 - \alpha^2 x_2\right).$$

If b is zero, b' is congruent with a^{-1}; therefore, in any case, I contains an operation S' of the form

$$(cx_1, dx_1 + c^{-1}x_2).$$

Again, Γ contains the operation T, where T denotes

$$(x_1, x_1 + x_2);$$

and I therefore contains $S'T^{-1}S'^{-1}T$, which is

$$(x_1, (1-c^2)x_1 + x_2).$$

Hence unless $1 - c^2 \equiv 0$, I must coincide with Γ. Now, when $p > 5$, c can always be chosen so that this congruence is not satisfied. If $p = 5$, the square of the above operation $\Sigma^{-1}S\Sigma S$, when unity is written for α, is

$$\left(x_1, \frac{4}{b}(b' + a)x_1 + x_2\right);$$

unless $b' + a \equiv 0$, this again requires that I coincides with Γ. If finally, the condition $b'+a \equiv 0$ is satisfied in S, it is not satisfied in $ST^{-1}S^{-1}T$, another operation belonging to I; and therefore again, in this case, I coincides with Γ.

Hence finally, if $n = 2$, the factor-group $\frac{\Gamma}{D}$ is simple, except when p is 2 or 3.

219. When n is greater than 2, it will be found that it is sufficient to deal in detail with the case $n = 3$, as the method will apply equally well for any greater value of n. Suppose here again that Γ has a self-conjugate sub-group I which contains D; and let S, denoting

$$(ax_1 + bx_2 + cx_3,\ a'x_1 + b'x_2 + c'x_3,\ a''x_1 + b''x_2 + c''x_3),$$

be one of the operations of I which is not contained in D. S cannot be permutable with all operations of the form $(x_1, x_2, x_3 + x_1)$, as it would then be permutable with every operation of Γ. We may therefore suppose without loss of generality that S and T are not permutable, T denoting $(x_1, x_2, x_3 + x_1)$. Then $T^{-1}STS^{-1}$ is an operation, distinct from identity, belonging to I. Now a simple calculation shews that this operation, say U, is of the form

$$\left(x_1 - cX, x_2 - c'X, Ax_1 + Bx_2 + Cx_3\right),$$

where X is the symbol with which S^{-1} replaces x_1.

If c and c' are both different from zero, Γ will contain an operation V of the form

$$\left(x_1 - \frac{c}{c'}x_2, x_2, x_3\right);$$

and I contains $V^{-1}UV$, which is of the form

$$(x_1, \alpha x_1 + \beta x_2 + \gamma x_3, \alpha' x_1 + \beta' x_2 + \gamma' x_3).$$

Moreover, if either c or c' is zero, the operation U itself leaves one symbol unaltered. Hence I always contains operations by which one symbol is unaltered.

This process may now be repeated to shew that I necessarily contains operations of the form

$$(x_1, x_2, \alpha'' x_1 + \beta'' x_2 + \gamma'' x_3);$$

and, since the determinant of the operation is unity, γ'' is necessarily congruent to unity. But it has been seen that the group Γ is generated from the last operation and the operations conjugate to it. Hence finally, if n is greater than 2, the factor-group $\frac{\Gamma}{D}$ is simple for all values of p.

220. The composition-series of G is now, except as regards the constitution of the simple group $\frac{\Gamma}{D}$, perfectly definite. It has, in fact, been seen that $\frac{G}{\Gamma}$ and D are cyclical groups of orders $p-1$ and d; and therefore if α, β, γ, ... are distinct primes whose product is $p-1$, and if α', β', γ', ... are distinct primes whose product is d; the successive factor-groups of G are first, a series of simple groups of prime orders α, β, γ, ...: then a simple group of composite order $\frac{N}{(p-1)d}$: and lastly, a series of simple groups of prime orders α', β', γ', ...

The sequence in which the set of simple groups of orders α, β, γ, ... are taken in the composition-series may be clearly any whatever, and the same is true of the set of factor-groups of orders α', β', γ', ...; but it is to be noticed that, when d is not equal to $p-1$, the composition-series is capable of further modifications. In this case, $\{A, \Gamma\}$ is a self-conjugate sub-group of G of order $\frac{N}{d}$, which has a maximum self-conjugate sub-group $\{A\}$ of order $p-1$. The successive composition-factors of G may therefore be taken in the sequence

$$\alpha',\ \beta',\ \gamma',\ \ldots :\ \frac{N}{(p-1)d}\ :\ \alpha,\ \beta,\ \gamma,\ \ldots :$$

and their arrangement may be yet further changed by considering the self-conjugate sub-group $\{A^m, \Gamma\}$, where m is a factor of $p-1$ less than $\frac{p-1}{d}$.

221. For every value of p^n, except 2^2 and 3^2, it thus appears that the linear group may be regarded as defining a simple group of composite order. We shall now proceed to a discussion of the constitution of the simple groups thus defined when $n = 2$, p being greater than 3*. In this case, the group Γ is defined by the congruences

$$\begin{aligned} y_1 &\equiv \alpha x_1 + \beta x_2, \\ y_2 &\equiv \gamma x_1 + \delta x_2, \quad \text{(mod. } p); \\ \alpha\delta - \beta\gamma &\equiv 1, \end{aligned}$$

and since $p - 1$ is divisible by 2 when p is an odd prime, d is equal to 2. Hence the self-conjugate operations of Γ are

$$(x_1, x_2) \quad \text{and} \quad (-x_1, -x_2).$$

The order of Γ is $p(p^2 - 1)$, and therefore the order of the simple group, H, which it defines is $\frac{1}{2}p(p^2 - 1)$. Suppose now, if possible, that Γ contains a sub-group g simply isomorphic with H. If S is any operation of Γ, not contained in g, the whole of the operations of Γ are contained in the two sets

$$g, \ Sg.$$

Now $(x_2, -x_1)$, whose square $(-x_1, -x_2)$ is a self-conjugate operation, cannot be contained in the simple group g. Hence both $(x_2, -x_1)$ and $(-x_1, -x_2)$ are contained in Sg, an obvious contradiction. Therefore Γ contains no sub-group simply isomorphic with H.

For a discussion of the properties of H, some concrete representation of the group itself is necessary; this may be obtained in the following way. Instead of the pair of homogeneous congruences that define each operation of Γ, let us, as in § 113, consider the single non-homogeneous congruence

$$y \equiv \frac{\alpha x + \beta}{\gamma x + \delta}, \ \text{(mod. } p),$$

*For the case $n = 3$ the reader may consult a paper by the author "On a class of groups defined by congruences," *Proc. L. M. S.* Vol. XXVI, (1895), pp. 58–106.

where

$$\alpha\delta - \beta\gamma \equiv 1.$$

Corresponding to every operation

$$(\alpha x_1 + \beta x_2,\ \gamma x_1 + \delta x_2)$$

of Γ, there will be a single operation of this new set; namely that in which α, β, γ, δ have respectively the same values. But since the operations

$$y \equiv \frac{\alpha x + \beta}{\gamma x + \delta} \quad \text{and} \quad y \equiv \frac{-\alpha x - \beta}{-\gamma x - \delta}$$

are identical, two operations

$$(\alpha x_1 + \beta x_2,\ \gamma x_1 + \delta x_2) \quad \text{and} \quad (-\alpha x_1 - \beta x_2,\ -\gamma x_1 - \delta x_2)$$

of Γ will correspond to each operation

$$y \equiv \frac{\alpha x + \beta}{\gamma x + \delta}$$

of the new set; the two self-conjugate operations

$$(x_1, x_2) \quad \text{and} \quad (-x_1, -x_2),$$

in particular, corresponding to the identical operation of the new set. Moreover, direct calculation immediately verifies that, to the product of any two operations of Γ, corresponds the product of the two corresponding operations of the new set. Hence the new set of operations forms a group of order $\frac{1}{2}p(p^2-1)$, with which Γ is multiply isomorphic; the group of order 2 formed by the self-conjugate operations of Γ corresponding to the identical operation of the new group.

The simple group H, of order $\frac{1}{2}p(p^2-1)$, which we propose to discuss, can therefore be represented by the set of operations

$$y \equiv \frac{\alpha x + \beta}{\gamma x + \delta}, \ (\text{mod. } p);$$

where

$$\alpha\delta - \beta\gamma \equiv 1,$$

α, β, γ, δ being integers reduced to modulus p.

222. Since the order of H is divisible by p and not by p^2, the group must contain a single conjugate set of sub-groups of order p. Now the operation

$$y \equiv x + 1,$$

or $(x + 1)$ as we will write it in future, is clearly an operation of order p: for its nth power is $(x + n)$, and p is the smallest value of n for which this is the identical operation. If $(x + 1)$ and $\left(\frac{\alpha x + \beta}{\gamma x + \delta}\right)$ are represented by P and S, then

$$S^{-1}PS = \left(\frac{(1 - \alpha\gamma)x + \alpha^2}{-\gamma^2 x + 1 + \alpha\gamma}\right).$$

This is identical with P, only if

$$\gamma \equiv 0, \quad \alpha^2 \equiv 1;$$

and therefore P is permutable with no operations except its own powers. On the other hand, if

$$\gamma \equiv 0,$$

then

$$S^{-1}PS = P^{a^2};$$

and therefore every operation, for which $\gamma \equiv 0$, transforms the sub-group $\{P\}$ into itself. These operations therefore form a sub-group: a result that may also be easily verified directly. The order of this sub-group is the number of distinct operations $\left(\frac{\alpha x + \beta}{\delta}\right)$ for which $\alpha\delta \equiv 1$. The ratio $\frac{\alpha}{\delta}$ must be a quadratic residue, while β may have any value whatever. Hence the order of the sub-group is $\frac{1}{2}(p - 1)p$; and H therefore contains $p + 1$ sub-groups of order p. Since H is a simple group, it follows (§ 125) that it can be represented as a transitive substitution group of degree $p+1$.

This representation of the group can be directly derived, as in § 113, from the congruences already used to define it. Thus if, in

$$y \equiv \frac{\alpha x + \beta}{\gamma x + \delta},$$

we write for x successively 0, 1, 2, ..., $p-1$, ∞, the $p+1$ values obtained for y, when reduced (mod. p), will be the same $p+1$ symbols in some other sequence. For if

$$\frac{\alpha x_1 + \beta}{\gamma x_1 + \delta} \equiv \frac{\alpha x_2 + \beta}{\gamma x_2 + \delta},$$

then

$$(\alpha\delta - \beta\gamma)(x_1 - x_2) \equiv 0,$$

and therefore $x_1 \equiv x_2$.

Each operation of H gives therefore a distinct substitution performed on the symbols 0, 1, ..., $p-1$, ∞; and the complete set of substitutions thus obtained gives the representation of H as a transitive substitution group of degree $p+1$. Since H contains operations of order p, this substitution group must be doubly transitive. That this is the case may also be shewn directly. Thus

$$\frac{y-a}{y-b} \equiv m\frac{x-a'}{x-b'}$$

is an operation changing a' into a and b' into b. This operation may be written

$$y \equiv \frac{k(bm-a)x + k(ab' - ma'b)}{k(m-1)x + k(b' - ma')},$$

and its determinant is

$$k^2 m(b-a)(b'-a').$$

If now $(b-a)(b'-a')$ is a quadratic residue (or non-residue) (mod. p), m may be any quadratic residue (or non-residue); and k can always be chosen so that the determinant is unity. There are therefore $\frac{1}{2}(p-1)$ substitutions in the group, changing any two symbols a', b' into any other two given symbols a, b. Further, if the operation $\left(\dfrac{\alpha x+\beta}{\gamma x+\delta}\right)$ keeps x unchanged in the substitution group, x must satisfy the congruence

$$x \equiv \frac{\alpha x+\beta}{\gamma x+\delta},$$

that is

$$\gamma x^2 + (\delta - a)x - \beta \equiv 0.$$

Such a congruence cannot have more than two roots; and therefore every substitution displaces all, all but one, or all but two, of the $p+1$ symbols.

223. The substitutions, which keep either one or two symbols fixed, must therefore be regular in the remaining p or $p-1$ symbols. Hence the order of every substitution which keeps just one symbol fixed must be p; and the order of every substitution that keeps two symbols fixed must be equal to or be a factor of $p-1$. Now it was seen in the last paragraph that the order of the sub-group that keeps two symbols fixed is $\frac{1}{2}(p-1)$. Moreover, if z is a primitive root (mod. p), the sub-group that keeps a and b fixed contains the operation

$$\frac{y-a}{y-b} = z^2\frac{x-a}{x-b},$$

and the order of this operation is $\frac{1}{2}(p-1)$. Hence, the sub-group that keeps any two symbols fixed is a cyclical group of order $\frac{1}{2}(p-1)$; and every operation that keeps two symbols fixed is some power of an operation of order $\frac{1}{2}(p-1)$. Since the group is a doubly transitive group of degree $p+1$, there must be $\frac{1}{2}(p+1)p$ sub-groups which keep two symbols fixed; and these must form a conjugate set. Each is therefore self-conjugate in a sub-group of order $p-1$. To determine the type of this sub-group, we may consider the sub-group keeping 0 and ∞ fixed: this is generated by Q, where Q denotes $\left(\dfrac{zx}{z^{-1}}\right)$. If $\left(\dfrac{\alpha x+\beta}{\gamma x+\delta}\right)$ is represented by S, then

$$S^{-1}QS = \left(\frac{(z\alpha\delta - z^{-1}\beta\gamma)x - (z-z^{-1})\alpha\beta}{(z-z^{-1})\gamma\delta x + z^{-1}\alpha\delta - z\beta\gamma}\right),$$

which can be a power of Q only if

$$\alpha\beta \equiv 0, \quad \gamma\delta \equiv 0.$$

Hence either

$$\beta \equiv \gamma \equiv 0,$$

in which case S is a power of Q: or

$$\alpha \equiv \delta \equiv 0, \quad \gamma \equiv -\beta^{-1}.$$

In the latter case, we have

$$S = \left(\frac{-\beta}{\beta^{-1}x}\right),$$

which is an operation of order 2; and then

$$S^{-1}QS = \left(\frac{z^{-1}x}{z}\right) = Q^{-1}.$$

The group of order $p-1$, which contains self-conjugately a cyclical sub-group of order $\frac{1}{2}(p-1)$ that keeps two symbols fixed, is therefore a group of dihedral (§ 202) type. Moreover, if t is any factor of $p-1$, this investigation shews that $\{S, Q\}$ is the greatest sub-group that contains $\{Q^t\}$ self-conjugately.

224. A substitution that changes all the symbols must either be regular in the $p+1$ symbols, or must be such that one of its powers keeps two symbols fixed. The latter case however cannot occur; for we have just seen that, if Q is an operation, of order $\frac{1}{2}(p-1)$, which keeps two symbols fixed, the only operations permutable with Q^r are the powers of Q. Hence the substitutions that change all the symbols must be regular in the $p+1$ symbols, and their orders must be equal to or be factors of $p+1$.

Suppose now that i is a primitive root of the congruence

$$i^{p^2-1} - 1 \equiv 0 \pmod{p},$$

so that i and i^p are the roots of a quadratic congruence with real coefficients; and consider the operation K, denoting

$$\frac{y-i^k}{y-i^{kp}} \equiv i^{2(p-1)}\frac{x-i^k}{x-i^{kp}},$$

where k is not a multiple of $p+1$. On solving with respect to y, K is expressed in the form

$$y \equiv \frac{\dfrac{i^{\left(\frac{k}{2}+1\right)(p-1)} - i^{-\left(\frac{k}{2}+1\right)(p-1)}}{i^{\frac{k}{2}(p-1)} - i^{-\frac{k}{2}(p-1)}}x - \dfrac{i^{(p-1)} - i^{-(p-1)}}{i^{\frac{k}{2}(p-1)} - i^{-\frac{k}{2}(p-1)}}i^{\frac{k}{2}(p+1)}}{\dfrac{i^{(p-1)} - i^{-(p-1)}}{i^{\frac{k}{2}(p-1)} - i^{-\frac{k}{2}(p-1)}}xi^{-\frac{k}{2}(p+1)} + \dfrac{i^{\left(\frac{k}{2}-1\right)(p-1)} - i^{-\left(\frac{k}{2}-1\right)(p-1)}}{i^{\frac{k}{2}(p-1)} - i^{-\frac{k}{2}(p-1)}}},$$

an operation of determinant unity. It will be found, on writing i^p for i in the coefficients of this operation, that they remain unaltered; therefore, since they are symmetric functions of i and i^p, they must be real numbers. The operation therefore belongs to H. The nth power of this operation is given by

$$\frac{y-i^k}{y-i^{kp}} \equiv i^{2n(p-1)}\frac{x-i^k}{x-i^{kp}},$$

and therefore, since the first power of i which is congruent to unity, (mod. p), is the (p^2-1)th, the order of the operation is $\frac{1}{2}(p+1)$. If we write kp for k in the operation K, the new operation is K^{-1}; but if k is replaced by any other number k', which is not a multiple of $p+1$, the new operation K', given by

$$\frac{y-i^{k'}}{y-i^{k'p}} \equiv i^{2(p-1)}\frac{x-i^{k'}}{x-i^{k'p}},$$

generates a new sub-group of order $\frac{1}{2}(p+1)$, which has no operation except identity in common with $\{K\}$. Now there are p^2-p numbers less than p^2-1 which are not multiples of $p+1$; therefore H contains $\frac{1}{2}(p^2-p)$ cyclical sub-groups of order $\frac{1}{2}(p+1)$, no two of which have a common operation except identity. The corresponding substitutions displace all the symbols.

225. A simple enumeration shews that the operations of the cyclical sub-groups of orders $\frac{1}{2}(p-1)$, p and $\frac{1}{2}(p+1)$, exhaust all the operations of the group. Thus there are, omitting identity from each sub-group:

(i) $\frac{1}{2}p(p+1)$ sub-groups of order $\frac{1}{2}(p-1)$, containing $\frac{1}{4}p(p^2-1)-\frac{1}{2}p^2-\frac{1}{2}p$ distinct operations;

(ii) $\frac{1}{2}p(p-1)$ sub-groups of order $\frac{1}{2}(p+1)$, containing $\frac{1}{4}p(p^2-1)-\frac{1}{2}p^2+\frac{1}{2}p$ distinct operations;

(iii) $p+1$ sub-groups of order p, containing p^2-1 distinct operations; and the sum of these numbers, with 1 for the identical operation, gives $\frac{1}{2}p(p^2-1)$, which is the order of the group.

Every operation that displaces all the symbols is therefore the power of an operation of order $\frac{1}{2}(p+1)$.

226. We shall now further shew that the $\frac{1}{2}p(p-1)$ sub-groups of order $\frac{1}{2}(p+1)$ form a single conjugate set, and that each is contained self-conjugately in a dihedral group of order $p+1$. Let S be any operation of H, which is permutable with $\{K\}$ and replaces i^k by some other symbol j. Then $S^{-1}KS$ is an operation which leaves j unaltered; it may therefore be expressed in the form

$$\frac{y-j}{y-j'} = m\frac{x-j}{x-j'}.$$

This can belong to the sub-group generated by K, only if j and j' are the same pair as i^k and i^{kp}. Hence j must be either i^k or i^{kp}; and similarly, if S replaces i^{kp} by j', the latter must be either i^{kp} or i^k. Hence either S must keep both the symbols i^k and i^{kp} unchanged or it must interchange them; and conversely, every operation which either keeps both the symbols unchanged or interchanges them, must transform $\{K\}$ into itself. If S keeps both of them unchanged, it is a power of K. If S interchanges them, it is of the form

$$\frac{y-i^k}{y-i^{kp}} \equiv m\frac{x-i^{kp}}{x-i^k};$$

and a simple calculation shews that

$$S^{-1}KS = K^{-1}.$$

If we take $m = 1$, S becomes

$$x + y \equiv i^k + i^{kp},$$

an operation belonging to H. Hence the cyclical sub-group $\{K\}$ is contained self-conjugately in the sub-group of $\{S, K\}$ which is of dihedral type. If there were any other operation S', not contained in $\{S, K\}$, which transformed K into its inverse, then SS' would be an operation permutable with K and not contained in $\{K\}$. It has just been seen that no such operation exists. Hence $\{S, K\}$, of order $p + 1$, is the greatest sub-group that contains $\{K\}$ self-conjugately; and $\{K\}$ must be one of $\frac{1}{2}p(p-1)$ conjugate sub-groups.

227. The distribution of the operations of H in conjugate sets is now known. A sub-group of order p is contained self-conjugately in a group of order $\frac{1}{2}p(p-1)$, while an operation of order p is permutable only with its own powers. There are therefore two conjugate sets of operations of order p, each set containing $\frac{1}{2}(p^2-1)$ operations. Again, each of the operations of a cyclical sub-group of order $\frac{1}{2}(p-1)$ or $\frac{1}{2}(p+1)$ is conjugate to its own inverse and to no other of its powers. Hence if $\frac{1}{2}(p+1)$ is even and therefore $\frac{1}{2}(p-1)$ odd, there are $\frac{1}{4}(p-3)$ conjugate sets of operations whose orders are factors of $\frac{1}{2}(p-1)$, each set containing p^2+p operations; $\frac{1}{4}(p-3)$ conjugate sets of operations whose orders are factors of $\frac{1}{2}(p+1)$, other than the factor 2, each set containing p^2-p operations; and a single set of operations of order 2, containing $\frac{1}{2}(p^2-p)$ operations. If $\frac{1}{2}(p-1)$ is even and $\frac{1}{2}(p+1)$ odd, there are $\frac{1}{4}(p-1)$ conjugate sets of operations whose orders are factors of $\frac{1}{2}(p+1)$, each containing p^2-p operations; $\frac{1}{4}(p-5)$ conjugate sets whose orders are factors of $\frac{1}{2}(p-1)$, other than the factor 2, each set containing p^2+p operations; and a single set of $\frac{1}{2}(p^2+p)$ conjugate operations of order 2. In either case, the group contains, exclusive of identity, $\frac{1}{2}(p+3)$ conjugate sets of operations.

228. Since $p-1$ and $p+1$ can have no common factor except 2, it follows that, if q^m denote the highest power of an odd prime, other than p, which divides the order of H, q^m must be a factor of $\frac{1}{2}(p-1)$ or of $\frac{1}{2}(p+1)$; and the sub-groups of order q^m must be cyclical. Moreover, since no two cyclical sub-groups of order $\frac{1}{2}(p-1)$, or $\frac{1}{2}(p+1)$, have a common operation except identity, the same must be true of the sub-groups of order q^m.

If 2^m is the highest power of 2 that divides $\frac{1}{2}(p-1)$ or $\frac{1}{2}(p+1)$, 2^{m+1} will be the highest power of 2 that divides the order of H. Moreover, a sub-group of order 2^{m+1} must contain a cyclical sub-group of order 2^m self-conjugately, and it must contain an operation of order 2 that transforms every operation of this cyclical sub-group into its own inverse; in other words, the sub-groups of order 2^{m+1} are of dihedral type.

Suppose now that two sub-groups of order 2^{m+1} $(m > 1)$ have a common sub-group of order 2^r $(r > 2)$. Such a sub-group must be either cyclical or dihedral: in the latter case, it contains self-conjugately a single cyclical sub-group of order 2^{r-1}. Hence, on the supposition made, a cyclical sub-group of order 4 at least would be contained self-conjugately in two distinct cyclical sub-groups of order 2^m. It has been seen that this is not the case; and therefore the greatest sub-group, that two sub-groups of order 2^{m+1} can have in common, must be a sub-group of order 4, whose operations, except identity, are all of order 2. Now every group of order 2^{m+1} contains one self-conjugate operation of order 2, and 2^m operations of order 2 falling into 2 conjugate sets of 2^{m-1} each. Moreover, the group of order $p \pm 1$, which has a cyclical sub-group of order 2^m and contains the operation A of order 2 of this cyclical group self-conjugately, has $\frac{1}{2}(p \pm 1)$ other operations of order 2; and therefore it contains $\dfrac{p \pm 1}{2^{m+1}}$ sub-groups of order 2^{m+1}, each of which has A for its self-conjugate operation. If now B is any operation of order 2 of this sub-group of order $p \pm 1$, and if it is distinct from A, then B enters into a sub-group of order 2^{m+1} that contains A self-conjugately. But since A is permutable with B, A must belong to the sub-group of order $p \pm 1$, which contains B self-conjugately; hence A enters into a sub-group of order 2^{m+1} which contains B self-conjugately. The sub-group $\{A, B\}$ is therefore common to two distinct sub-groups of order 2^{m+1}. Now no group of order 2^r $(r > 2)$ can be common to two sub-groups of order 2^{m+1}; and therefore $\{A, B\}$ must (§ 80) be permutable with some operation S whose order s is prime to 2. If s is not 3, S must be permutable with A and B: and then $\{A, S\}$ and $\{B, S\}$ would be two distinct sub-groups of orders $2s$, whose operations are permutable with each other. It has been seen that H does not contain such sub-groups. Hence $s = 3$; and S transforms A, B and AB cyclically, or $\{S, A, B\}$ is a sub-group of tetrahedral type (§ 202).

The number of quadratic* sub-groups contained in H may be directly enumerated. A group of order 2^{m+1} contains 2^{m-1} such sub-groups, which fall into 2 conjugate sets of 2^{m-2} each; a single group of order 8 containing each quadratic group self-conjugately. The quadratic groups, contained in the $\frac{p \pm 1}{2^{m+1}}$ sub-groups of order 2^{m+1} of a sub-group of order $p \pm 1$, are clearly all distinct, and each quadratic group belongs to just 3 groups of order $p \pm 1$; thus $\{A, B\}$ belongs to the 3 groups which contain A, B and AB respectively as self-conjugate operations. Hence the total number of quadratic groups contained in H is

$$\tfrac{1}{2}p(p \mp 1)\frac{p \pm 1}{2^{m+1}} 2^{m-1} \tfrac{1}{3} = \frac{\tfrac{1}{2}p(p^2 - 1)}{12}.$$

229. The greatest sub-group of a group of order 2^{m+1}, that contains a quadratic group self-conjugately, is a group of order 8 and dihedral type; and it has been shewn that 3 is the only factor, prime to 2, that occurs in the order of the sub-group containing a quadratic group self-conjugately. Hence finally, the order of the greatest group containing a quadratic group self-conjugately is 24, and the $\frac{\frac{1}{2}p(p^2 - 1)}{12}$ quadratic groups fall into two conjugate sets of $\frac{\frac{1}{2}p(p^2 - 1)}{24}$ each. The group of order 24, that contains a quadratic group self-conjugately, contains also a self-conjugate tetrahedral sub-group, while the sub-groups of order 8 are dihedral. Hence (§ 84) this group must be of octohedral type.

Since every tetrahedral sub-group of H contains a quadratic sub-group self-conjugately, and every octohedral sub-group contains a tetrahedral

*A non-cyclical group of order 4 is called a *quadratic* group.

sub-group self-conjugately, there must also be two conjugate sets of tetrahedral sub-groups and two conjugate sets of octohedral sub-groups, the number in each set being $\frac{\frac{1}{2}p(p^2-1)}{24}$.

230. We have hitherto supposed $m > 1$, or what is the same thing, $p \equiv \pm 1 \pmod{8}$. If now $m = 1$, so that $p \equiv \pm 3 \pmod{8}$, the highest power of 2 that divides the order of H is 2^2; and, since 2^2 is not a factor of $\frac{1}{2}(p \pm 1)$, the sub-groups of order 2^2 are quadratic. Moreover, since 2^2 is the highest power of 2 dividing the order of H, the quadratic sub-groups form a single conjugate set. Each sub-group of order $p \pm 1$, which has a self-conjugate operation of order 2, contains $\frac{1}{4}(p \pm 1)$ sub-groups of order 4, and each of the latter belongs to 3 of the former. The total number is as before

$$\frac{\frac{1}{2}p(p^2-1)}{12},$$

and since they form a single conjugate set, each quadratic group is self-conjugate in a group of order 12. Also, for the same reason as in the previous case, this sub-group is of tetrahedral type.

Finally, since every sub-group of H of tetrahedral type must contain a quadratic sub-group self-conjugately, H must contain a single conjugate set of $\frac{\frac{1}{2}p(p^2-1)}{12}$ tetrahedral sub-groups. In this case, the order of H is not divisible by 24, and therefore the question of octohedral sub-groups does not arise.

231. The group H always contains tetrahedral sub-groups; when its order is divisible by 24, it contains also octohedral sub-groups. Now if $p \equiv \pm 1 \pmod{5}$, the order of H is divisible by 60: it may be shewn as follows that, in these cases, H contains sub-groups of icosahedral type.

Let us suppose, first, that $p \equiv 1 \pmod{5}$; and let j be a primitive root of the congruence

$$j^5 \equiv 1 \pmod{p}.$$

Then $\left(\frac{jx}{j^{-1}}\right)$, which we will denote by A, is an operation of order 5. The operations of order 2 of H are all of the form B, where B denotes

$\left(\dfrac{\alpha x + \beta}{\gamma x - \alpha}\right)$, since each is its own inverse. Now

$$AB = \left(\frac{\alpha j x + \beta j^{-1}}{\gamma j x - \alpha j^{-1}}\right);$$

and (§ 203) if A and B generate an icosahedral group,

$$(AB)^3 = 1.$$

A simple calculation shews that, if this condition is satisfied, then

$$\alpha^2(j - j^{-1})^2 \equiv 1.$$

Also, since the determinant of B is unity,

$$\alpha^2 + \beta\gamma \equiv -1.$$

These two congruences have just $p - 1$ distinct solutions, the solutions α, β, γ and $-\alpha$, $-\beta$, $-\gamma$ being regarded as identical. There are therefore $p - 1$ operations of order two in H, namely the operations

$$\left(\frac{\dfrac{1}{j - j^{-1}} x + \beta}{\gamma x - \dfrac{1}{j - j^{-1}}}\right),$$

where

$$\beta\gamma \equiv -1 - \frac{1}{(j - j^{-1})^2},$$

which with A generate an icosahedral sub-group.

The group generated by

$$\left(\frac{jx}{j^{-1}}\right) \quad \text{and} \quad \left(\frac{\dfrac{1}{j - j^{-1}} x + \beta_0}{\gamma_0 x - \dfrac{1}{j - j^{-1}}}\right),$$

contains 5 of the $p-1$ operations of order 2 of the form

$$\left(\frac{\dfrac{1}{j-j^{-1}}x+\beta}{\gamma x-\dfrac{1}{j-j^{-1}}}\right),$$

viz. those for which

$$\beta \equiv \beta_0 j^n, \quad \gamma \equiv \gamma_0 j^{-n}, \quad (n=0,1,2,3,4).$$

Hence the sub-group $\{A\}$, of order 5, belongs to $\frac{1}{5}(p-1)$ distinct icosahedral sub-groups. Now each icosahedral sub-group has 6 sub-groups of order 5; and H contains $\frac{1}{2}p(p+1)$ sub-groups of order 5 forming a single conjugate set. The number of icosahedral sub-groups in H is therefore

$$\tfrac{1}{6}\frac{p-1}{5}\,\tfrac{1}{2}p(p+1)=\frac{\frac{1}{2}p(p^2-1)}{30}.$$

The group of isomorphisms of the icosahedral group is the symmetric group of degree 5 (§ 173). Now H can contain no sub-group simply isomorphic with the symmetric group of degree 5. For if it contained such a sub-group, an operation of order 5 would be conjugate to its own square; and this is not the case.

Hence (§ 165), if an icosahedral sub-group K of H is contained self-conjugately in a greater sub-group L, then L must be the direct product of K and some other sub-group. This also is impossible; for the greatest sub-group of H in which any cyclical sub-group, except those of order p, is contained self-conjugately, is of dihedral type. Hence L must coincide with K, and K must be one of $\dfrac{\frac{1}{2}p(p^2-1)}{60}$ conjugate sub-groups. The icosahedral sub-groups of H therefore fall into two conjugate sets of $\dfrac{\frac{1}{2}p(p^2-1)}{60}$ each.

In a similar manner, when $p\equiv -1 \pmod{5}$, we may take as a typical operation A, of order 5,

$$\frac{y-i}{y-i^p}\equiv i^{\frac{p^2-1}{5}}\frac{x-i}{x-i^p};$$

and it may be shewn, the calculation being rather more cumbrous than in the previous case, that there are just $p+1$ operations B, of the form $\left(\frac{\alpha x+\beta}{\gamma x-\alpha}\right)$, such that

$$(AB)^3 = 1,$$

and that five of these belong to the icosahedral group generated by A and any one of them. It follows, exactly as in the previous case, that H contains $\frac{\frac{1}{2}p(p^2-1)}{30}$ icosahedral sub-groups, which fall into two conjugate sets, each set containing $\frac{\frac{1}{2}p(p^2-1)}{60}$ groups.

232. Finally, we proceed to shew that H has no other sub-groups than those which have been already determined. Suppose, first, that a sub-group h of H contains two distinct sub-groups of order p. These must, by Sylow's theorem, form part of a set of $kp+1$ sub-groups of order p conjugate within h. Now H contains only $p+1$ sub-groups of order p, and therefore k must be unity and h must contain all the sub-groups of order p; or since H is simple, h must contain and therefore coincide with H. Hence the only sub-groups of H, whose orders are divisible by p, are those that contain a sub-group of order p self-conjugately. They are of known types.

Suppose next that g is a sub-group of H, whose order n is not divisible by p, and let S_1 be an operation of g whose order q_1 is not less than the order of any other operation of g. In H the sub-group $\{S_1\}$ is self-conjugate in a dihedral group of order $p\pm 1$; and the greatest sub-group of this group, which contains no operation of order greater than q_1, is a dihedral group of order $2q_1$. Hence in g the sub-group $\{S_1\}$ is self-conjugate in a group of order q_1 or $2q_1$, and therefore it forms one of $\frac{n}{q_1}$ or of $\frac{n}{2q_1}$ conjugate sub-groups. Moreover, no two of these sub-groups contain a common operation except identity; and they therefore contain, excluding identity, $\frac{n(q_1-1)}{\epsilon_1 q_1}$ distinct operations, where ϵ_1 is either 1 or 2.

Of the remaining operations of g, let S_2 be one whose order q_2 is not less than that of any of the others. The operation S_2 cannot be permutable with any of the $\dfrac{n(q_1-1)}{\epsilon_1 q_1}$ operations already accounted for, since S_2 is not a power of any one of these operations. Hence, exactly as before, $\{S_2\}$ must form one of $\dfrac{n}{\epsilon_2 q_2}$ conjugate sub-groups in g, ϵ_2 being either 1 or 2; and these sub-groups contain $\dfrac{n(q_2-1)}{\epsilon_2 q_2}$ operations which are distinct from identity, from each other, and from those of the previous set. This process may be continued till the identical operation only remains. Hence, finally, n being the total number of operations of g, we must have

$$n = 1 + \sum_\nu \frac{n(q_\nu - 1)}{\epsilon_\nu q_\nu}$$

or

$$\frac{1}{n} = 1 - \sum_\nu \frac{q_\nu - 1}{\epsilon_\nu q_\nu}.$$

233. In this equation, let r of the ϵ's be 1 and s of them be 2, so that $r+s$ is their total number, say m. Then

$$\begin{aligned}\frac{1}{n} &= 1 - \sum\left(1 - \frac{1}{q_\lambda}\right) - \sum\left(\tfrac{1}{2} - \frac{1}{2q_\mu}\right)\\ &= 1 - r + \sum\frac{1}{q_\lambda} - \tfrac{1}{2}s + \sum\frac{1}{2q_\mu}\\ &\leqslant 1 - r + \tfrac{1}{2}r - \tfrac{1}{2}s + \tfrac{1}{4}s\\ &\leqslant 1 - \tfrac{1}{2}r - \tfrac{1}{4}s\\ &\leqslant 1 - \tfrac{1}{4}m.\end{aligned}$$

Hence, since n is a positive integer, there cannot be more than three terms under the sign of summation. Moreover, since

$$\frac{1}{n} \leqslant 1 - \tfrac{1}{2}r - \tfrac{1}{4}s,$$

r cannot be greater than 1, and therefore not more than one of the ϵ's can be unity. Also, when one of the ϵ's is unity, we have

$$\begin{aligned}\frac{1}{n} &= \frac{1}{q} - \tfrac{1}{2}s + \sum \frac{1}{2q_\mu} \\ &\leqslant \tfrac{1}{2} - \tfrac{1}{2}s + \tfrac{1}{4}s \\ &\leqslant \tfrac{1}{4}(2-s),\end{aligned}$$

so that, in this case, s cannot be greater than unity. The solutions are now easily obtained by trial.

(i) For one term in the sum, the only possible solution is

$$\epsilon_1 = 1, \quad n = q_1,$$

and the corresponding group is cyclical.

(ii) For two terms in the sum, the solutions are

$$\begin{aligned}&(\alpha)\quad \epsilon_1 = \epsilon_2 = 2, \quad n = \frac{2q_1q_2}{q_1+q_2};\\ &(\beta)\quad \epsilon_1 = 2, \quad \epsilon_2 = 1, \quad q_2 = 2, \quad n = 2q_1;\\ &(\gamma)\quad \epsilon_1 = 1, \quad \epsilon_2 = 2, \quad q_1 = 3, \quad q_2 = 2, \quad n = 12.\end{aligned}$$

To the solution (α) there corresponds no sub-group; for $n < 2q_1$, and the values $q_1 = q_1$, $\epsilon_1 = 2$ imply that g has a sub-group of order $2q_1$.

To the solution (β) correspond the sub-groups of order $2q_1$ of dihedral type, for which q_1 is odd, so that the operations of order 2 form a single conjugate set.

To the solution (γ) corresponds a sub-group of order 12 containing 8 operations of order 3 and 3 operations of order 2, i.e. a tetrahedral sub-group.

(iii) For three terms in the sum, the solutions are

$$\begin{aligned}&(\alpha)\quad \epsilon_1 = \epsilon_2 = \epsilon_3 = 2, \quad && q_2 = 2, \quad q_3 = 2, \quad n = 2q_1;\\ &(\beta)\quad \text{"} \qquad \text{"} \quad q_1 = 3, && q_2 = 3, \quad q_3 = 2, \quad n = 12;\\ &(\gamma)\quad \text{"} \qquad \text{"} \quad q_1 = 4, && q_2 = 3, \quad q_3 = 2, \quad n = 24;\\ &(\delta)\quad \text{"} \qquad \text{"} \quad q_1 = 5, && q_2 = 3, \quad q_3 = 2, \quad n = 60.\end{aligned}$$

To the solution (α) correspond the sub-groups of order $2q_1$ of dihedral type, in which q_1 is even, so that the operations of order 2, which do not belong to the cyclical sub-group of order q_1, fall into two distinct conjugate sets.

To the solution (β) would correspond a group of order 12 containing 3 operations of order 2 and 4 sub-groups of order 3 which fall into two conjugate sets of 2 each. Sylow's theorem shews that such a group cannot exist; and therefore there is no sub-group of H corresponding to this solution.

Solution (γ) gives a group of order 24, with 3 conjugate cyclical sub-groups of order 4, 4 conjugate cyclical sub-groups of order 3, and 6 other operations of order 2 forming a single conjugate set. No operation of this group is permutable with each of the 4 sub-groups of order 3; and therefore, if the group exists, it can be represented as a transitive group of 4 symbols. On the other hand, the order of the symmetric group of 4 symbols, which (§ 203) is simply isomorphic with the octohedral group, is 24; and its cyclical sub-groups are distributed as above. Hence to this solution there correspond the octohedral sub-groups of H.

Solution (δ) gives a group of order 60, with 6 conjugate sub-groups of order 5, 10 conjugate sub-groups of order 3, and a conjugate set of 15 operations of order 2. It has been shewn, in § 85, that there is only one type of group of order 60 that has 6 sub-groups of order 5; viz. the alternating group of degree 5: and that, in this group, the distribution of sub-groups in conjugate sets agrees with that just given. Moreover, the alternating group of degree 5 is simply isomorphic with the icosahedral group. Hence to this solution there correspond the icosahedral sub-groups of H.

234. When $p > 11$, then $\frac{1}{2}p(p-1) > 60$; and, when $p > 3$, $\frac{1}{2}p(p-1) > p+1$. Hence when $p > 11$, the order of the greatest sub-group of H is $\frac{1}{2}p(p-1)$, and the least number of symbols in which H can be expressed as a transitive substitution group is $p+1$.

When p is 5, 7 or 11, however, H can be expressed as a transitive substitution group of p symbols*.

*This is another of the results stated in the letter of Galois referred to in the footnote on p. 215.

For, when $p = 5$, H contains a tetrahedral sub-group of order 12, forming one of 5 conjugate sub-groups; therefore H can be expressed as a transitive group of 5 symbols. It is to be noticed that in this case H is an icosahedral group.

When $p = 7$, H contains an octohedral sub-group of order 24, which is one of 7 conjugate sub-groups; and H can therefore be expressed as a transitive group of 7 symbols. Similarly, when $p = 11$, H contains an icosahedral sub-group of order 60, which is one of 11 conjugate sub-groups; and the group can be expressed transitively in 11 symbols.

235. The simple groups, of the class we have been discussing in the foregoing sections, are self-conjugate sub-groups of the triply transitive groups of degree $p+1$, defined by

$$y \equiv \frac{\alpha x + \beta}{\gamma x + \delta}, \text{ (mod. } p),$$

the existence of which was demonstrated in § 113. In fact, since $\left(\dfrac{\alpha x + \beta}{\gamma x + \delta}\right)$ and $\left(\dfrac{k\alpha x + k\beta}{k\gamma x + k\delta}\right)$ represent the same transformation, the determinant, $\alpha\delta - \beta\gamma$, of any transformation may always be taken as either unity or a given non-residue; and it follows at once that the transformations of determinant unity form a self-conjugate sub-group of the whole group of transformations.

If, as in § 113, α, β, γ, δ, are powers of i, where i is a primitive root of the congruence

$$i^{p^n-1} \equiv 1, \text{ (mod. } p),$$

the triply transitive group G of degree $p^n + 1$, which is defined by the transformations, has again, when p is an odd prime, a self-conjugate sub-group H of order $\frac{1}{2}p^n(p^{2n} - 1)$, which is given by the transformations of determinant unity. It follows from Theorem IX, § 134, that G, being a triply transitive group of degree $p^n + 1$, must have, as a self-conjugate sub-group, a doubly transitive simple group; and it is easy to shew that H is this sub-group.

In fact, if a simple group h is a self-conjugate sub-group of G it must be contained in H. Also, since h is a doubly transitive group of degree p^n+1, it must contain every operation of order p that occurs in G. Now we may shew that these operations generate H. Thus $\left(\dfrac{2x+1}{-x}\right)$ and $(x+2-i-i^{-1})$ are operations of order p belonging to G. Therefore $\left(\dfrac{(i+i^{-1})x+1}{-x}\right)$ belongs to h. But this operation is transformed into $\left(\dfrac{ix}{i^{-1}}\right)$ by $\left(\dfrac{x+i^{-1}}{x+i}\right)$. Hence $\left(\dfrac{ix}{i^{-1}}\right)$ belongs to h; and a sub-group of h which keeps one symbol unchanged is the group of order $\frac{1}{2}p^n(p^n-1)$ generated by $(x+1)$ and $\left(\dfrac{ix}{i^{-1}}\right)$. The order of h therefore is not less than $\frac{1}{2}p^n(p^{2n}-1)$; in other words h is identical with H.

When $p=2$, every power of i is a quadratic residue, and the determinant of every transformation is unity. In this case it may be shewn, by an argument similar to the above, that the group G of order $2^n(2^{2n}-1)$ is itself a simple group.

We are thus led to recognize the existence of a doubly-infinite series of simple groups of orders $2^n(2^{2n}-1)$ and $\frac{1}{2}p^n(p^{2n}-1)$, which are closely analogous to the groups of order $\frac{1}{2}p(p^2-1)$ already discussed. For an independent proof of the existence of these simple groups and for an investigation of their properties, the reader is referred to the memoirs mentioned below*.

236. We will now return to the linear homogeneous group G of transformations of n symbols, taken to a prime modulus p; and consider it more directly as the group of isomorphisms of an Abelian group of order p^n and type $(1, 1, \ldots, \text{to } n \text{ units})$. As in § 156, it may be expressed in the form of a substitution group performed on the p^n-1 symbols of the operations, other than identity, of the Abelian group. In this form it is clearly transi-

*Moore: "On a doubly-infinite series of simple groups," *Chicago Congress Mathematical Papers*, (1893); Burnside: "On a class of groups defined by congruences," *Proc. L. M. S.* Vol. XXV, (1894), pp. 113–139.

tive, since there are isomorphisms changing any operation of the Abelian group into any other operation. If P is any operation of the Abelian group, an isomorphism which changes any one of the $p-1$ operations

$$P,\ P^2,\ \ldots,\ P^{p-1},$$

into any other, will certainly interchange the set among themselves. Hence, when expressed as a group of degree $p^n - 1$, G is imprimitive; and the symbols forming an imprimitive system are those of the operations, other than identity, of any sub-group of order p of the Abelian group. If

$$P_1,\ P_2,\ \ldots,\ P_n$$

are a set of generating operations of the Abelian group, an isomorphism, which changes each of the sub-groups

$$\{P_1\},\ \{P_2\},\ \ldots,\ \{P_n\}$$

into itself, must be of the form

$$\begin{pmatrix} P_1\ ,\ P_2\ ,\ldots,\ P_n \\ P_1^{\alpha_1}, P_2^{\alpha_2}, \ldots, P_n^{\alpha_n} \end{pmatrix}.$$

This isomorphism changes P_1P_2 into $(P_1P_2^{\frac{\alpha_2}{\alpha_1}})^{\alpha_1}$; therefore it will only transform the sub-group $\{P_1P_2\}$ into itself when $\alpha_1 \equiv \alpha_2$, (mod. p). If then the given isomorphism changes every sub-group of order p into itself, we must have

$$\alpha_1 \equiv \alpha_2 \equiv \ldots \equiv \alpha_n, \text{ (mod. } p).$$

Hence the only operations of G, which interchange the symbols of each imprimitive system among themselves, are those given by the powers of

$$\begin{pmatrix} P_1, P_2, \ldots, P_n \\ P_1^{\alpha}, P_2^{\alpha}, \ldots, P_n^{\alpha} \end{pmatrix},$$

where α is a primitive root of p. This operation is the same as that denoted by A in § 217. It follows immediately that the factor-group $\dfrac{G}{\{A\}}$ can be represented as a transitive group in $\frac{p^n - 1}{p - 1}$ symbols. In fact, the operations of $\{A\}$ are the only operations of G which transform each of the $\frac{p^n - 1}{p - 1}$ sub-groups of order p in itself; and these $\frac{p^n - 1}{p - 1}$ sub-groups must be permuted among themselves by every operation of G. The substitution group thus obtained is doubly transitive; for if P and P' are any two operations of the Abelian group such that P' is not a power of P, and if Q and Q' are any other two operations of the Abelian group subject to the same condition, there certainly exists an isomorphism of the form

$$\begin{pmatrix} P, P', \dots \\ Q, Q', \dots \end{pmatrix},$$

and this isomorphism changes the sub-groups $\{P\}$ and $\{P'\}$ into the sub-groups $\{Q\}$ and $\{Q'\}$.

These results will still hold if, instead of considering G the total group of isomorphisms, we take Γ the group of isomorphisms of determinant unity. Thus the determinant of

$$\begin{pmatrix} P, P', \dots \\ Q^\alpha, Q', \dots \end{pmatrix}$$

is α times the determinant of

$$\begin{pmatrix} P, P', \dots \\ Q, Q', \dots \end{pmatrix}.$$

It is therefore possible always to choose α so that the determinant of

$$\begin{pmatrix} P, P', \dots \\ Q^\alpha, Q', \dots \end{pmatrix}$$

shall be unity; and this isomorphism still changes the sub-groups $\{P\}$ and $\{P'\}$ into $\{Q\}$ and $\{Q'\}$ respectively.

The lowest power of A contained in Γ is (§ 217) $A^{\frac{p-1}{d}}$. Hence the group $\frac{\Gamma}{\{A^{\frac{p-1}{d}}\}}$ can be represented as a doubly transitive group of degree $\frac{p^n - 1}{p - 1}$.

This group is (§ 220) simply isomorphic with the simple group of order $\frac{N}{(p-1)d}$, which is defined by the composition-series of G.

We may sum up these results as follows:—

THEOREM. *The homogeneous linear group of order*

$$N = (p^n - 1)(p^n - p) \dots (p^n - p^{n-1})$$

when p^n is neither 2^2 nor 3^2, defines, by its composition-series, a simple group of order $\frac{N}{(p-1)d}$, where d is the greatest common factor of $p-1$ and n. This simple group can be represented as a doubly transitive group of degree $p^{n-1} + p^{n-2} + \dots + p + 1$.

237. The $\frac{p^n - 1}{p - 1}$ symbols, permuted by one of these doubly transitive simple groups, may be regarded as the sub-groups of order p of an Abelian group of order p^n and type $(1, 1, \dots, \text{to } n \text{ units})$. Now every pair of sub-groups of such an Abelian group enters in one, and only in one, sub-group of order p^2; and every sub-group of order p^2 contains $p+1$ sub-groups of order p. Hence from the $\frac{p^n - 1}{p - 1}$ symbols permuted by the doubly transitive group, $\frac{p^n - 1 \cdot p^{n-1} - 1}{p - 1 \cdot p^2 - 1}$ sets of $p+1$ symbols each may be formed, such that every pair of symbols occurs in one set and no pair in more than one set, while the sets are permuted transitively by the operations of the group. These groups therefore belong to the class of groups referred to in § 148. The sub-group, that leaves two symbols unchanged, permutes the remaining symbols in two transitive systems of $p-1$ and $p^{n-1} + p^{n-2} + \dots + p^2$; and the sub-group, that leaves unchanged each of the symbols of one of the sets of $p+1$, is contained self-conjugately in a sub-group whose order is $(p+1)p$ times that of a sub-group leaving two symbols unchanged. This latter sub-group permutes the symbols in two transitive systems of $p+1$ and $p^{n-1} + p^{n-2} + \dots + p^2$. It may be pointed out that, when n is 3, such a sub-group is simply isomorphic with, but is not conjugate to, the sub-groups that leave one symbol unchanged: this may be seen at once by noticing that an Abelian group, of order p^3 and type $(1, 1, 1)$, has the same number of sub-groups of orders p and p^2.

238. Some special cases may be noticed. First, when $p = 2$, both $p - 1$ and d are unity, and the homogeneous linear group is itself a simple group.

If $n = 3$, then $N = 168$; so that the group of isomorphisms of a group of order 8, whose operations are all of order 2, is the simple group of order 168 (§ 146).

If $n = 4$, then $N = 2^6 \cdot 3^2 \cdot 5 \cdot 7$. This is the order of the alternating group of 8 symbols; and it may be shewn that this group is simply isomorphic with the group of isomorphisms.

The Abelian group of order 16 contains 35 sub-groups of order 4; and it may be shewn that, from these 35 sub-groups, sets of 5 can be formed in 56 distinct ways, so that each set of 5 contains every operation of order 2 of the Abelian group once, and once only. If

$$P_1,\ P_2,\ P_3,\ P_4$$

are a set of generating operations of the Abelian group, such a set of 5 sub-groups of order 4 is given by

$$\{P_1, P_2\},\ \{P_3, P_4\},\ \{P_1P_3, P_2P_4\},\ \{P_1P_4, P_1P_2P_3\},\ \{P_2P_3, P_1P_3P_4\}.$$

Now

$$\begin{pmatrix} P_1, & P_2\ , & P_3, & P_4 \\ P_1, & P_2P_3, & P_4, & P_2P_4 \end{pmatrix}$$

is an isomorphism of order 7 of the Abelian group, and P_1 is the only operation of the group, except identity, which is left unchanged by this isomorphism. It may be directly verified that the 7 sets of 5 groups of order 4, into which the given set is transformed by the powers of this isomorphism, contain every sub-group of order 4 of the Abelian group. The 7 sets, being interchanged among themselves by this isomorphism of order 7 which leaves only P_1 unchanged, must be interchanged among themselves by isomorphisms of order 7 which leave any other single operation of the Abelian group unchanged. There are therefore at least 15 isomorphisms of order 7 which interchange the 7 sets among themselves. Now the isomorphisms, which interchange the 7 sets among themselves, form a sub-group of the group of isomorphisms, which is isomorphic with a group of degree 7; and the only groups of degree 7, which contain at least 15 operations of order 7, are the symmetric and the alternating groups. The group of isomorphisms must therefore contain a sub-group which is isomorphic with the symmetric or with the alternating group of degree 7. Hence at once, since the group of isomorphisms is simple, it must contain a sub-group which is simply isomorphic with the alternating group of degree 7. Since this must be one of 8 conjugate sub-groups, the group of substitutions itself is simply isomorphic with the alternating group of degree 8.

If $p^n = 3^3$, then $p^{n-1} + \ldots + p + 1 = 13$, $d = 1$, and $N = 2^4 \cdot 3^3 \cdot 13$. There is therefore a doubly transitive simple group of degree 13 and order $2^4 \cdot 3^3 \cdot 13$ (§§ 145, 149).

239. The homogeneous linear group may be generalized by taking for the coefficients powers of a primitive root of

$$i^{p^\nu - 1} \equiv 1, \text{ (mod. } p),$$

instead of powers of a primitive root of

$$i^{p-1} \equiv 1, \text{ (mod. } p).$$

When the coefficients are thus chosen, the order of the group $G_{p,n,\nu}$, defined by all sets of transformations

$$\left.\begin{array}{l} x_1^1 \equiv a_{11}x_1 + a_{12}x_2 + \ldots + a_{1n}x_n, \\ x_2^1 \equiv a_{21}x_1 + a_{22}x_2 + \ldots + a_{2n}x_n, \\ \ldots\ldots\ldots\ldots\ldots\ldots\ldots\ldots\ldots\ldots \\ x_n^1 \equiv a_{n1}x_1 + a_{n2}x_2 + \ldots + a_{nn}x_n, \end{array}\right\} \text{(mod. } p),$$

whose determinant differs from zero, may be shewn, as in § 172, to be

$$N - (p^{n\nu} - 1)(p^{n\nu} - p^\nu)(p^{n\nu} - p^{2\nu}) \ldots (p^{n\nu} - p^{(n-1)\nu});$$

and the order of the sub-group Γ, formed of those transformations whose determinant is unity, is $\dfrac{N}{p^\nu - 1}$. The only self-conjugate operations of Γ are the operations of the sub-group generated by $(ix_1, ix_2, \ldots, ix_n)$, which are contained in Γ. If δ is the greatest common factor of $p^\nu - 1$ and n, these self-conjugate operations of Γ form a cyclical sub-group γ of order δ.

Finally, the argument of § 219 maybe repeated to shew that $\frac{\Gamma}{\gamma}$ is a simple group.

The homogeneous linear group $G_{p,n,\nu}$, when values of ν greater than unity are admitted, thus defines a triply infinite system of simple groups; it may be proved that these groups can, for all values of ν, be expressed as doubly transitive groups of degree $\frac{p^{n\nu}-1}{p^{\nu}-1}$.

240. We may shew, in conclusion, that the group $G_{p,n,\nu}$ is simply isomorphic with a sub-group of $G_{p,n\nu,1}$. For this purpose, we consider the group defined by

$$x_1^1 \equiv x_1 + i^{r_1}, \quad x_2^1 \equiv x_2 + i^{r_2}, \quad \dots, \quad x_n^1 \equiv x_n + i^{r_n},$$
$$(r_1, r_2, \dots, r_n = 0, 1, 2, \dots, p^{\nu} - 1);$$

the congruences being taken to modulus p. This is an Abelian group of order $p^{n\nu}$ and type $(1, 1, \dots, \text{to } n\nu \text{ units})$. Moreover, the operation

$$x_1^1 \equiv a_{11}x_1 + \dots + a_{1n}x_n,$$
$$\dots\dots\dots\dots\dots\dots\dots\dots$$
$$x_n^1 \equiv a_{n1}x_1 + \dots + a_{nn}x_n,$$

of $G_{p,n,\nu}$ transforms the given operation of the Abelian group into

$$x_1^1 \equiv x_1 + i^{s_1}, \quad x_2^1 \equiv x_2 + i^{s_2}, \quad \dots, \quad x_n^1 \equiv x_n + i^{s_n};$$

where

$$i^{s_1} \equiv a_{11}i^{r_1} + a_{12}i^{r_2} + \dots + a_{1n}i^{r_n},$$
$$\dots\dots\dots\dots\dots\dots\dots\dots\dots\dots$$
$$i^{s_n} \equiv a_{n1}i^{r_1} + a_{n2}i^{r_2} + \dots + a_{nn}i^{r_n}.$$

Every operation of $G_{p,n,\nu}$ as defined in § 239, is therefore permutable with the Abelian group, and gives a distinct isomorphism of it; or in other words, as stated above, $G_{p,n,\nu}$ is simply isomorphic with a sub-group of $G_{p,n\nu,1}$.

Further, the sub-group

$$x_1^1 \equiv x_1 + i^r, \quad x_2^1 \equiv x_2, \quad \ldots, \quad x_n^1 \equiv x_n,$$
$$(r = 0, 1, 2, \ldots, p^\nu - 1),$$

is transformed by the given operation of $G_{p,n,\nu}$ into the sub-group

$$x_1^1 \equiv x_1 + a_{11} i^r, \quad x_2^1 \equiv x_2 + a_{21} i^r, \quad \ldots, \quad x_n^1 \equiv x_n + a_{n1} i^r,$$
$$(r = 0, 1, 2, \ldots, p^\nu - 1).$$

If

$$a_{21} \equiv a_{31} \equiv \ldots \equiv a_{n1} \equiv 0,$$

the two sub-groups are identical; but if these conditions are not satisfied, they have no operation in common except identity. Moreover,

$$a_{11}, \ a_{21}, \ \ldots, \ a_{n1}$$

may each have any value from 0 to $i^{p^\nu - 1}$, simultaneous zero values alone excluded. Hence the sub-group of order p^ν defined by

$$x_1^1 \equiv x_1 + i^r, \quad x_2^1 \equiv x_2, \quad \ldots, \quad x_n^1 \equiv x_n,$$
$$(r = 0, 1, \ldots, p^\nu - 1),$$

is one of $\dfrac{p^{n\nu} - 1}{p^\nu - 1}$ conjugate sub-groups in the group formed by combining the Abelian group with $G_{p,n,\nu}$; and no two of these conjugate sub-groups have a common operation except identity.

The $p^{n\nu} - 1$ operations, other than identity, of an Abelian group of order $p^{n\nu}$ and type $(1, 1, \ldots, \text{to } n\nu \text{ units})$, can therefore be divided into $\dfrac{p^{n\nu} - 1}{p^\nu - 1}$ sets of $p^\nu - 1$ each, such that each set, with identity, forms a sub-group of order p^ν; and the group $G_{p,n,\nu}$ is isomorphic with a group of isomorphisms of the Abelian group which permutes among themselves such a set of $\dfrac{p^{n\nu} - 1}{p^\nu - 1}$ sub-groups of order p^ν.

Ex. 1. Shew that the $\dfrac{p^{n\nu} - 1 \cdot p^{n\nu} - p \ldots p^{n\nu} - p^\nu}{p - 1 \cdot p^2 - 1 \ldots p^\nu - 1}$ sub-groups of order p^ν of an Abelian group of order $p^{n\nu}$ and type $(1, 1, \ldots, \text{to } n\nu \text{ units})$ can be divided into sets of $\dfrac{p^{n\nu}-1}{p^{\nu}-1}$ each, such that each set contains every operation of the group, other than identity, once and once only; and discuss in how many distinct ways such a division may be carried out.

Ex. 2. Shew that the simple group, defined by the group of isomorphisms of an Abelian group of order p^n and type $(1, 1, \ldots, 1)$, admits a class of contragredient isomorphisms, which change the operations of the simple group, that correspond to isomorphisms leaving a sub-group of order p of the Abelian group unaltered, into operations that correspond to isomorphisms leaving a sub-group of order p^{n-1} of the Abelian group unaltered.

CHAPTER XV.

ON SOLUBLE AND COMPOSITE GROUPS.

241. THE most general problem of pure group-theory, so far as it is concerned with groups of finite order, is the determination and analysis of all distinct types of group whose order is a given integer. The solution of this problem clearly involves the previous determination of all types of simple groups whose orders are factors of the given integer.

Now there is no known criterion by means of which we can say whether, corresponding to an arbitrarily given composite integer N as order, there exists a simple group or not. For certain particular forms of N, this question can be answered in the affirmative. For instance, when $N = \frac{1}{2}n!$, previous investigation enables us to state that there is a simple group of order N; but even in these cases we cannot, in general, say how many distinct types of simple group of order N there are.

On the other hand, we have seen that, for certain forms of N, it is possible to state that there is no simple group of order N. Thus, when N is the power of a prime, or is the product of two distinct primes, there is no simple group whose order is N; and further, every group of order N is soluble. Again, if N is divisible by a prime p, and contains no factor of the form $1 + kp$, a group of order N cannot be simple; for, by Sylow's theorem, it must contain a self-conjugate sub-group having a power of p for its order.

We propose, in the present Chapter, to prove a series of theorems which enable us to state, in a considerable variety of cases, that a group, which has a number of given form for its order, is either soluble or composite. If the results appear fragmentary, it must be remembered that this branch of the subject has only recently received attention: it should be regarded rather as a promising field of investigation than as one which is thoroughly explored.

The symbols $p_1, p_2, p_3, \ldots$ will be used throughout the Chapter to denote distinct primes in ascending order of magnitude; while distinct primes, without regard to their magnitude, will be denoted by $p, q, r, \ldots$.

242. THEOREM I. *If H is one of n conjugate sub-groups of a* *group G of order N, and if N is not a factor of $n!$, G cannot be simple.*

The group G is isomorphic with the group of substitutions given on transforming the set of n conjugate sub-groups by each of the operations of G. This is a transitive group of degree n, and its order therefore is equal to or is a factor of $n!$. Hence, if N is not a factor of $n!$, the isomorphism cannot be simple; G is therefore not simple.

Corollary*. If a prime p divides N, and if m is the greatest factor of N which is congruent to unity (mod. p), G will be composite unless N is a factor of $m!$.

There cannot, in fact, be more than m conjugate sub-groups whose order is the highest power of p that divides N; so that this result follows from the theorem.

In dealing with a given integer as order, it may happen that, though no single application of this theorem will prove the corresponding group to be composite, a repeated application of Sylow's theorem, on which the theorem depends, will lead to that result. As an example, let us consider groups whose order is $3^2 \cdot 5 \cdot 11$. Any such group must contain either 1 or 45 sub-groups of order 11, and either 1 or 11 sub-groups of order 5. Hence, if the group is simple, it must contain 45 sub-groups of order 11, and 11 sub-groups of order 5; and each one of the latter must be contained self-conjugately in a sub-group of order 45. Now the 45 sub-groups of order 11 would contain 450 distinct operations of order 11, leaving only 45 others; and therefore a sub-group of order 45, if the group contain such a sub-group, must in this case be self-conjugate. This is, however, in direct contradiction to the assumption that the group contains 11 sub-groups of order 5; therefore the group cannot be simple.

243. THEOREM II. *Every group whose order is the power of a prime is soluble.*

This has already been proved in § 54.

THEOREM III†. *A group G whose order is $p^\alpha q^\beta$, where α is less than $2m$, m being the index to which p belongs* (mod. q), *is soluble.*

---

*Hölder, *Math. Ann.* Vol. XL, (1892), p. 57.

†Frobenius, *Berliner Sitzungsberichte* 1895, p. 190, and Burnside, *Proc. L. M. S.* Vol. XXVI, (1895), p. 209; for the case where $\alpha \not> m$.

If $\alpha < m$, G contains a self-conjugate sub-group H of order q^β.

If $\alpha = m$, G contains either 1 or p^m sub-groups of order q^β. In the latter case, if q^r is the order of the greatest sub-group common to two groups of order q^β, and if h is such a sub-group, h must (§ 80) be contained self-conjugately in a sub-group k of order $p^x q^{r+s}$, $x > 0$, $s > 0$; and this sub-group must contain more than one sub-group of order q^{r+s}. Hence x must be m; and therefore h must be common to all the p^m sub-groups of order q^β, so that h is self-conjugate, and s is $\beta - r$. If r is zero, no two sub-groups of order q^β have a common operation except identity, and the p^m sub-groups contain $p^m(q^\beta - 1)$ distinct operations, so that a sub-group of order p^m is self-conjugate. Hence when α is equal to m, G contains either a self-conjugate sub-group of order q^r $(r > 0)$ or else one of order p^m.

If $m < \alpha < 2m$, G contains 1 or p^m sub-groups of order q^β. If there is only one, it is self-conjugate.

If there are p^m sub-groups of order q^β, and if q^r $(r > 0)$ is the order of the greatest group common to any two of them, it may be shewn, exactly as in the preceding case, that G contains a self-conjugate sub-group of order q^r.

If $r = 0$, the $q^\beta - 1$ operations, other than identity, of each of the p^m sub-groups of order q^β are all distinct. Now G is isomorphic with a group of degree p^m. If the isomorphism is multiple, G must have a self-conjugate sub-group whose order is a power of p. If the isomorphism is simple, we may regard G as a transitive group of degree p^m; and every operation, except identity, of a sub-group of order q^β will leave one symbol only unchanged. Hence q^β is equal to or is a factor of $p^m - 1$. Moreover, when G is thus regarded, the sub-groups of order p^α must be transitive; for all the sub-groups of order q^β will be given on transforming any one of them by the operations of a sub-group of order p^α.

Consider now a sub-group of order $p^{\alpha-m}q^\beta$, which keeps one symbol fixed and contains a sub-group of order q^β self-conjugately. It will contain q^γ $(\gamma \leqslant \beta)$ sub-groups of order $p^{\alpha-m}$. Any one of these must keep p^x symbols fixed; for it is the sub-group of a transitive group of order p^α and degree p^m, which keeps one symbol fixed. If $\gamma < \beta$, there are operations of order q which transform one of these sub-groups into itself, and therefore interchange among themselves the p^x symbols unchanged by the sub-group. Hence $p^x - 1$ must be divisible by q. This requires that $x = m$, which is impossible. Hence $\gamma = \beta$, and the sub-group of order $p^{\alpha-m}q^{\beta}$ contains q^{β} sub-groups of order $p^{\alpha-m}$. It is obvious that no two of these can enter in the same sub-group of order p^{α}; for if they did, they would generate a group of order $p^{\alpha-m+x}$ $(x > 0)$, a sub-group of this order cannot enter in a group of order $p^{\alpha-m}q^{\beta}$. Therefore G contains q^{β} sub-groups of order p^{α}.

Since the sub-groups of G, of order p^{α}, are transitive, their self-conjugate operations must (§ 106) displace all the symbols, and they cannot therefore be permutable with any operation whose order is a power of q. Suppose now that every operation of a sub-group Q, of order q^x, is permutable with an operation P of order p^y, and that there is no sub-group of order q^{x+1} of which this is true. If S is a self-conjugate operation of a sub-group of order p^{α} to which P belongs, P is transformed into itself and Q into a new sub-group Q' by S. Hence the sub-group which contains P self-conjugately has two, and therefore p^m, sub-groups of order q^x. No two of these can belong to the same sub-group of order q^{β}, for if they did they would generate a group of order $q^{x+x'}$ $(x' > 0)$. Now since P is permutable with Q, it must leave unchanged the symbol which Q leaves unchanged. Hence P must leave every symbol unchanged; i.e. it must be the identical operation. The order of every operation of G is therefore either a power of p or a power of q.

Suppose now that p^r is the order of the greatest sub-group that is common to any two sub-groups of order p^{α}, and that h is such a sub-group. Then (§ 80) h must be permutable with an operation of order q. Since no operation of h is permutable with any operation of order q, it follows that $p^r - 1$ is divisible by q, and therefore that $r = m$.

Let k, of order $p^{m+t}q^{\gamma}$, be the greatest sub-group that contains h self-conjugately. If $\gamma = \beta$, h is common to each of the q^{β} sub-groups of order p^{α}, and it is therefore a self-conjugate sub-group of G.

If $\gamma < \beta$, let Q be a sub-group, of order q^{γ}, of k. In G there must be a group, whose order is divisible by $q^{\gamma+1}$, containing Q self-conjugately. Hence there must be operations in G, which transform Q into itself and h into a conjugate sub-group h'. If h and h' have a common sub-group, it must be transformed into itself by every operation of Q. This is impossible, since $p^s - 1$ is not divisible by q when $s < m$.

In the group $\{h', Q\}$, Q is one of p^m conjugate sub-groups, and therefore no operation of h' can transform Q into itself. Hence if an operation P' of h' transforms h into itself, it must transform Q into a group Q', which is conjugate to Q in k. Hence, k' being the greatest sub-group that contains h' self-conjugately, $\{Q, Q'\}$ must be common to k and k'. But $\{Q, Q'\}$ containing two sub-groups of order q^γ, must contain p^m such sub-groups; and therefore the p^m sub-groups of order q^γ that enter in k are identical with those that enter in k'. This is impossible if H and H' are distinct; for the p^m sub-groups of k, of order q^γ, generate $\{h, Q\}$. Hence no operation of h', except identity, transforms h into itself, and the number of sub-groups in the conjugate set to which h belongs must not be less than p^m; so that

$$p^m \not> p^{\alpha-m-t} q^{\beta-\gamma}.$$

Now we have seen that

$$p^m \equiv 1 \ (\text{mod. } q^\beta),$$
$$\text{and} \quad q^\beta \equiv 1 \ (\text{mod. } p^{\alpha-m});$$

from these congruences, and the inequality $\alpha - m < m$, it follows that

$$p^m \not< p^{\alpha-m} q^\beta - q^\beta + 1.$$

This inequality is inconsistent with the previous one, which follows directly from the supposition that γ is less than β. Hence $\gamma = \beta$, and h is a self-conjugate sub-group of G.

It follows therefore that, on every possible supposition, G must have a self-conjugate sub-group. If this sub-group be represented by G_1, the same reasoning may be repeated with respect to the groups $\dfrac{G}{G_1}$ and G_1. Hence G is soluble.

Corollary. All groups whose orders are $p_1 p_2^\alpha$, $p_1^2 p_2^\alpha$, $p_1^3 p_2^\alpha$, $p_1^4 p_2^\alpha$, $p_1^5 p_2^\alpha$, $p_1^\alpha p_2$ are soluble.

Since the congruence

$$p_1^2 \equiv 1 \ (\text{mod. } p_2)$$

is satisfied only by $p_1 = 2$, $p_2 = 3$, the results stated follow immediately from the theorem, except for the cases $2^4 3^\alpha$ and $2^5 3^\alpha$.

If in these cases there are 16 sub-groups of order 3^α $(\alpha > 1)$, there must (§ 78) be sub-groups of order $3^{\alpha-1}$ common to two sub-groups of order 3^α; and, in the groups of order $2^4 3^\alpha$ or $2^5 3^\alpha$, such a sub-group, if not self-conjugate, must be one of either 4 or 8 conjugate sub-groups. The groups must therefore be isomorphic with groups of degree 4 or 8; from this it follows immediately that they must be soluble (§ 146).

244. THEOREM IV. *Groups of order $p_1^\alpha p_2^2$ are soluble.*

If a group G, of order $p_1^\alpha p_2^2$, contains only p_2 sub-groups of order p_1^α, it cannot be simple, for a group of order p_2^2 cannot be expressed as a substitution-group of p_2 symbols. Similarly, if G contains a single group of order p_1^α, it is not simple.

If G contains p_2^2 sub-groups of order p_1^α, and if these sub-groups have no common operations, except identity, there are $p_2^2(p_1^\alpha - 1)$ operations in G whose orders are powers of p_1; and therefore a group of order p_2^2 is self-conjugate. If the operations of the p_2^2 sub-groups of order p_1^α are not all distinct, let p_1^r be the order of the greatest sub-group common to any two of them; and let h be such a sub-group. Then (§ 80) h must be self-conjugate in a group k of order $p_1^{r+s} p_2^\beta$ $(s > 0, \beta = 1 \text{ or } 2)$. If $\beta = 2$, h is self-conjugate in G; and if $\beta = 1$, k must contain p_2 sub-groups of order p_1^{r+s}, and therefore

$$p_2 \equiv 1 \text{ (mod. } p_1).$$

Now (§ 78)

$$p_2^2 \equiv 1 \text{ (mod. } p_1^{\alpha-r}),$$

and therefore

$$p_2 \equiv 1 \text{ (mod. } p_1^{\alpha-r}),$$

unless $p_1 = 2$; but, if $p_1 = 2$, then

$$p_2 \equiv \pm 1 \text{ (mod. } p_1^{\alpha-r-1}).$$

Again, if $s = \alpha - r$, h is one of p_2 conjugate sub-groups; as before, G cannot then be simple.

Suppose now that H is a sub-group of order p_1^α, and that the self-conjugate operations of H form a sub-group h. This must be one of 1, p_2 or p_2^2 conjugate sub-groups; in the first two cases, G cannot be simple. Moreover, if any two sub-groups of the conjugate set to which h belongs have a common sub-group, it must be self-conjugate in a group containing more than one sub-group of order p_1^α; again, G cannot be simple.

Let

$$H,\ H_1,\ \ldots,\ H_{p_2-1}$$

be the p_2 sub-groups (§ 80) of order p_1^α that contain h; and suppose that h is not contained in h. Every operation of each of the p_2 sub-groups

$$\mathrm{h},\ \mathrm{h}_1,\ \ldots,\ \mathrm{h}_{p_2-1},$$

is permutable with every operation of h. But these sub-groups, since they occur in k and not in h, generate a sub-group of k, whose order is divisible by p_2. Hence there must be an operation Q, of order p_2, which is permutable with every operation of h. This operation would be permutable with every operation of a group of order $p_1^r p_2^2$ at least, and it would therefore be one of $p_1^{\alpha-r}$ conjugate operations at most. Now, if $p_1 > 2$, $p_1^{\alpha-r} < p_2$; and as a group of order p_2^2 cannot be represented in terms of less than p_2 symbols, G would in this case have a self-conjugate sub-group. If $p_1 = 2$, then either $p_1^{\alpha-r}$ or $p_1^{\alpha-r} = 2(p_2+1)$. A group of order p_2^2 cannot be expressed in less than $2p_2$ symbols; a cyclical group of order p_2^2 cannot be expressed in $2(p_2+1)$ symbols; and a group of degree $2(p_2+1)$, which contains a non-cyclical sub-group of order p_2^2, must (§ 141) contain the alternating group. Hence again, in this case, G must have a self-conjugate sub-group.

Finally, suppose that h contains the p_2 sub-groups

$$\mathrm{h},\ \mathrm{h}_1,\ \ldots,\ \mathrm{h}_{p_2-1}.$$

If h contains further sub-groups of the conjugate set to which h belongs, they must be permuted among themselves in sets of p_2 when h is transformed by an operation Q, of order p_2, belonging to k. Hence we may assume that

$$\begin{array}{llll} \mathrm{h}, & \mathrm{h}_1, & \ldots, & \mathrm{h}_{p_2-1}, \\ \mathrm{h}_{p_2}, & \mathrm{h}_{p_2+1}, & \ldots, & \mathrm{h}_{2p_2-1}, \\ \cdots & \cdots & \cdots & \cdots \\ \mathrm{h}_{(x-1)p_2}, & \mathrm{h}_{(x-1)p_2+1}, & \ldots, & \mathrm{h}_{xp_2-1}, \end{array}$$

is the complete set of sub-groups of the conjugate set to which h belongs, contained in h; and that, when transformed by Q, those in each line are permuted among themselves. If $\{Q, Q'\}$ is a sub-group* of order p_2^2, the p_2^2 sub-groups

$$\mathrm{h},\ \mathrm{h}_1,\ \ldots,\ \mathrm{h}_{p_2^2-1},$$

are permuted transitively among themselves, when transformed by the operations of $\{Q, Q'\}$. Hence if Q' change h into h_{yp_2}, it must change the set

$$\mathrm{h},\ \mathrm{h}_1,\ \ldots,\ \mathrm{h}_{p_2-1},$$

into the set

$$\mathrm{h}_{yp_2},\ \mathrm{h}_{yp_2+1},\ \ldots,\ \mathrm{h}_{(y+1)p_2-1};$$

and therefore

$$H_{yp_2},\ H_{yp_2+1},\ \ldots,\ H_{(y+1)p_2-1},$$

is a set of p_2 sub-groups of order p_2^{α}, which have the common sub-group $Q'^{-1}hQ'$.

Now, since by supposition h is not a self-conjugate sub-group of H, there must (§ 55) be in H a sub-group h' conjugate to and permutable with h. Let this be the sub-group of order p_1^r, common to

$$H,\ H_1',\ \ldots,\ H_{p_2-1}'.$$

No one of the sets

$$H_{yp_2},\ H_{yp_2+1},\ \ldots,\ H_{(y+1)p_2-1},\quad (y = 0, 1, \ldots, x-1),$$

can have more than one group in common with this set, as otherwise p_1^r would not be the order of the greatest group common to two groups of order p_1^{α}. Hence, of the set of groups

$$\mathrm{h},\ \mathrm{h}_1',\ \ldots,\ \mathrm{h}_{p_2-1}',$$

*If this sub-group is cyclical, then $Q = Q'^{p}$.

at least $p_2 - x$ are distinct from the groups, conjugate to h, which enter in h. Every group of order p_1^{r+s} occurring in k will have a similar set of $p_2 - x$ sub-groups of the set h, which do not occur in h. Hence, since the operations of h are the only operations common to two sub-groups of order p_1^{r+s} in k, there must be in k not less than $p_2(p_2 - x) + p_2 x$, i.e. p_2^2, sub-groups of the set to which h belongs.

In other words, k must contain a self-conjugate sub-group of G.

On every supposition that can be made, we have shewn that G must have a self-conjugate sub-group. If this sub-group is G_1, and if the order of either G_1 or $\frac{G}{G_1}$ contains p_2^2, we may repeat the same reasoning; while if the orders of $\frac{G}{G_1}$ and G_1 are both divisible by p_2, we have already seen that they must be soluble. Finally, then, G itself must be soluble.

245. THEOREM V. *A group of order $p_1^\alpha p_2^\beta$, in which the sub-groups of orders p_1^α and p_2^β are Abelian, is soluble.*

Suppose, first, that there are p_2^γ $(\gamma < \beta)$ sub-groups of order p_1^α. The group is then isomorphic with a substitution group of degree p_2^γ. Now the operations of a sub-group of order p_2^β transform the p_2^γ sub-groups of order p_1^α transitively among themselves; and therefore, in the isomorphism between the given group and the substitution group of degree p_2^γ, there corresponds to an Abelian sub-group of order p_2^β a transitive substitution group of degree p_2^γ. But (§ 124) an Abelian group can only be represented as a transitive substitution group of a number of symbols equal to its order. Hence the isomorphism between the given group and the substitution group is multiple; and the given group is not simple.

Suppose, next, that there are p_2^β sub-groups of order p_1^α, and therefore (§ 81) $p_1^\alpha - 1$ distinct sets of conjugate operations whose orders are powers of p_1. Let P be an operation whose order is a power of p_1; and let H, of order $p_1^\alpha p_2^\delta$, be the greatest group which contains P self-conjugately. Then H contains at least p_2^δ operations whose orders are powers of p_2, and therefore at least p_2^δ distinct operations of the form PQ, where P and Q are permutable and

$$Q^{p_2^\delta} = 1.$$

Now P is one of $p_2^{\beta-\delta}$ conjugate operations; corresponding to each of these, there is a similar set of p_2^{δ} operations of the form PQ; while no two such operations can be identical. The group therefore contains p_2^{β} operations of the form PQ; and similarly, it contains an equal number for each set of conjugate operations whose order is a power of p_1. Hence, finally, the group contains $p_2^{\beta}(p_1^{\alpha}-1)$ operations whose orders are divisible by p_1. There is therefore in this case a self-conjugate sub-group of order p_2.

The group therefore always contains a self-conjugate sub-group. A repetition of the same reasoning shews that it is soluble.

246. THEOREM VI. *A group of order $p_1^{\alpha_1}p_2^{\alpha_2}\dots p_n^{\alpha_n}$, in which the sub-groups of order $p_1^{\alpha_1}, p_2^{\alpha_2}, \dots, p_{n-1}^{\alpha_{n-1}}$ are cyclical, is soluble**.

If the number of operations of the group, whose orders divide $p_r^{\alpha_r}p_{r+1}^{\alpha_{r+1}}\dots p_n^{\alpha_n}$, is exactly equal to this number, it follows from Theorem VII, Cor. I, § 87, that the same is true for the number of operations of the group whose orders divide $p_{r+1}^{\alpha_{r+1}}\dots p_n^{\alpha_n}$. Now, when $r=1$, this relation obviously holds; and therefore it is true for all values of r. The group therefore contains just $p_n^{\alpha_n}$ operations whose orders divide $p_n^{\alpha_n}$; hence a sub-group of order $p_n^{\alpha_n}$ is self-conjugate. Any sub-group of order $p_{n-1}^{\alpha_{n-1}}$ must transform this self-conjugate sub-group of order $p_n^{\alpha_n}$ into itself, so that the group must contain a sub-group of order $p_{n-1}^{\alpha_{n-1}}p_n^{\alpha_n}$. If there were more than one sub-group of this order, the group would contain more than $p_{n-1}^{\alpha_{n-1}}p_n^{\alpha_n}$ operations whose orders divide this number. Hence the group contains a single sub-group of order $p_{n-1}^{\alpha_{n-1}}p_n^{\alpha_n}$, and this sub-group must be self-conjugate. By continuing this reasoning it may be shewn that, for each value of r, the group contains a single sub-group of order $p_r^{\alpha_r}p_{r+1}^{\alpha_{r+1}}\dots p_n^{\alpha_n}$, such a sub-group being necessarily self-conjugate. The group is therefore soluble; for we have already seen that a group whose order is the power of a prime is always soluble.

The self-conjugate sub-group of order $p_r^{\alpha_r}\dots p_n^{\alpha_n}$ must contain the complete set of sub-groups of order $p_r^{\alpha_r}$. If then there are m of these sub-groups, m must be a factor of $p_{r+1}^{\alpha_{r+1}}\dots p_n^{\alpha_n}$; and a sub-group of order $p_r^{\alpha_r}$ must be

* *Proc. L. M. S.* Vol. XXVI, (1895), p. 199.

contained self-conjugately in a sub-group of order

$$p_1^{\alpha_1}\dots p_r^{\alpha_r}\frac{p_{r+1}^{\alpha_{r+1}}\dots p_n^{\alpha_n}}{m}.$$

Hence if i and j are any two indices between 1 and n, the group contains sub-groups of order $p_i^{\alpha_i}p_j^{\alpha_j}$. Now the above sub-group, which contains a sub-group of order $p_r^{\alpha_r}$ self-conjugately, is of the same nature as the original group. Hence, i and j being any two indices less than r, it contains a sub-group of order $p_i^{\alpha_i}p_j^{\alpha_j}p_r^{\alpha_r}$. This process may be continued to shew that, if μ is any factor of $p_1^{\alpha_1}\dots p_n^{\alpha_n}$ which is relatively prime to $\dfrac{p_1^{\alpha_1}\dots p_n^{\alpha_n}}{\mu}$, then the group contains sub-groups of order μ.

Ex. Shew that, when the groups of order $p_n^{\alpha_n}$ are also cyclical, there are sub-groups whose orders are any factors whatever of the order of the group.

247. A special case of the class of groups under consideration is that in which the order contains no repeated prime factor. These groups have formed the subject of a memoir by Herr O. Hölder*. He shews that they are capable of a specially simple form of representation, which we shall now consider.

Let

$$N = p_1p_2\dots p_n$$

be the order of a group G. It has been seen above that a sub-group of order p_n is self-conjugate. Suppose then that $p_l, p_m, \dots, p_n$ are the orders of those cyclical sub-groups of prime order which are contained self-conjugately in G. They generate a self-conjugate sub-group H of order $p_lp_m\dots p_n$, say μ. Each of the cyclical sub-groups $\{P_l\}, \{P_m\}, \dots, \{P_n\}$ of H, generated by an operation of prime order, is permutable with all the rest; and therefore (§ 34) the operations $P_l, P_m, \dots, P_n$ are all permutable. Hence, since their orders are distinct primes, the sub-group H is cyclical. Let now K be a sub-group of G of order ν, where $N = \mu\nu$. No self-conjugate sub-group of G can be contained in K. For if it contained such a sub-group of order $p_\alpha p_\beta\dots p_\gamma$ $(\alpha < \beta < \dots < \gamma)$, this sub-group would contain a single cyclical sub-group of order p_γ; and G would contain a self-conjugate sub-group of order p_γ, contrary to supposition. Since

*"Die Gruppen mit quadratfreier Ordnungszahl" *Göttingen Nachrichten*, 1895: pp. 211–229. Compare also Frobenius, *Berliner Sitzungsberichte*, 1895, pp. 1043, 1044.

then K contains no self-conjugate sub-group of G, it follows (§ 123) that G can be expressed transitively in μ symbols; when it is so expressed, an operation M of order μ, that generates H, will be a circular substitution of μ symbols. The only substitutions performed on the μ symbols, which are permutable with M, are its own powers; and therefore no operation of G is permutable with M except its own powers. Every operation of K must therefore give a distinct isomorphism of $\{M\}$. Now (§ 169) the group of isomorphisms of a cyclical group is Abelian. Hence K must be Abelian; and since its order contains no repeated prime factor, it must be cyclical. The group G is therefore completely defined by the relations

$$M^{\mu} = 1, \quad N^{\nu} = 1, \quad N^{-1}MN = M^{\alpha},$$

where α belongs to index ν, (mod. μ).

248. The theorem of § 246 is a particular case of the following more general one, due to Herr Frobenius*, from which a number of interesting and important results may be deduced.

THEOREM VII. *If G is a group, of order $N = mn$, where m and n are relatively prime, and where*

$$m = p^{\alpha} q^{\beta} \dots r^{\gamma};$$

and if sub-groups $P, Q, \dots, R$, of orders $p^{\alpha}, q^{\beta}, \dots, r^{\gamma}$, are Abelian, while $\theta(P)$ and $\frac{N}{p^{\alpha}}$, $\theta(Q)$ and $\frac{N}{p^{\alpha}q^{\beta}}$, $\dots$, $\theta(R)$ and $\frac{N}{m}$, are relatively prime; then G contains (i) *exactly n operations whose orders divide n, and* (ii) *a sub-group of order m which has a self-conjugate sub-group of order r^{γ}.*

The theorem will be proved inductively, by shewing that, if it is true for any similar group of order $m'n'$, where m' is a factor of m, it is also true for G.

Suppose that the greatest group, which contains R self-conjugately, is a group H' of order $p^{\alpha'} q^{\beta'} \dots r^{\gamma} n'$, where n' is a factor of n. Let N' be any operation of this sub-group, whose order ν' is a factor of n'. Then (§ 177) since $\theta(R)$ and ν' are relatively prime, N' is permutable with every operation of R. Hence every operation of H', whose order divides n', is permutable with every operation of R.

*"Über auflösbare Gruppen, II": *Berliner Sitzungsberichte*, 1895, p. 1035.

Let now S be any operation of R; and suppose that the order of K, the greatest sub-group which contains S self-conjugately, is $p^{\alpha_1}q^{\beta_1}\dots r^{\gamma}n_1$. If the order of S is r^{δ}, the order of $\frac{K}{\{S\}}$ is $p^{\alpha_1}q^{\beta_1}\dots r^{\gamma-\delta}n_1$, say m_1n_1. We have seen in § 176 that, if P_1 is a sub-group of P, then $\theta(P_1)$ is equal to or is a factor of $\theta(P)$. Hence if P_1 is a sub-group of order p^{α_1} of $\frac{K}{\{S\}}$, $\theta(P_1)$ and $\frac{N}{p^{\alpha}}$ are relatively prime; therefore, a fortiori, $\theta(P_1)$ and $\frac{m_1n_1}{p^{\alpha_1}}$ are relatively prime. Similarly $\theta(Q_1)$ and $\frac{m_1n_1}{p^{\alpha_1}q^{\beta_1}}$ are relatively prime, and so on. Moreover m_1 is a factor of m. We may therefore assume that the theorem holds for $\frac{K}{\{S\}}$. Hence this group contains exactly n_1 operations whose orders divide n_1, and it also contains a sub-group of order m_1 in which a sub-group of order $r^{\gamma-\delta}$ is self-conjugate. Therefore K contains a sub-group of order $p^{\alpha_1}q^{\beta_1}\dots r^{\gamma}$ in which a sub-group of order r^{γ} is self-conjugate, and exactly n_1 operations of the form SN, where S and N are permutable and the order of N divides n.

Now S is one of $p^{\alpha-\alpha_1}q^{\beta-\beta_1}\dots\frac{n}{n_1}$ conjugate operations in G; corresponding to each of these, there is such a set of n_1 operations of the form SN, while (§ 16) no two of these operations can be identical. Hence G contains $p^{\alpha-\alpha_1}q^{\beta-\beta_1}\dots n$ operations of the form SN, arising from the conjugate set to which S belongs. If then we sum for the distinct conjugate sets of operations of G whose orders are powers of r, the number of operations of G, whose orders divide $r^{\gamma}n$ and do not divide n, is $n\sum p^{\alpha-\alpha_1}q^{\beta-\beta_1}\dots$.

Now since H' is the greatest group that contains R self-conjugately, the greatest common sub-group of K and H' is the greatest sub-group of K that contains R self-conjugately. The order of this group certainly contains $p^{\alpha_1}q^{\beta_1}\dots r^{\gamma}$ as a factor; and if the order is $p^{\alpha_1}q^{\beta_1}\dots r^{\gamma}\nu$, then S is one of $p^{\alpha'-\alpha_1}q^{\beta'-\beta_1}\dots\frac{n'}{\nu}$ conjugate operations in H'. Every operation of H' however, whose order divides n', is permutable with every operation of R; and therefore $\nu = n'$. Hence the number of operations of H', excluding identity, whose orders are powers of r, is $\sum' p^{\alpha'-\alpha_1} q^{\beta'-\beta_1} \ldots$, where the summation is extended to all the distinct conjugate sets of operations of H' whose orders are powers of r. It has been shewn (Theorem IV, Corollary, § 81) that the number of distinct conjugate sets of operations in G, whose orders are powers of r, is the same as the number in H'. Hence the symbols $\sum$ and $\sum'$ represent summations of the same number of terms.

Finally, in accordance with the induction we are using, we may assume that the number of operations of G whose orders divide $r^\gamma n$ is exactly equal to $r^\gamma n$. For the order of G may be separated into the factors

$$m' = p^\alpha q^\beta \ldots,$$

and

$$n' = r^\gamma n,$$

where m' is a factor of m, while the conditions of the theorem hold for this separation.

Hence the number of operations of G whose orders divide n, is equal to

$$n\left(r^\gamma - \sum p^{\alpha-\alpha_1} q^{\beta-\beta_1} \ldots\right),$$

while at the same time

$$r^\gamma - 1 = \sum p^{\alpha'-\alpha_1} q^{\beta'-\beta_1} \ldots.$$

Unless $\alpha' = \alpha$, $\beta' = \beta$, $\ldots$, the former of these numbers is negative, which is impossible; these conditions must therefore be satisfied. The number of operations, whose orders divide n, is therefore exactly equal to n; and the order of the greatest group H' that contains R self-conjugately is $p^\alpha q^\beta \ldots r^\gamma n'$. The factor-group $\dfrac{H'}{R}$ has for its order $p^\alpha q^\beta \ldots n'$, and it satisfies the same conditions as G. Hence it contains a sub-group of order $p^\alpha q^\beta \ldots$; and H' therefore contains a sub-group of order $p^\alpha q^\beta \ldots r^\gamma$.

The free use of the inductive process that has been made in the preceding proof may possibly lead the reader to doubt its validity. He will find it instructive to verify the truth of the theorem directly in the simpler cases. In fact, the direct verification for the case in which m is a power of a prime is essential to the formal proof; it has been omitted for the sake of brevity.

Corollary I. If $m = m_1 m_2$, where m_1 and m_2 are relatively prime, G has a single conjugate set of sub-groups of order m_1.

That G contains groups of order m_1 may be shewn inductively at once. For since it is clearly true when m contains only one, or two, distinct prime factors, we may assume it true when m_1 does not contain r^γ as a factor. Now G has a sub-group g of order $p^\alpha q^\beta \dots r^\gamma$, which contains a sub-group R of order r^γ self-conjugately. The factor-group $\frac{g}{R}$ contains then a sub-group of order m_1' where m_1' is any factor of $\frac{m}{r^\gamma}$ which is relatively prime to $\frac{m}{m_1' r^\gamma}$; and therefore g contains a sub-group of order $m_1' r^\gamma$.

If now I and I' are two sub-groups of G of order $m_1' r^\gamma$, they may be assumed to contain the same sub-group R of order r^γ. For the sub-groups of G of order r^γ form a single conjugate set; and therefore, if I' contained a sub-group R' of this set, $S^{-1}I'S$ would contain $S^{-1}R'S$, which may be taken to be R. Now R is a self-conjugate sub-group of both I and I', and therefore the factor-groups $\frac{I}{R}$ and $\frac{I'}{R}$ of order m_1' are sub-groups of the factor-group $\frac{H'}{R}$. Hence if the result is true when m_1 does not contain the factor r^γ, it is also true when m_1 does contain r^γ. But when m_1 is equal to p^α, the result is obviously true; and therefore it is true generally.

Corollary II. If the sub-groups $P, Q, \dots, R$ of G contain characteristic sub-groups $P_0, Q_0, \dots, R_0$ of orders $p^{\alpha_0}, q^{\beta_0}, \dots, r^{\gamma_0}$; then G contains a sub-group of order $m_0 = p^{\alpha_0} q^{\beta_0} \dots r^{\gamma_0}$, which has a self-conjugate sub-group of order r^{γ_0}.

Assuming the truth of this statement when m_0 does not contain the factor r, the factor-group $\frac{H'}{R}$ must contain a sub-group of order $\frac{m_0}{r^{\gamma_0}}$; and therefore G contains a sub-group J of order $p^{\alpha_0} q^{\beta_0} \dots r^\gamma$ which has R for a self-conjugate sub-group. Hence, since R_0 is a characteristic sub-group of R, J contains a sub-group of order $p^{\alpha_0} q^{\beta_0} \dots$ and a self-conjugate sub-group of order r^{γ_0}. It therefore contains a sub-group of order $p^{\alpha_0} q^{\beta_0} \dots r^{\gamma_0}$, which has a self-conjugate sub-group of order r^{γ_0}.

249. The reader will have no difficulty in seeing, as has already been stated, that Theorem VI is a direct result of the theorem proved in the last paragraph. We shall proceed at once to further applications of it.

THEOREM VIII. *A group G of order*

$$N = p_1^{\alpha_1} p_2^{\alpha_2} \dots p_n^{\alpha_n},$$

in which the sub-groups of every order $p_r^{\alpha_r}$ $(r = 1, 2, \dots, n-1)$ *are Abelian, with either one or two generating operations, is generally soluble; the special case* $p_1 = 2$, $p_2 = 3$, *may constitute an exception**.

If an Abelian sub-group P_r of order $p_r^{\alpha_r}$ is generated by a single operation, it is cyclical and $\theta(P_r)$ is (§ 176) equal to $p_r - 1$. If P_r is generated by two independent and permutable operations, $\theta(P_r)$ is equal to $(p_r - 1)(p_r^2 - 1)$. Now no prime greater than p_r can divide $(p_r - 1)(p_r^2 - 1)$ unless $p_r + 1$ be a prime; and this is only possible when p_r is equal to 2. Hence unless $P_1 = 2$, $p_2 = 3$, $\theta(P_r)$ and $\dfrac{N}{p_1^{\alpha_1} p_2^{\alpha_2} \dots p_r^{\alpha_r}}$ are relatively prime for all values of r from 1 to $n-1$. Omitting for the present this exceptional case, the conditions of Theorem VII are satisfied by G; it therefore contains exactly $p_n^{\alpha_n}$ operations whose orders are divisible by p_n. Therefore G has a self-conjugate sub-group P_n of order $p_n^{\alpha_n}$, and hence also one or more sub-groups of order $p_{n-1}^{\alpha_{n-1}} p_n^{\alpha_n}$. There are however only $p_{n-1}^{\alpha_{n-1}} p_n^{\alpha_n}$ operations in G whose orders are divisible by no prime smaller than p_{n-1}, and therefore the sub-group of order $p_{n-1}^{\alpha_{n-1}} p_n^{\alpha_n}$ must be self-conjugate. This process clearly may be continued to shew that G has a self-conjugate sub-group of every order $p_r^{\alpha_r} p_{r+1}^{\alpha_{r+1}} \dots p_n^{\alpha_n}$ $(r = 2, 3, \dots, n)$; from which it follows immediately that G is soluble.

Returning now to the exceptional case, suppose that a group of order 2^{α_1} is contained self-conjugately in a maximum group of order $2^{\alpha_1} 3^{\beta} m$. If every operation of the group of order 2^{α_1} is self-conjugate within this sub-group, G contains $2^{\alpha_1} - 1$ distinct conjugate sets of operations whose orders are powers of 2; hence it follows that there are just $\dfrac{N}{2^{\alpha_1}}$ operations

*Frobenius, *loc. cit.*, p. 1041.

in G whose orders are not divisible by 2. In this case, the conditions of Theorem VII are satisfied by G; and it is still soluble.

If, lastly, the operations of the sub-groups of order $2^{\alpha_1}3^{\beta}m$, whose orders are powers of 2, are not all self-conjugate, there must be an operation B, whose order is a power of 3, in this sub-group which is not permutable with every operation of the sub-group of order 2^{α_1}. Let now

$$A,\ A_1,\ \ldots,\ A_r,\ A_{r+1},\ \ldots$$

be a characteristic series of the group A of order 2^{α_1}; and suppose that A_{r+1} is the greatest of these groups with every one of whose operations B is permutable. Then (§ 175) B is not permutable with every operation of $\frac{A_r}{A_{r+1}}$. Hence $\frac{A_r}{A_{r+1}}$ is a quadratic group; and its three operations of order 2 must be permuted cyclically when transformed by B. It follows that $\{A_r, B\}$ is multiply isomorphic with a tetrahedral group. Hence finally, under the conditions of the theorem, G is certainly soluble unless it contains a sub-group which is isomorphic with a tetrahedral group.

Corollary. A group of order $p_1^{\alpha_1}p_2^{\alpha_2}\ldots p_n^{\alpha_n}$ in which no one of the indices $\alpha_1,\ \alpha_2,\ \ldots,\ \alpha_{n-1}$ is greater than 2, is soluble unless it has a sub-group which is isomorphic with a tetrahedral group. For a group of order p_r^2 is necessarily an Abelian group which is generated by either one or two independent operations.

250. We shall next consider certain groups of even order in which the operations of odd order form a self-conjugate sub-group. If G is a group of order N, where

$$N = 2^{\alpha}n, \quad (n \text{ odd}),$$

we shall suppose, in this and the following paragraphs, that a sub-group $\mathcal{Q}$ of order 2^{α} is Abelian, and that $\theta(\mathcal{Q})$ and n are relatively prime. Let us take first the case in which $\mathcal{Q}$ is cyclical, so that $\theta(\mathcal{Q})$ is unity and the latter condition is satisfied for all values of n. If such a group is represented in regular form as a substitution group of $2^{\alpha}n$ letters, the substitution corresponding to an operation of order 2^{α} will consist of n cycles of 2^{α} symbols each. This is an odd substitution; and therefore G has a self-conjugate sub-group of order $2^{\alpha-1}n$. In this sub-group there are cyclical operations of order $2^{\alpha-1}$. Hence the same reasoning will apply to it, and it contains a self-conjugate sub-group of order $2^{\alpha-2}n$. Now (Theorem VII, Cor. I. § 87) G contains exactly $2^{\alpha-2}n$ operations whose orders are not divisible by $2^{\alpha-1}$. Hence the sub-group of order $2^{\alpha-2}n$ must be self-conjugate in G. Proceeding thus, it may be shewn that G contains self-conjugate (and characteristic) sub-groups of every order $2^{\alpha-r}n$ $(r = 1, 2, \ldots, \alpha)$*.

251. Suppose, secondly, that, under the same conditions, G contains a self-conjugate sub-group $\mathscr{Q}$ of order 2^α. Since $\theta(\mathscr{Q})$ and n are relatively prime, every operation of $\mathscr{Q}$ is permutable with every operation of G (§ 177). If $\mathscr{Q}'$ is a sub-group of $\mathscr{Q}$ of order $2^{\alpha-1}$, $\mathscr{Q}'$ must therefore be self-conjugate in G. Now the order of the factor group $\dfrac{G}{\mathscr{Q}'}$ is $2n$; hence, by the preceding result, it has a self-conjugate sub-group of order n. It follows that G has a self-conjugate sub-group of order $2^{\alpha-1}n$. This group, again, has a self-conjugate sub-group of order $2^{\alpha-2}n$, and so on. Hence G has a sub-group of order n. Now by Theorem VII (§ 248), G has just n operations whose divide n. Hence the sub-group H, of order n, must be self-conjugate; and G is the direct product of the two groups $\mathscr{Q}$ and H.

252. Suppose next that $\mathscr{Q}$ is not self-conjugate in G; and let I, of order $2^\alpha n'$, be the greatest group that contains $\mathscr{Q}$ self-conjugately. Every operation of I is permutable with every operation of $\mathscr{Q}$; and therefore (§ 81) G contains $2^{\alpha-1}$ distinct sets of conjugate operations whose orders are powers of 2. The case, in which $\mathscr{Q}$ is cyclical, has been already dealt with and will now be excluded; we may thus assume that $\mathscr{Q}$ contains $2^\beta-1$ $(\beta \nless 2)$ operations of order 2, and that G contains an equal number of sets of conjugate operations of order 2.

If possible, let no two sub-groups of order 2^α have a common operation except identity. Then if two operations S and T, of order 2, are chosen, belonging to distinct conjugate sets and to different sub-groups of order 2^α, they cannot be permutable with each other. They will therefore generate a dihedral sub-group of order $2m$. If m were odd, S and T would be conjugate operations. Hence m must be even; and the dihedral sub-group

*Frobenius, "Über auflösbare Gruppen, II" *Berliner Sitzungsberichte*, 1895, p. 1039.

must contain a self-conjugate operation U of order 2. Since U is permutable with both S and T, it must occur in at least two different sub-groups of order 2^α; and therefore the supposition, that no two sub-groups of order 2^α have a common operation except identity, is impossible.

Let now $\mathcal{Q}'$ of order $2^{\alpha'}$ be a sub-group common to $\mathcal{Q}$ and $\mathcal{Q}_1$, two sub-groups of order 2^α; and suppose that no two sub-groups of order 2^α have a common sub-group which contains $\mathcal{Q}'$ and is of greater order. Since every operation of $\mathcal{Q}$ is self-conjugate in I, $\mathcal{Q}'$ must be self-conjugate in I. If then J, of order $2^\alpha n'r$, is the greatest sub-group of G that contains $\mathcal{Q}'$ self-conjugately, J contains r sub-groups of order 2^α. Hence the factor-group $\frac{J}{\mathcal{Q}'}$ contains r sub-groups of order $2^{\alpha-\alpha'}$, and no two of these have a common sub-group; for if they had, some two sub-groups of order 2^α contained in J would have a common sub-group greater than and containing $\mathcal{Q}'$, contrary to supposition. The r sub-groups of order $2^{\alpha-\alpha'}$ of $\frac{J}{\mathcal{Q}'}$ must therefore be cyclical.

253. We will apply the results of the last paragraph to the case in which all the operations of $\mathcal{Q}$ are of order 2. Then $\mathcal{Q}'$ must be of order $2^{\alpha-1}$, since $\frac{\mathcal{Q}}{\mathcal{Q}'}$ is cyclical. Suppose now that R is an operation of J of odd order, which is permutable with $\mathcal{Q}'$ but not with $\mathcal{Q}$. If R were self-conjugate in a sub-group whose order is divisible by 2^α, this sub-group would contain $\mathcal{Q}'$ and therefore one or more sub-groups of order 2^α containing $\mathcal{Q}'$. But R is not permutable with any sub-group of order 2^α that contains $\mathcal{Q}'$; and therefore the highest power of 2 that divides the order of the group in which R is self-conjugate is $2^{\alpha-1}$, so that R is one of 2μ (μ odd) conjugate operations. If now A is an operation of order 2 that belongs to $\mathcal{Q}$ and not to $\mathcal{Q}'$, no operation conjugate to R can be permutable with A. Hence in the substitution group of degree 2μ, that results on transforming the set of operations conjugate to R among themselves by all the operations of G, the substitution corresponding to A is an odd substitution. This substitution group has therefore a self-conjugate sub-group whose order is half its own, and therefore G has a self-conjugate sub-group of order $2^{\alpha-1}n$. In the same way it may be shewn that this sub-group has a self-conjugate sub-group of order $2^{\alpha-2}n$; and so on. Hence G has a sub-group H of order n. But it follows from Theorem VII, § 248, that G has exactly n operations whose orders divide n; and therefore H is a self-conjugate sub-group. Moreover, since $\frac{G}{H}$ is an Abelian group of order 2^α, it must contain self-conjugate sub-groups of every order 2^r $(r = 1, 2, \ldots, \alpha - 1)$; and G therefore has self-conjugate sub-groups of every order $2^r n$. In general however these sub-groups are not characteristic, as is the case when $\mathcal{Q}$ is cyclical.

254. We will consider next the case where $\mathcal{Q}$ is generated by two operations A and B, of orders $2^{\alpha-1}$ and 2. There are in $\mathcal{Q}$ three operations of order 2, namely, $A^{2^{\alpha-2}}$, $A^{2^{\alpha-2}}B$, and B; and in $\mathcal{Q}$ there are no operations, of order greater than 2, of which the two latter are powers. Suppose that G contains an operation B', conjugate to B, with which $A^{2^{\alpha-2}}$ is not permutable. Then $\{A^{2^{\alpha-2}}, B'\}$ is a dihedral group, and if R is an operation of odd order of this group, no power of A is permutable with R. Since $A^{2^{\alpha-2}}$ and B' are not conjugate operations in G, there must be an operation of order 2, conjugate to $A^{2^{\alpha-2}}B$, and permutable with R. The sub-group, in which $\{R\}$ is self-conjugate, therefore contains representatives of each of the three sets of conjugate operations of order 2 that belong to G; and in this sub-group these representatives form three distinct conjugate sets. Hence no operation conjugate to $A^{2^{\alpha-2}}$ can be permutable with R; and therefore no power, except identity, of an operation conjugate to A is permutable with R. Hence R is one of $2^{\alpha-1}\mu$ (μ odd) conjugate operations, with none of which is A permutable. It follows that, in the substitution group of degree $2^{\alpha-1}\mu$, with which G is isomorphic, A is an odd substitution. Hence G contains a self-conjugate sub-group of order $2^{\alpha-1}n$; and since this is of the same type as G, the same reasoning may be applied to it. Exactly as before, G will contain self-conjugate sub-groups of every order $2^r n$ $(r = 0, 1, \ldots, \alpha - 1)$.

We have assumed that there is an operation B', conjugate to B, with which $A^{2^{\alpha-2}}$ is not permutable. If $A^{2^{\alpha-2}}$ were permutable with every operation that is conjugate to B, the sub-group in which $A^{2^{\alpha-2}}$ is permutable would contain a self-conjugate sub-group G' of G whose order is divisible by 2. In this case, we may deal with $\frac{G}{G'}$ exactly as we have been dealing with G.

The results of §§ 250–254 may be summed up in the following form:—

THEOREM IX. *If in a group G of order $2^\alpha n$, where n is odd, a sub-group $\mathscr{Q}$ of order 2^α is an Abelian group of type (α), $(\alpha-1, 1)$, or $(1, 1, \ldots, 1)$, and if $\theta(\mathscr{Q})$ and n are relatively prime, then G contains self-conjugate sub-groups of each order $2^\beta n$ $(\beta = 0, 1, \ldots, \alpha-1)$.*

255. Still representing the order of G by $2^\alpha n$, where n is odd, there are two cases, in which the sub-groups of order 2^α are not Abelian, where it may be shewn without difficulty that G contains a self-conjugate sub-group of order n.

THEOREM X. *If the order of G is $2^\alpha n$, where n is odd and not divisible by 3, and if a sub-group $\mathscr{Q}$ of order 2^α is of the type*

$$\text{(i)}\quad A^{2^{\alpha-1}} = 1,\quad B^2 = A^{2^{\alpha-2}},\quad B^{-1}AB = A^{-1},$$

$$\textit{or}\quad \text{(ii)}\quad A^{2^{\alpha-1}} = 1,\quad B^2 = 1,\qquad B^{-1}AB = A^{-1},$$

G contains a self-conjugate sub-group of each order $2^\beta n$ $(\beta = 0, 1, \ldots, \alpha-1)$.

For each of these types, $\vartheta(\mathscr{Q})$ is 3 or 1; and if $\mathscr{Q}'$ is any sub-group of $\mathscr{Q}$, $\vartheta(\mathscr{Q}')$ is either 3 or 1. Hence if any operation of G of odd order is permutable with $\mathscr{Q}$ or $\mathscr{Q}'$, it must be permutable with every operation of $\mathscr{Q}$ or $\mathscr{Q}'$. In each type, the self-conjugate operations form a cyclical sub-group of order 2, namely $\{A^{2^{\alpha-2}}\}$.

We will consider first the case where $\mathscr{Q}$ is of type (i). In this case, $\{A^{2\alpha-3}\}$ and $\{B\}$ are sub-groups of G of order 4, the former being self-conjugate in $\mathscr{Q}$ while the latter is not. If they are conjugate sub-groups in G, we have seen (§ 82) that G must contain an operation of odd order S, such that the sub-groups $S^{-n}\{B\}S^n$ $(n = 0, 1, 2, \ldots)$ are permutable with each other. This is impossible, since every operation that is permutable with a sub-group $\mathscr{Q}'$ is permutable with all its operations. Hence $\{A^{2^{\alpha-3}}\}$ and $\{B\}$ are not conjugate sub-groups; and $A^{2^{\alpha-3}}$ and B are not conjugate operations.

Now $A^{2^{\alpha-3}}$ is self-conjugate in $\{A\}$ and is one of two conjugate operations in $\mathcal{Q}$; therefore in G it must be one of 2μ conjugate operations, where μ is odd. When these 2μ conjugate operations are transformed by $A^{2^{\alpha-3}}$, two only, namely, $A^{2^{\alpha-3}}$ and $A^{-2^{\alpha-3}}$, remain unchanged. Let us suppose that, in the resulting substitution, there are x cycles of 2 symbols and y cycles of 4 symbols.

When the 2μ conjugate operations are transformed by B, none can remain unchanged; and we may suppose that the resulting substitution has x' cycles of 2 symbols and y' cycles of 4 symbols.

Now since

$$A^{2^{\alpha-2}} = B^2,$$
$$y = y',$$

and therefore

$$1 + x = x'.$$

Hence one of the two substitutions $A^{2^{\alpha-3}}$ and B must be odd; G has therefore a self-conjugate sub-group of order $2^{\alpha-1}n$. The groups of order $2^{\alpha-1}$ contained in this self-conjugate sub-group are either of type (i), with $\alpha - 1$ written for α: or they are cyclical; for, like $\mathcal{Q}$, they can only contain a single operation of order 2. If they are cyclical, the reasoning of § 250 may be applied; and if they are of type (i) the same reasoning will apply to the self-conjugate sub-group of order $2^{\alpha-1}n$ that has been used for G. Finally, then, G must contain sub-groups of orders n, $2n$, 2^2n, ..., $2^{\alpha-1}n$, each of which is contained self-conjugately in the next; and therefore the sub-group H of order n must be self-conjugate.

If $\mathcal{Q}$ is of type (ii), A is one of 2μ conjugate operations in G, where μ is odd. It may be shewn, as in the previous case, that B and $A^{2^{\alpha-2}}$ cannot be conjugate in G, so that no one of the operations conjugate to A has B for one of its powers. If now B were permutable with any one of these 2μ conjugate operations, the group would contain an Abelian sub-group of order 2^α, which is not the fact. Hence the substitution, given on transforming the 2μ operations by B, consists of μ transpositions and is an odd substitution. Therefore G contains a self-conjugate sub-group of order $2^{\alpha-1}n$, in which the sub-groups of order $2^{\alpha-1}$ are cyclical. Hence, again, there is a self-conjugate sub-group of order n.

256. The only non-Abelian groups of order 2^3 are those of types (i) and (ii) of § 255, when $\alpha = 3$. The Abelian groups of order 2^2 and 2^3 are all included in the types considered in Theorem IX, § 254. For an Abelian group $\mathcal{Q}$, of order 2^3 and type $(2, 1)$, $\theta(\mathcal{Q})$ is 3; and for one of order 2^3 and type $(1, 1, 1)$, $\theta(\mathcal{Q})$ is 21. Hence Theorems IX and X shew that, if 2, 2^2 or 2^3 divide the order of a group, but not 2^4, then the operations of odd order form a self-conjugate sub-group, with possible exceptions when 12 or 56 is a factor of the order. Hence:—

THEOREM XI. *A group of even order cannot be simple unless* 12, 16 *or* 56 *is a factor of the order**.

257. In further illustration of the methods of the preceding paragraphs, we will deal with another case in which the sub-groups of order p^α are not Abelian.

Let G be a group of order $p^\alpha m$, where m is relatively prime to $p(p-1)\ldots(p^\alpha-1)$; and suppose that a sub-group H of order p^α is such that within it every operation is either self-conjugate or one of p conjugate operations†. Let h be a sub-group of G whose order is a power of p; and let S be an operation of G whose order is not divisible by p. Then since m is relatively prime to $p(p-1)\ldots(p^\alpha-1)$, if S is permutable with h, it is permutable with every operation of h. Hence it follows from § 82 that no operation of H, which is not self-conjugate in H, can be conjugate in G to a self-conjugate operation of H. Suppose now that S_1 and S_2 are two operations of H, each of which in H is one of p conjugate operations; and that while S_1 and S_2 are conjugate in G, they are not conjugate in I, the greatest group that contains H self-conjugately. Let H_1 and H_2 be the sub-groups of H, of orders $p^{\alpha-1}$, which contain S_1 and S_2 self-conjugately. Since S_1 and S_2 are conjugate in G, all the sub-groups of order $p^{\alpha-1}$ which contain either of them self-conjugately belong to the same conjugate set. If then H_1 and H_2 are identical, there must be an operation, of order prime to p, which will transform S_1 into S_2 and H_1 into itself. This, we have seen, is impossible. If H_1 and H_2 are not identical, there must be

*Burnside, *Proc. L. M. S.* Vol. XXVI, (1895), p. 332.

†The reader will easily verify that, when this condition is satisfied, $\frac{H}{\mathrm{h}}$ is Abelian, h being the sub-group formed by the self-conjugate operations of H.

an operation which will transform S_1 into S_2 and H_1 into H_2. Now H_1 and H_2 both contain the sub-group h formed of the self-conjugate operations of H; and since no self-conjugate operation of H is conjugate in G with any operation of H which is not self-conjugate, the operation in question, which transforms S_1 into S_2 and H_1 into H_2, must transform h into itself. Hence S_1 and S_2 are conjugate operations in that sub-group, G', of G which contains h self-conjugately. Now in $\frac{G'}{\mathrm{h}}$ the operations, that correspond to S_1 and S_2, are self-conjugate operations of $\frac{H}{\mathrm{h}}$. Since then these operations are conjugate in $\frac{G'}{\mathrm{h}}$, they must (§ 82) be conjugate in $\frac{I}{\mathrm{h}}$. This however is impossible, since every operation of $\frac{I}{\mathrm{h}}$ whose order is a power of p is self-conjugate. Finally, then, no two operations of H, which are not conjugate in H, can be conjugate in G; and the number of distinct sets of conjugate operations in G, whose orders are powers of p, is equal to the number in H. From this it follows, as in previous cases, that the number of operations of G whose orders divide m is equal to m.

258. THEOREM XII. *The only simple groups, whose orders are the products of four or of five primes, are groups of orders* 60, 168, 660, *and* 1092*: and no group, whose order contains less than four prime factors, is simple.*

Groups, whose orders are p_1, p_1p_2, p_1^2, p_1^3, $p_1^2p_2$, $p_1p_2^2$ or $p_1p_2p_3$ are all proved to be soluble by previous theorems in the present chapter.

If the order of a group contains four prime factors, it must be of one of the forms p_1^4, $p_1^3p_2$, $p_1^2p_2^2$, $p_1^2p_2p_3$, $p_1p_2^3$, $p_1p_2^2p_3$, $p_1p_2p_3^2$ or $p_1p_2p_3p_4$. If p_1 is an odd prime, groups of any one of these orders have already been shewn to be soluble; while if p_1 is 2, the only case which can give a simple group is $2^2p_2p_3$. If a group of this order is simple, it follows from Theorem VIII, Cor. (§ 249), that p_2 must be 3; and the order of the group $12p_3$. A cyclical sub-group of order p_3 must be one of $1+kp_3$ conjugate sub-groups. Hence $1+kp_3$ must be a factor of 12, so that p_3 is either 5 or 11. If p_3 were 11, the 12 conjugate sub-groups of order 11 would contain 120 distinct operations of order 11, and the tetrahedral sub-group, which the group (if simple) must contain, would be self-conjugate. Hence p_3 must be 5 and the order of the group is 60. We have already seen that a simple group of order 60 actually exists, namely, the icosahedral group; and that there is only one type for such a group (§ 85).

If the order of a group contains five prime factors, it must be of one of the forms:—

$$p_1^5,\ p_1^4p_2,\ p_1^3p_2^2,\ p_1^3p_2p_3,\ p_1^2p_2^3,\ p_1^2p_2^2p_3,\ p_1^2p_2p_3^2,\ p_1^2p_2p_3p_4,\ p_1p_2^4,$$
$$p_1p_2^3p_3,\ p_1p_2^2p_3^2,\ p_1p_2^2p_3p_4,\ p_1p_2p_3^3,\ p_1p_2p_3^2p_4,\ p_1p_2p_3p_4^2,\ p_1p_2p_3p_4p_5.$$

If p_1 is an odd prime, it follows from previous theorems that none of these forms, except $p_1^3p_2p_3$ and $p_1p_2^3p_3$, can give simple groups.

Taking first the order $p_1p_2^3p_3$, let us suppose (without limitation to the particular case) that a group of order $p_1p_2^mp_3$ is simple. It must contain p_3 or p_1p_3 conjugate sub-groups of order p_2^m. In either case, the operations of these sub-groups must be all distinct, or else they must all have a common sub-group, for the order of the group contains no factor congruent to unity (mod. p_2), except p_3 or p_1p_3. Now the group contains (§ 248) just $p_2^mp_3$ operations whose orders are not divisible by p_1; and if there are p_3 sub-groups of order p_2^m whose operations are all distinct, there are $(p_2^m-1)p_3$ operations whose orders are powers of p_2. Hence, in this case, a sub-group of order p_3 is self-conjugate. On the other hand, if there are p_1p_3 sub-groups of order p_2^m, their operations cannot be all distinct; and the group has a self-conjugate sub-group whose order is a power of p_2. A repetition of this reasoning shews that a group of order $p_1p_2^mp_3$ is always soluble.

We consider, next, a group whose order is $p_1^3p_2p_3$. If p_1 is odd, p_1^2 cannot be congruent to unity, (mod. p_2) or (mod. p_3). The same is true, when $p_1=2$, if p_2 is not equal to 3. We shall therefore first deal with a group of order $p_1^3p_2p_3$ on the supposition that p_2 is not 3.

If neither p_2 nor p_3 divides p_1^3-1, it follows from §§ 248, 257 that there are just p_2p_3 operations in the group whose orders divide p_2p_3. When this is the case, the group clearly cannot be simple.

If p_3 divides p_1^3-1, and if there are more than p_2p_3 operations whose orders divide p_2p_3, a sub-group of order p_1^3 must be self-conjugate in a sub-group of order $p_1^3p_3$, so that the group is isomorphic with a group of degree p_2. In this case, again, the group cannot be simple.

Finally, then, we have only to deal with the case in which a sub-group of order p_1^3 is self-conjugate in a sub-group of order $p_1^3p_2$, while in this sub-group an operation of order p_2 is one of p_1^3 conjugate operations. Now the congruences

$$p_1^3 \equiv 1 \ (\text{mod. } p_2),$$
$$p_2 \equiv 1 \ (\text{mod. } p_1),$$

are inconsistent; therefore, if a sub-group of order p_2 is self-conjugate in a sub-group of order $p_1^\alpha p_2$, the latter must be Abelian. Hence the group contains $p_1^3(p_2 - 1)p_3$ operations whose orders are divisible by p_2. Now if the sub-groups of order p_1^3 have a common sub-group, it must be common to all of the sub-groups of order p_1^3, and it is a self-conjugate sub-group. If, however, no two have a common sub-group, the group contains $(p_1^3 - 1)p_3$ operations of order p_1; and there remain only p_3 operations whose orders divide p_3. The group is therefore in any case composite.

If now $p_2 = 3$, a group of order $2^3 3p_3$ has, by Sylow's theorem, a self-conjugate sub-group of order p_3 unless p_3 is 5, 7, 11 or 23. If $p_3 = 23$, the group (if simple) would have just 24 operations whose orders divide 24; it is thence easily seen to be non-existent. If $p = 11$, the group (if simple) could be expressed as a doubly transitive group of degree 12. In this form, however, each operation, which transforms an operation of order 11 into its own inverse, would be a product of 5 transpositions and therefore an odd substitution. This group therefore cannot be simple. If $p_3 = 5$, the group could be expressed as a doubly transitive group of degree 6. The sub-group of order 4 which transforms a sub-group of order 5 into itself would be cyclical; and the corresponding substitution being odd, the group could not be simple.

Hence, finally, the only possibility is a group of order 168. That a simple group of this order actually exists is shewn in § 146; also, there is only one type of such group.

When p_1 is equal to 2, the only case, that requires discussion in addition to those we have dealt with, is $p_1^2p_2p_3p_4$. A group of this order can only be simple (§ 249) when it contains a tetrahedral sub-group. In this case, the operations of order 2 form a single conjugate set, and the group contains just $3p_3p_4$ operations whose orders are divisible by 2. If a sub-group of order $2^\alpha 3q$ is the greatest that contains a sub-group of order 3 self-conjugately, the group will contain $2^{2-\alpha}(3-1)p_3p_4$ operations whose orders are divisible by 3 and not by 2. Hence the group must contain either p_3p_4, $5p_3p_4$, or $7p_3p_4$, operations whose orders divide p_3p_4. On the other hand, the number of these operations may be expressed in the form

$$x(p_3-1)p_4+y(p_4-1)+1,$$

where x is a factor of 12, and y is a factor of $12p_3$ which is congruent to unity, (mod. p_4).

Hence

$$x(p_3-1)p_4+y(p_4-1)=zp_3p_4-1, \quad (z=1,5,7).$$

The case $z=1$ leads to $y=1$, so that the sub-group of order p_4 is self-conjugate.

The case $z=5$ gives no solution; but when $z=7$, it will be found that there are two solutions, namely,

$$x=6,\quad y=12,\quad p_3=5,\quad p_4=11,$$
$$\text{and}\quad x=6,\quad y=14,\quad p_3=7,\quad p_4=13.$$

That simple groups actually exist corresponding to the two orders 660 and 1092 thus arrived at, is shewn in § 221. There is also in each case a single type; the verification of this statement is left to the reader.

259. As has already been stated at the beginning of the present chapter, the solution of the general problem of pure group-theory, namely, the determination of all possible types of group of any given order, depends essentially on the previous determination of all possible simple groups. A complete solution of this latter problem is not to be expected; but for orders, which do not exceed some given limit, the problem may be attacked directly. The first determination of this kind was due to Herr O. Hölder*, who examined all possible orders up to 200: he proved that none of them, except 60 and 168, correspond to a simple group. Mr Cole† continued the

* *Math. Ann.* Vol. XL, (1892), pp. 55–88.

† *American Journal of Mathematics*, Vol. XV, (1893), pp. 303–315.

investigation, by examining all orders from 201 to 660, with the result of shewing that in this interval the only orders, which have simple groups corresponding to them, are 360, 504 and 660. The existence of a simple group of order 504 had not been recognised before Mr Cole's investigation.

The author* has carried on the examination from 661 to 1092, with the result of shewing that 1092 is the only number in this interval which is the order of a simple group.

As the limit of the order is increased, such investigations as these rapidly become more laborious, since a continually increasing number of special cases have to be dealt with. There is little doubt however but that, with the aid of the theorems proved in the present chapter, the investigation might be continued without substantial difficulty up to 2000.

We shall here be content with verifying the result of Herr Hölder's and Mr Cole's investigations for orders up to 660. The method used in dealing with particular numbers may suggest to the reader how the determination might be continued.

260. It follows from Theorem XII that, if an odd number can be the order of a simple group, it must be the product of at least 6 prime factors. Moreover, by previous theorems, we have seen that 3^6, 3^5p and 3^4p^2 cannot be orders of simple groups. Hence certainly no odd number less than $3^4 \cdot 5 \cdot 7$ or 2835† can be the order of a simple group. Therefore, by Theorem XI, we need only examine numbers less than 660 which are divisible by 12, 16, or 56. When each number less than 660 which is divisible by 12, 16, or 56 is written in the form $p_1^{\alpha_1} p_2^{\alpha_2} \ldots$, it will be found that, with eleven exceptions, Theorems II to XII of the present chapter immediately shew that there are no simple groups corresponding to them. The exceptions are 60, 168, 240, 336, 360, 480, 504, 528, 540, 560, 660. We will first deal with such of these numbers as do not actually correspond to simple groups.

---

* *Proc. London Math. Soc.*, Vol. XXVI, (1895), pp. 333–338.

† It is not, of course, here suggested that 2835 is the order of a simple group. It may, in fact, be shewn with little difficulty that no number of the form 3^4pq, 3^3p^2q, or 3^2p^3q, where p and q are odd primes, can be the order of a simple group. A much higher limit than 2835 may therefore be given for the order, if odd, of a simple group.

$$240 = 2^4 \cdot 3 \cdot 5.$$

A simple group of this order would contain 6 or 16 sub-groups of order 5. If it contained 6 sub-groups of order 5, the group could be expressed as a transitive substitution group of degree 6; and there is no group of degree 6 and order 240. If there were 16 sub-groups of order 5, each would be self-conjugate in a group of order 15, which must be cyclical. The group then would be a doubly transitive group of degree 16 and order $16 \cdot 15$. Such a group (§ 105) contains a self-conjugate sub-group of order 16.

$$336 = 2^4 \cdot 3 \cdot 7.$$

A simple group of this order would contain 8 sub-groups of order 7, each self-conjugate in a group of order 42. We have seen (§ 146) that there is no simple group of degree 8 and order $8 \cdot 7 \cdot 6$.

$$480 = 2^5 \cdot 3 \cdot 5.$$

Since 2^5 is not a factor of 5!, a group of order 480 must, if simple, contain 15 sub-groups of order 2^5. Now 2^5 is not congruent to unity, (mod. 4), and therefore (§ 78) some two sub-groups of order 2^5 must have a common sub-group of order 2^4. Such a sub-group, of order 2^4, must (§ 80) either be self-conjugate, or it must be contained self-conjugately in a sub-group of order $2^5 \cdot 3$ or $2^5 \cdot 5$. In either case the group is composite.

$$528 = 2^4 \cdot 3 \cdot 11.$$

There must be 12 sub-groups of order 11, each being self-conjugate in a group of order 44. A group of order 44 necessarily contains operations of order 22, and such an operation cannot be represented as a substitution of 12 symbols. The group is therefore not simple.

$$540 = 2^2 \cdot 3^3 \cdot 5.$$

There must be 10 sub-groups of order 3^3, each self-conjugate in a group of order $2 \cdot 3^3$. There must also be 36 sub-groups of order 5, each self-conjugate in a group of order 15. Now when the group is expressed as transitive in 10 symbols, the substitutions of order 5 consist of 2 cycles of 5 symbols each, and no such substitution can be permutable with a substitution of order 3. Hence the group is not simple.

$$560 = 2^4 \cdot 5 \cdot 7.$$

There must be 8 sub-groups of order 7. A transitive group of degree 8 and order $8 \cdot 7 \cdot 10$ does not exist (§ 146).

That there are actually simple groups of orders 60, 168, 360, and 660, has already been seen; the verification that there is only a single type of simple group of order 360 may be left to the reader. It remains then to consider the order 504.

$$504 = 2^3 \cdot 3^2 \cdot 7.$$

A simple group of this order must contain 8 or 36 sub-groups of order 7. There is no group of degree 8 and order 504 (§ 146). Hence there must be 36 sub-groups of order 7.

Again, there must be 7 or 28 sub-groups of order 3^2; and since there is no group of degree 7 and order 504 (*l.c.*), there are 28 sub-groups of order 3^2. If two of these have a common operation P of order 3, it must (§ 80) be self-conjugate in a group containing more than one sub-group of order 3^2. If the order of this sub-group is $2^2 {\cdot} 3^2$, it must contain a sub-group of order 2^2 self-conjugately; and therefore this sub-group of order 2^2 must be self-conjugate in a group of order $2^3 {\cdot} 3^2$ at least. Such a group would be one of 7 conjugate sub-groups, and this is impossible. Similarly, P cannot be self-conjugate in a group whose order is greater than $2^2 \cdot 3^2$. Hence no two sub-groups of order 3^2 have a common operation, except identity.

There are therefore in the group 216 operations of order 7, and 224 operations whose orders are powers of 3, leaving just 64 other operations.

Suppose now that A is an operation of order 2, self-conjugate in a sub-group of order 2^3. If A is self-conjugate in a group of order 2^3x (x being a factor of $3^2 \cdot 7$), this group must contain at least x operations of odd order and therefore at least x operations of the form AS, where S is an operation of odd order permutable with A. There are also x operations of this form, corresponding to each of the $\frac{63}{x}$ operations conjugate to A; so that the whole group contains 63 operations of the form AS, where S is of odd order and permutable with A, while A is one of a set of conjugate operations of order 2, each of which is self-conjugate in a sub-group of order 2^3. Hence taking the identical operation with these 63 operations of even order, all the operations of the group are accounted for; and there are therefore no operations whose orders are powers of 2 except those of the conjugate set to which A belongs. The sub-groups of order 2^3 are therefore Abelian groups whose operations are all of order 2; and since the 7 operations of order 2, in a group of order 2^3, are all conjugate in the group of order 504, they must (§ 81) be conjugate in the sub-group within which the group of order 2^3 is self-conjugate. Hence, finally, there must be 9 sub-groups of order 2^3, each self-conjugate in a group of order $2^3 \cdot 7$, in which the 7 operations of order 2 form a single conjugate set.

A simple group of order 504, if it exists, must therefore be expressible as a triply transitive group of degree 9, in which the sub-group of order 56 that keeps one symbol fixed is of known type. From this, several inferences can be drawn. Firstly, all the 63 operations of order 2 of the 9 sub-groups of order 2^3 must be distinct, since each operation of order 2 keeps just one symbol fixed. There is therefore a conjugate set of 63 operations of order 2, and there are no other operations of even order in the group. Secondly, the 224 operations whose orders are powers of 3 cannot all be operations of order 3. For, if they were, there would be 8 operations of order 3 containing any three given symbols in a cycle; and therefore the group would contain operations other than identity which keep 3 symbols fixed. This is not the fact; the sub-groups of order 3^2 are therefore cyclical.

Conversely, we may now shew that such a triply transitive group of degree 9, if it exists, is certainly simple. The distribution of its operations in conjugate sets has, in fact, been determined as follows. There are:—

3	conjugate sets of operations of order	7	each containing	72	operations
3	"	9	"	56	"
1	"	3	"	56	"
1	"	2	"	63	"

A self-conjugate sub-group, that contains a single operation of order 7, must contain all of them; and such a sub-group, that contains a single operation of order 9, must contain all the operations of orders 3 and 9. Hence if a self-conjugate sub-group contains operations of order 9, its order must be of the form

$$1 + 224 + 216x + 63y;$$

and if it contains no operations of order 9, its order must be of the form

$$1 + 56x' + 216y' + 63z';$$

each of the symbols x, y, x', y', z' being either zero or unity. The order must at the same time be a factor of 504. A very brief consideration will shew that all these conditions cannot be satisfied, and therefore that no self-conjugate sub-group exists.

That there is a triply transitive group of degree 9 and order $9 \cdot 8 \cdot 7$, has already been seen in § 113; and the actual formation of the substitutions generating this group will verify that it satisfies the conditions just obtained.

We may however very simply construct the group by the method of § 109; and this process has the advantage of shewing at the same time that there is only one type.

There is no difficulty in constructing the sub-group of order 56 as a doubly transitive group of 8 symbols. If we denote by s and t the two substitutions

$$(1254673) \quad \text{and} \quad (12)(34)(56)(78),$$

$\{s, t\}$ is, in fact, a group of the desired type. If now A is any substitution of order 2 in the symbols 1, 2, ..., 8, 9 which contains the transposition (89), and satisfies the conditions

$$AsA = S_1, \quad AtA = S_2AS_3,$$

where S_1, S_2, S_3 belong to $\{s, t\}$: then (§ 109) it follows that $\{A, s, t\}$ is a triply transitive group of degree 9 and order 504.

Now every operation of order 2 in the group, which interchanges 8 and 9, must transform s into its inverse. Also from the 9 symbols only 7 operations of order 2, satisfying these conditions, can be formed; and if one of them belongs to the group, they must all belong to it. Hence there cannot be more than one type of group satisfying the conditions.

Finally, if

$$A = (89)(23)(46)(57),$$

then

$$AsA = s^{-1},$$

and AtA can be expressed in the form

$$(1635842)A(1685437),$$

where the two operations (1635842) and (1685437) belong to $\{s, t\}$. It follows, then, that $\{A, s, t\}$ is a triply transitive group of degree 9 and order 504, which satisfies all the conditions, and that there is only a single type of such group.

261. The order of a non-soluble group must be divisible by the order of some non-cyclical simple group. Moreover, if N is the order of a non-cyclical simple group and if n is any other number, there is always at least one non-soluble group of order Nn; for, G being a simple group of order N and H any group of order n, the direct product of G and H is a non-soluble group of order Nn. Herr Hölder* has determined all distinct types of non-soluble groups whose orders are less than 480. He shews that, besides the three simple groups of orders 60, 168, and 360, there are 22 non-soluble groups whose orders are given by the following table.

120	180	240	300	336	360	420
3	1	8	1	3	5	1

For the proof of all these results except the first, and for the very interesting and suggestive methods that lead to them, the reader is referred to Herr Hölder's memoir. The case of a non-soluble group of order 120 is susceptible of simple treatment; and we will consider it here as exemplifying

*"Bildung zusammengesetzter Gruppen," *Math. Ann.* Vol. XLVI., (1895), pp. 321–422; in particular, p. 420.

how composite groups may be constructed when their factor-groups and the isomorphisms of the latter are completely known.

262. The composition-factors of a non-soluble group G of order 120 are 2 and 60. If these may be taken in either order, then G contains a self-conjugate sub-group of order 2 and a self-conjugate sub-group of order 60. The latter, being simple, must be an icosahedral group and cannot contain the former. Hence, in this case, G must (§ 34) be the direct product of an icosahedral group and a group of order 2.

Next, suppose that the composition-factors can only be taken in the order 2, 60. Then G contains a self-conjugate sub-group H of icosahedral type and no self-conjugate sub-group of order 2. Hence if S is any operation of G which is not contained in H, the isomorphism of H which arises on transforming its operations by S must (§ 165) be contragredient. It follows that G is simply isomorphic with a group of isomorphisms of H. Now H is isomorphic with the alternating group of degree five, and its group of isomorphisms is therefore (§ 173) simply isomorphic with the symmetric group of degree five. Hence, in this case, G is simply isomorphic with the symmetric group of degree five.

Lastly, suppose that the composition-factors can only be taken in the order 60, 2. Then G has a self-conjugate sub-group of order 2 and no self-conjugate sub-group of order 60. Hence (§ 35) G has no sub-group of order 60. Suppose that A is the self-conjugate operation of order 2, and arrange the operations of G in the sets

$$1,\ A;\quad S_1,\ S_1A;\quad S_2,\ S_2A;\ \ldots;\quad S_{59},\ S_{59}A.$$

Since $\dfrac{G}{\{A\}}$ is simply isomorphic with an icosahedral group, it must (§ 203) be possible to choose three sets

$$S',\ S'A;\quad S'',\ S''A;\quad S''',\ S'''A;$$

such that either operation of the first, multiplied by either operation of the second, belongs to the third set; while at the same time the cube of either operation of the first set, the fifth power of either operation of the second set, and the square of either operation of the third set, belong to the set 1, A. Now if

$$S'^3 = 1,$$

then

$$(S'A)^3 = A.$$

Hence we may assume, without loss of generality, that

$$S'^3 = 1, \quad S''^5 = 1, \quad S'S'' = S''' \quad \text{or} \quad S'''A.$$

If S''' is of order 2, so also is $S'''A$, and G would then contain an icosahedral sub-group. Hence

$$S'''^4 = 1,$$

and

$$S'''^2 = A.$$

Now

$$S_1^3 = 1, \quad S_2^5 = 1, \quad (S_1S_2)^2 = 1,$$

is a complete set of defining relations for the icosahedral group; so that, from the symbols S_1 and S_2, exactly 60 distinct products can be formed. Hence from S', S'', and A, where

$$S'^3 = 1, \quad S''^5 = 1, \quad (S'S'')^2 = A, \quad A^2 = 1,$$

A being permutable both with S' and S'', exactly 120 distinct products can be formed. It follows that the relations just given are the defining relations of a non-soluble group of order 120, which has a self-conjugate sub-group of order 2 and no sub-group of order 60; and that there is only one type of such group. We have already seen (§ 221) that there must be at least one such type of group; viz. the group defined by

$$\begin{aligned} x_1' &\equiv \alpha x_1 + \beta x_2, \\ x_2' &\equiv \gamma x_1 + \delta x_2, \quad (\text{mod. } 5). \\ \alpha\delta - \beta\gamma &\equiv 1, \end{aligned}$$

Ex. If p is an odd prime, shew that a group, whose composition-factors are 60 and p, must be the direct product of an icosahedral group and a group of order p. (Hölder.)

263. Let G be a composite group; and suppose that the factor-groups of the composition-series of G are the two non-cyclical simple groups H' and H in the order given. Then (§ 165) if H_1 is the group formed of all the operations of G which are permutable with every operation of H, the direct product $\{H, H_1\}$ of H and H_1 is a self-conjugate sub-group of G. Now the composition-factors of $\{H, H_1\}$ are composition-factors of G. Hence H_1 must either be isomorphic with H', or it must reduce to the identical operation. The latter alternative can only occur if $\frac{L}{H}$ has a sub-group simply isomorphic with H', L being the group of isomorphisms of H. In the cases of the simple groups whose groups of isomorphisms have been investigated in Chapter XI, i.e. the alternating groups, the doubly transitive groups of degree $p^n + 1$ and order $\frac{1}{2}p^n(p^{2n} - 1)$, and the triply transitive groups of degree $2^n + 1$ and order $2^n(2^{2n} - 1)$, $\frac{L}{H}$ was found to be an Abelian group. (These groups include all the simple groups whose orders do not exceed 660.) If H and H' belong to these classes of groups, then G must be the direct product of simple groups simply isomorphic with H and H'. For instance, a group of order 3600, whose composition-factors are 60 and 60, must be the direct product of two icosahedral groups.

This result may clearly be extended to the case of a group, the factor-groups of whose composition-series are a number of non-cyclical simple groups $H_1, H_2, \ldots, H_n$, such that no two of them are of the same type, while for each of them $\frac{L}{H}$ is soluble. Such a group must be the direct product of n simple groups which are isomorphic with $H_1, H_2, \ldots, H_n$.

If, however, several of these groups are of the same type, the inference is no longer necessarily true. Thus if G is the direct product of five icosahedral groups, and if L is the group of isomorphisms of G, the order of $\frac{L}{G}$ is $2^5 \cdot 5!$, and this group contains icosahedral sub-groups. A group of order $(60)^6$, the factor-groups of whose composition-series are all of order 60, is not therefore necessarily the direct product of 6 icosahedral groups.

Note to § 257.

The property stated in the footnote on p. 405 may be proved as follows. Let S_1 and S_2 be two operations of the group which are not permutable with each other; and suppose that

$$S_2^{-1} S_1 S_2 = S_1 \Sigma.$$

The sub-group of order p^{n-1}, which contains S_1 self-conjugately, also contains $S_1\Sigma$ self-conjugately (§ 55); therefore it contains Σ self-conjugately. Similarly the sub-group, which contains S_2 self-conjugately, contains $S_2\Sigma^{-1}$ and therefore Σ self-conjugately. Hence Σ, being contained self-conjugately in two distinct sub-groups of order p^{n-1}, is a self-conjugate operation. From this it follows at once that $\frac{H}{\mathrm{h}}$ is Abelian.

Note to § 258.

The statement, on p. 408, that the congruences $p_1^3 \equiv 1$ (mod. p_2) and $p_2 \equiv 1$ (mod. p_1) are inconsistent, is subject to an exception if $p_1^2 + p_1 + 1$ is a prime. When this is the case and when $p_2 = p_1^2 + p_1 + 1$, a group of order $p_1^2 p_2 p_3$ can only be simple, under the conditions assumed in the text, if it contains $(p_1^3 - 1)p_3$ operations of order p_1, $p_1^2(p_2 - 1)p_3$ operations of order p_2, $p_1^2 p_2(p_3 - 1)$ operations of order p_3, and no other operation except identity. The relation

$$p_1^3 p_2 p_3 = (p_1^3 - 1)p_3 + p_1^2(p_2 - 1)p_3 + p_1^2 p_2(p_3 - 1) + 1$$

cannot however be satisfied; the group is therefore composite.

Note to § 260.

No simple group of odd order is at present known to exist. An investigation as to the existence or non-existence of such groups would undoubtedly lead, whatever the conclusion might be, to results of importance; it may be recommended to the reader as well worth his attention. Also, there is no known simple group whose order contains fewer than three different primes. This suggests that Theorems III and IV, §§ 243, 244, may be capable of generalisation. Investigation in this direction is also likely to lead to results of interest and importance.

APPENDIX.

THE technical phraseology that has been used in this book is borrowed almost entirely from French or German; and far the greater number of important memoirs on the subject of finite groups are written in one or the other of those languages. To enable the reader to refer, with as little trouble as possible, to the writings of foreign mathematicians, a table is here given of the French and the German equivalents of the more important technical terms. Even abroad, the phraseology of the subject has not yet arrived at that settled state in which every writer uses a technical term in the same sense; but the variations of usage are not very serious. In his recently published *Lehrbuch der Algebra*, however, Herr Weber has introduced or adopted several deviations from ordinary usage; the chief of these are noted, by the addition of his name after the term, in the subjoined table.

Group	Groupe
	Gruppe
Abelian group	Groupe des opérations échangeables
	Abel'sche Gruppe
Alternating group	Groupe alterné
	Alternierende Gruppe
Complete group	Vollkommene Gruppe
Composite group	Groupe composé
	Zusammengesetzte Gruppe
Factor-group	Groupe facteur
	Factorgruppe
Primitive or imprimitive group	Groupe primitif ou non-primitif
	Primitive oder imprimitive Gruppe
Transitive or intransitive group	Groupe transitif ou intransitif
	Transitive oder intransitive Gruppe
Simple group	Groupe simple
	Einfache Gruppe
Soluble group	Groupe résoluble
	Auflösbare Gruppe
	Metacyclische Gruppe* (*Weber*)

*This term is used by other German writers to denote the holomorph of a group of prime order.

Substitution group	Groupe des substitutions *or* système des substitutions conjuguées
	Substitutionengruppe
Symmetric group	Groupe symétrique
	Symmetrische Gruppe
Group of isomorphisms	Gruppe der Isomorphismen
Sub-group	Sousgroupe
	Untergruppe
	Theiler (*Weber*)
Conjugate sub-groups	Sousgroupes conjugués
	Gleichberechtigte Untergruppen
	Conjugirte Theiler (*Weber*)
Self-conjugate sub-group	Sousgroupe invariant
	Ausgezeichnete Untergruppe *or* invariante Untergruppe
	Normaltheiler (*Weber*)
Maximum self-conjugate sub-group	Sousgroupe invariant maximum
	Ausgezeichnete *or* invariante Maximaluntergruppe
	Grösster Normaltheiler (*Weber*)
Characteristic sub-group	Charakteristische Untergruppe
Composition-series	Suite de composition
	Reihe der Zusammensetzung
	Compositionsreihe (*Weber*)
Composition factors	Facteurs de composition
	Factoren der Zusammensetzung
Chief series	Hauptreihe
Characteristic series	Lückenlose Reihe charakteristischer Untergruppen
Simple isomorphism	Isomorphisme holoédrique
	Holoëdrischer *or* einstufiger Isomorphismus
Multiple isomorphism*	Isomorphisme meriédrique

*It is to be noticed that, if G is multiply isomorphic with H, then H is "meriédriquement isomorphe" with G.

	Meriëdrischer *or* mehrstufiger Isomorphismus
Isomorphism of a group with itself	Isomorphismus der Gruppe in sich
Cogredient or contragredient isomorphism	Cogredient oder contragredient Isomorphismus
Degree (of a group)	Degré
	Grad
Order " "	Ordre
	Ordnung
	Grad (*Weber*)
Genus " "	Geschlecht
Substitution	Substitution
	Substitution *or* Buchstabenvertauschung
Circular substitution	Substitution circulaire
	Cirkularsubstitution
Even and odd substitutions	Substitutions positives et négatives
	Gerade und ungerade Substitutionen
Regular substitution	Substitution régulière
	Reguläre Substitution
Similar substitutions	Substitutions semblables
	Ähnliche Substitutionen
Transposition	Transposition
	Transposition

One other term, introduced by Herr Weber, for which no simple equivalent is used in this book, must be mentioned. The greatest sub-group which is common to two groups he calls the "Durchschnitt" of the two groups.

The ratio of the order of a sub-group H, to the order of the group G containing it, is called by French writers the "indice," by German the "Index," of H in G. The phrase is most commonly used of a substitution group in relation to the symmetric group of the same degree.

The smallest number of symbols displaced by any substitution, except identity, of a substitution group is called by French writers the "classe," by Germans the "Klasse," of the group.

INDEX.

(The numbers refer to pages.)

www.ingramcontent.com/pod-product-compliance
Ingram Content Group UK Ltd.
Pitfield, Milton Keynes, MK11 3LW, UK
UKHW040235300826
14791UKWH00001B/153

9 789393 971814